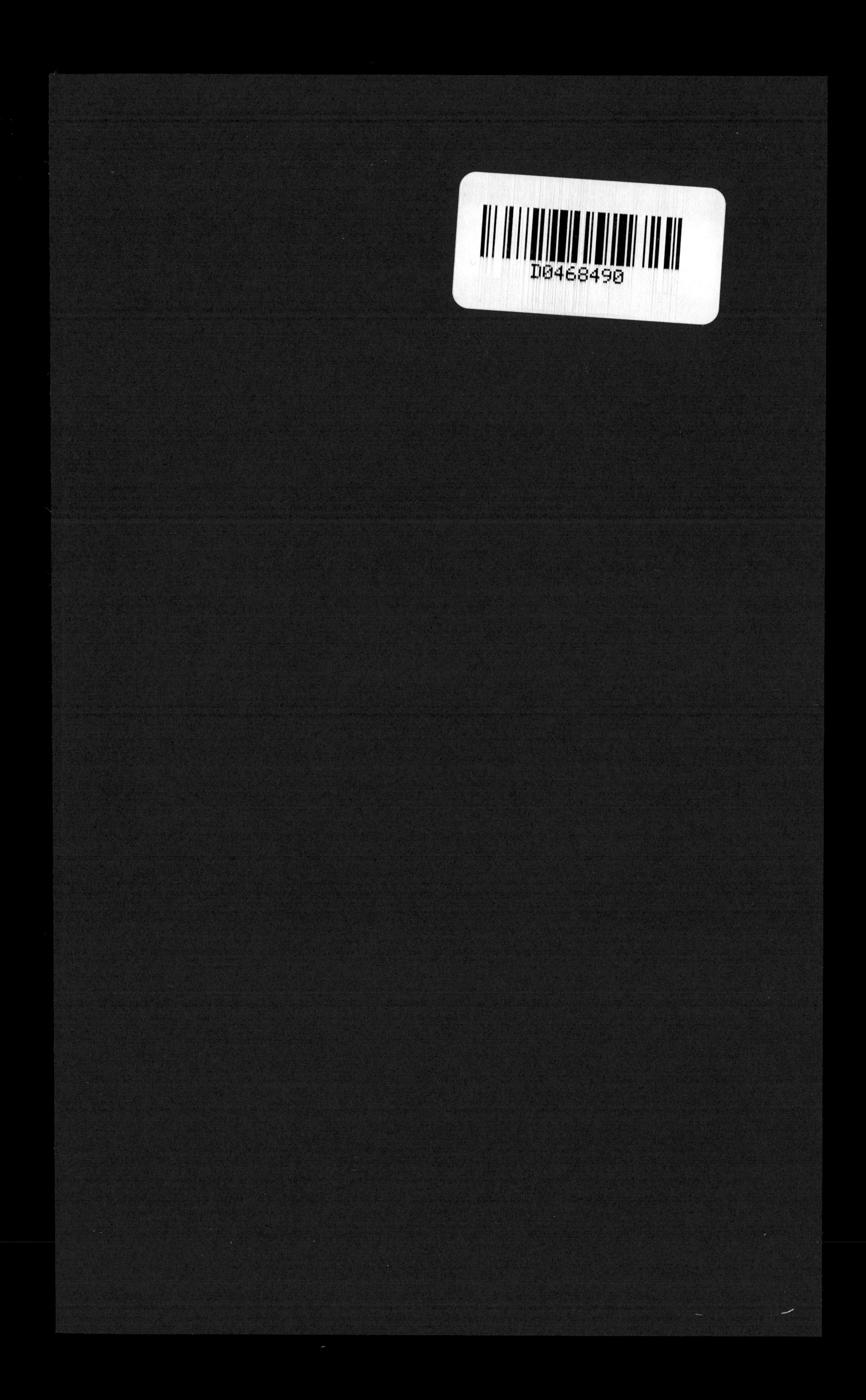
D0468490

# THE SPIRIT OF INQUIRY

*How one extraordinary society shaped modern science*

SUSANNAH GIBSON

*With a Foreword by Simon Conway Morris FRS*

Great Clarendon Street, Oxford, OX2 6DP,
United Kingdom

Oxford University Press is a department of the University of Oxford.
It furthers the University's objective of excellence in research, scholarship,
and education by publishing worldwide. Oxford is a registered trade mark of
Oxford University Press in the UK and in certain other countries



First Edition published in 2019
Impression: 1



Published in the United States of America by Oxford University Press
198 Madison Avenue, New York, NY 10016, United States of America

British Library Cataloguing in Publication Data
Data available

Library of Congress Control Number: 2018951266

ISBN 978–0–19–883337–6

Printed in Great Britain by
Clays Ltd, Elcograf S.p.A.

*To SLDF*

# CONTENTS

a glorious Phantom may
Burst, to illumine our tempestuous day.

*England in 1819*
Percy Bysshe Shelley

# ACKNOWLEDGEMENTS

I would like to thank the Cambridge Philosophical Society for being so generous with their resources and time, especially for making their archives so readily accessible and for allowing me access to their (many and varied) publications and private papers. The Society's council, its fellows, and its staff have been unfailingly helpful; thanks especially to Jim Woodhouse, Simon Conway Morris, Alan Blackwell, Beverley Larner, Janet Hujon, Sara Lees, and Wendy Cattell. Jim Secord, as always, has been a stalwart advisor and tireless reader of early drafts.

This book would not have been possible without the work of Joan Bullock-Anderson, who catalogued the Society's archive in 2014, nor without the support of the Whipple Library and its wonderful staff—Anna Jones, Dawn Moutrey, Agnieszka Lanucha, and Jack Dixon—where that archive currently resides. Yvonne Nobis and the staff of the Betty and Gordon Moore Library, Sandra Freshney of the Sedgwick Museum archives, and the staff of Cambridge University Library's Rare Books Room have also been extremely helpful.

Thanks to the many people who spoke to me of their experiences of the Society over the years, particularly the Henslow fellows (and especially Alex Liu and Emily Mitchell), the former executive secretary Judith Winton Thomas, and many former editors of the Society's journals, as well as former members of the Council.

Thanks to the following for sharing their expertise: Boris Jardine, Josh Nall, Stephen Courtney, Jack Tavener, and the anonymous reviewer; Catherine Clarke, for her early advice on the manuscript; and Ed Potten, for asking many pertinent questions over the last two years.

At Oxford University Press, I would like to thank Latha Menon and Jenny Nugee for their hard work, and also Rosanna van den Bogaerde,

Jonathan Rowley, Caroline Quinnell, and all involved in the book's production.

I am grateful to Girton College (who continue to lead the way on social innovation). But, most of all, I am grateful to my family— especially Seb, Amos, Ridley, and Oisín (my constant companion through much of the writing of this manuscript).

# FOREWORD

by Simon Conway Morris

There is much to be said for the understated, not least in the coinage of self-depreciation and getting things done without undue fuss. Such is the hallmark of the Cambridge Philosophical Society, whose history has remained largely, indeed unjustly, neglected. Now, thanks to the forensic skills of Susannah Gibson, the Society is placed in a remarkable new light, as both witness and participant in those momentous events that revealed entirely new worlds, those of organic evolution, deep time, and fundamental physics. So much is apparent from the Society's early days, when reports were received from Charles Darwin on his world-changing voyage, George Biddell Airy lectured on subjects as disparate as eye defects and pendulums, James Clerk Maxwell first addressed the Society at the age of twenty-two, J.J. Thomson spoke on the still mysterious cathode rays, and, to a packed house, Arthur Eddington outlined the sensational confirmation of Einstein's theories of space-time as the result of observations of how starlight was seen to be bent during a solar eclipse. This interval of little more than a century led to profound changes in our perspective. But it also opened some terrible possibilities. So it was that, as war spread across Europe, Rudolph Peierls, publishing in the Society's *Proceedings*, presented calculations that, with further refinement, led to the inexorable conclusion that a mass of enriched uranium smaller than a tennis ball could, in principle, annihilate a city.

From its establishment in 1819, the inspiration of its founders, John Stevens Henslow and Adam Sedgwick (along with Edward Daniel Clarke), was reflected in the Society, which was driven by a deep curiosity as to why the world is as it is, and how, with such knowledge, it might then be improved. With hindsight, this path to our modern world, where astounding discoveries are almost commonplace, reads as simple narrative—but it was not so. Susannah is adept at showing how intricate and braided the actual story was, and, as importantly, how the situation both in the University of Cambridge and across the country was, on occasion, very far from propitious.

Near the end of her book, Susannah remarks how the 'Society has [now] become just a small part of the vast landscape of Cambridge science—and that is the true mark of its success'; this is high praise indeed. Success, as is so often the case with English enterprises, lies in the genius of reinvention. In the case of the Society, it was sometimes involuntary. In the 1860s, looming financial catastrophe, largely the result of the egregious Mr Crouch, compelled a migration from its original home to a site that fortuitously made the Society a neighbour of the new Cavendish Laboratory. Here, as was repeatedly the case during its two-hundred-year history, the Society served as a vital catalyst in the epic developments of Cambridge science.

Susannah's account reveals science as not only exhilarating but dotted with events and characters. Think of the crowds watching co-founder Clarke launching a hot-air balloon from Jesus College, as well as his arranging volcanic eruptions in his packed lectures. And what about Michael Foster's distillation of noisome 'excrementitious fluids' in his laboratory that, strange to report, won him valuable bench space at the expense of his then neighbour, the Plumian Professor of Astronomy? In the wider field, the Society also repeatedly gained territory for the advancement of Cambridge science. Thus, it helped to win space for a new University Botanic Garden, provided the seedcorn for the now immense collections devoted to natural history, and, as importantly, provided the nucleus for the world's finest scientific library (where, until its shameful closure, I spent endless hours pursuing my own research into evolution). So too, while the mantle of experimental expertise necessarily passed to the University departments, the Society pioneered the study of anthropometrics, thereby laying the groundwork for statistical rigour in biology.

Nor has the Society rested on its laurels. Recently, it inaugurated a scheme of Henslow Fellowships for gifted young scientists, and Susannah mentions the ongoing contributions of one recent and one current Fellow, Alex Liu and Emily Mitchell, in their studies of the dawn of animal life. So, the Society looks forward with confidence, but so too it looks back at its own initiation in 1819 as a turning point in Cambridge science. It may be a cliché that all scientists stand on the shoulders of giants (though less often is it remembered that this conceit goes back to at least the time of Bernard of Chartres), but it is no accident that the official seal of the Society is in the form of Roubiliac's celebrated statue of Isaac Newton. Located in the antechapel of Trinity College, this foundation (where a memorial to Adam Sedgwick is also to be found) as well as my own College of St John's (once home to John Henslow) played their own parts in the nurturing of the Society. Here, we join Susannah in saluting a remarkable institution which did, indeed, play its small part in the making of Cambridge science, but played it very well.

**SIMON CONWAY MORRIS, FRS**
*President of the Cambridge Philosophical Society* (2018–2019)

# PREFACE

The Antikythera mechanism is one of the most extraordinary objects to survive from the world of ancient Greece. On this smallish fragment of metal, heavily corroded from centuries spent beneath the waves, the shape of a cross enclosed within a circle immediately catches the eye. It is clearly the work of a skilled craftsman; but what is it? It has been an object of fascination since its discovery in 1902, and modern scientific testing, using x-ray tomography and high-resolution surface scanning, have confirmed a long-held suspicion that the circles that appear over and over again in the body of the object were once an intricate system of gear wheels. There were more than thirty of these precisely cut brass wheels, and they were used to model the motions of the heavens. Whoever made this extraordinary device must have had a detailed knowledge of the movements of the celestial bodies, as well as astounding technical skill: the sophistication of the Antikythera mechanism was not matched for many centuries, when the first mechanical astronomical clocks were developed in China, and it was many more centuries before the technology reappeared in Europe.

The first meeting I ever attended of the Cambridge Philosophical Society was about this ancient machine. A packed lecture hall hung on every word of Mike Edmunds, Emeritus Professor of Astrophysics at the University of Cardiff, as he revealed the secrets of the device to us. His talk wove together archaeology, materials science, history, and cosmology. The Antikythera mechanism fundamentally changes the way we understand ancient technology; it can give us insights into the workshop of the Greek craftsman and into the mind of the Greek astronomer. Edmunds's study of it not only makes use of cutting-edge analytical techniques; it also provides an ideal example for explaining those techniques to a lay audience. It is, in short, the perfect topic for discussion at a meeting of a philosophical society.

But what exactly is a philosophical society? Should an audience member be surprised to hear about metallurgical imaging techniques rather than Plato's idealism or Kant's metaphysics? It has long been a refrain amongst the Society's staff when talking to prospective members that 'the Society is not involved in philosophy, but in natural philosophy, which is to say science'. That is a neat summation of a series of complex terms which have had multiple different meanings throughout history; perhaps too neat?

The phrase 'natural philosophy' has ancient roots. It is often associated with Aristotle and his holistic study of the natural world. But, as the centuries wore on and Aristotle's work was reimagined in line with new Christian doctrines, natural philosophy grew into something else: a way of understanding the natural world as it was created by God. The subjects studied by natural philosophers were incredibly varied and included the sciences of motion and mechanics, the properties and qualities of matter, the art of astronomy, and more esoteric notions such as change, chance, and causes.[1]

By the late seventeenth century, natural philosophy was evolving again. That is when Isaac Newton, working in Cambridge, published his most famous book—*Philosophiæ naturalis principia mathematica*, or, *The mathematical principles of natural philosophy*. This book did something quite unexpected: it merged abstract philosophical study with the precision of mathematics. For Newton, natural philosophy was still essentially a religious activity, one complementary to his strongly held views on theology, but the union of philosophy and mathematics would have a profound effect on the field. It made individual sciences such as astronomy broaden out, because now they did not simply deal with mathematical calculations, but were permitted to seek out the underlying causes of the phenomena they addressed; and it allowed philosophical arts such as studies of motion to become more precise as they turned from qualitative to quantitative.[2]

Some people speak of this period at the end of the seventeenth century as a 'scientific revolution', the point at which natural philosophy ceased and modern science began. But, in reality, natural philosophy lived on for many years. In the eighteenth century—the century of Enlightenment—natural philosophy remained a broad art. More than that, it began to draw in a broader spectrum of participants. The Enlightenment ideal of egalitarianism meant that ordinary people were being exposed to aspects of culture (be they artistic or scientific) that would previously have been closed to them. The seventeenth century had seen the creation of a new entity for the elites—the scientific society—and now, in the eighteenth century, there was growing demand for similar societies to cater to the general populace. In London, the Royal Society (founded in 1660) had been the centre of elite natural philosophy for over a hundred years but, from the late eighteenth century, two new kinds of organization began to grow up. First, based in London, were the specialist scientific institutions, which catered for single subjects like natural history, geology, and astronomy. Then there were the provincial societies.

These provincial societies, usually based in industrial towns such as Manchester, or fashionable spa towns such as Bath, often styled themselves as 'literary and philosophical', reflecting the societies' intention to introduce their members to a wide span of knowledge across the arts and sciences. But that was not all the societies did—the original concept of natural philosophy as a sort of spiritual or moral experience was as relevant as ever: for example, the founders of the Literary and Philosophical Society in the northern town of Halifax, a centre of wool manufacture, hoped that they would encourage 'a taste for scientific and other liberal pursuits, which may serve to elevate the intellectual and moral character, and thus to promote . . . the best interests of mankind'.[3] The societies were an instant hit and audiences flocked to their grand new lecture halls to hear about all the marvels of the era: electricity and steam power; the powerful new machines of the industrial age; advances in medicine; and explorations of new lands; but also about poetry, music, and fine paintings. Membership of the provincial societies grew rapidly and, alongside their new lecture halls, the societies built museums, libraries, and classrooms. The societies were true Enlightenment institutions.

The Cambridge Philosophical Society, which was founded in 1819, is part of this same tradition, and yet it stands apart. For one thing, it was the only such society to be founded in an English university town and, though it was officially independent, the Society and the University had an intimate relationship. For another, membership was only open to former students of the University of Cambridge, which meant (as we shall see) that every single fellow of the Society had a thorough grounding in mathematics, for no one was permitted to graduate without knowing his Newton. Perhaps this is what distinguished the Cambridge Philosophical Society from its sister societies across the country: while the provincial societies devoted the main part of their energies to natural history and the more descriptive sciences, the Cambridge Philosophical Society became increasingly mathematical in its early years. This meant that it aligned itself with the nineteenth-century trends of specialization and mathematization, especially in the physical sciences. While members of the provincial societies favoured talks on travel, were most likely to borrow novels from their libraries, and often focused on education over original research, the fellows of the Cambridge Philosophical Society relished talks on the calculus, built up one of the most impressive scientific libraries in the country, and actively sought to create *new* knowledge.[4] The Cambridge Philosophical Society was perfectly poised to be part of the campaign to create what we now think of as 'science'.

This history has been conjured from a small archive the Society has preserved, which was catalogued for the first time in 2014 as part of the preparations for the Society's bicentenary celebrations in 2019.[5] I am the first historian to systematically examine the archive since it was catalogued.[6] Not knowing where to begin when faced with a wall of identical grey archival boxes, I opened the uppermost box and found an old red Oxo tin, rusting around the edges and declaring itself the bringer of 'meatier meal times'. Tied to the outside of the box was a soot-blackened metal ladle, while inside rolled a handful of red wax balls, which were the size and shape of holly berries, along with the stump of a candle, a few spent matches, and the old seal of the Society. These simple objects instantly transport the viewer back to the early days of the Society. They are a rare material link to its past and a first step to unravelling its complex history.

From that archive, sometimes frustratingly incomplete, I have tried to tell a story which has often been lost in the bewilderingly rich history of Cambridge. It is a story that is much bigger than it may first appear, for understanding the history of the Cambridge Philosophical Society gives new insights into the history of science both in the city and, more widely, in Britain, Europe, and the rest of the world. It gives us a new understanding of how Cambridge came to be what it is and of how science developed from being a peripheral activity undertaken by a small number of wealthy gentlemen to being an enormously well-funded activity that can affect every aspect of our lives.

# 1

# THE FENLAND PHILOSOPHERS

From the tall chalk cliffs south of Alum Bay, you can see clear across the water to the northern shore, where coloured bands of stone stripe vertically above the beach. Reds, pinks, greys, and browns, veined occasionally with white: these ancient rocks jut out harshly beneath the green curves of the land above. Like a skeleton exposed, the sheer cliffs reveal the hidden workings that lie under ground, usually out of sight. You cannot understand the skin of the land without understanding its bones, its muscles. Scrambling down the rocks, one bright April day in 1819, two young geologists hoped to do just that—understand.

Adam Sedgwick and his friend John Stevens Henslow had come to the Isle of Wight to see these extraordinary vertical beds (see Figures 1 and 2). At the base of the cliffs, in the curve of the bay where the chalk gives way suddenly to darker rocks, they used their hammers to crack open the earth. The rocks here were rich in clay, and rich in fossils. Sedgwick and Henslow found ancient oysters, the long, elegant shells of *pinna*, two kinds of bivalve, and one of univalve. They examined the specimens carefully, chipping away at the matrix that had held them for so long, turning them over in their hands, comparing them to similar species. But what did these fossils mean? The same collection of fossilized shells was known to exist in another kind of rock, the so-called London Clay. Was this the same rock, far from the other known outcrops, or was it mere coincidence that the fossils matched? Sedgwick and Henslow tried to picture the world as it had been when the rocks were formed—perhaps millions of years earlier: they conjured a tranquil sea that was shallow, warm, and filled

**Figure 1** Adam Sedgwick in his 40s, wearing full academical dress in his role as Woodwardian Professor (1833).

**Figure 2** John Stevens Henslow in his 50s when he was Professor of Botany (1849).

with living things. Now, the only evidence for that earlier world was the rock face in front of them with its scattered fragments of preserved shells.[1]

This was Henslow's first foray into field geology, and Sedgwick's second. Geology was a young science; born in the second half of the eighteenth century, it was growing quickly in the first decades of the nineteenth. This new science aimed to unlock the secrets of the past by examining the present-day world. If Sedgwick and Henslow could understand the make-up of these rocks, they might be able to figure out how they had been created, what the landscape had been like at the time of their creation, and whether there had ever been a physical link between them and the London Clay. Their fieldwork was slow and methodical: partly to ensure that their data would be reliable and partly because both were just learning the methods of their new science. Henslow had lately completed an undergraduate degree at the University of Cambridge and was determined to learn as much about the natural sciences as possible.[2] Sedgwick had a more pressing reason for learning the techniques of geology: he had recently been appointed Woodwardian Professor of Geology at Cambridge but knew next to nothing of the subject.

Sedgwick was an intelligent and able man. He had been born in 1785 in the village of Dent, which was in the northern English county of Yorkshire and was where his father Richard Sedgwick was the local vicar and schoolmaster. Adam studied under his father until he was 16, and then attended the grammar school in nearby Sedbergh before finally spending a summer being tutored by the legendary John Dawson. Dawson had begun life as a shepherd in the Yorkshire Dales but somehow developed an interest in mathematics. He used wool spun from his flock to knit stockings, and he used the money earned from selling these to buy mathematical textbooks. Dawson was an entirely self-taught mathematician but he soon became competent enough to start teaching pupils of his own. Richard Sedgwick had been one of Dawson's earliest students, and so it was natural that Richard sent his son Adam to him to learn the mathematics needed to earn a place at Cambridge. Dawson had, since the 1780s, come to specialize in preparing boys for their studies at Cambridge, and a staggering number of his former pupils went on to be awarded top marks in the University's examinations.[3]

After a summer studying under Dawson, Adam Sedgwick won a place at Cambridge. In the early nineteenth century, Cambridge University was primarily a religious training ground which supplied clergy to the Church of England, though it also offered a broad education to boys who came from reasonably wealthy families or were clever enough to win scholarships. There were two parts to the system: the colleges (seventeen of them when Sedgwick arrived) which were autonomous bodies run by fellows with responsibility for the education, bodily welfare, and spiritual guidance of their students; and the University, which was a small administrative body mostly concerned with examinations and the awarding of degrees. At the time, there were only two universities in England: Cambridge and Oxford (compared to five in Scotland). Both of the English universities were conservative Anglican institutions that were closed to most of society: women, Catholics, Jews, and members of the dissenting Protestant churches were barred from entry.

Sedgwick was accepted into Trinity College in 1804 as a sizar—the name given to a poorer student who paid reduced fees in return for performing chores around the college. He travelled the couple of hundred miles from the rolling green hills of Yorkshire to the vast flatness of Cambridgeshire in a 'six-inside' stage coach, a journey that took three days and two nights and which Sedgwick later recalled as being utterly dismal.[4] This was the furthest Sedgwick had ever travelled from home. When he reached Cambridge and entered Trinity College—the largest, wealthiest, and most powerful of the colleges that made up the University—he found a world unlike anything he had known before. The black-gowned figures of Cambridge, the strange jargon they used, and the countless unspoken rules were a mystery to him. Many of his fellow students came from wealthy and privileged families; they had moved in fashionable circles, they had travelled, and they had cultivated the manners of gentlemen. For affluent students, Cambridge could be a place of great decadence: 'wine-parties' were popular student events at which 'pine-apples and preserved fruits of all sorts, and ices in varied columns' were served and 'frivolous youths seem to vie with one another in the multiplicity of their wines. Champagne and claret are now considered almost indispensable.'[5] To such students, Sedgwick appeared uncouth and provincial: he wore out-of-date clothes, displayed unsophisticated manners, and spoke with a thick

Yorkshire accent. But he had a quick mind and a boyish animation; he charmed his peers and soon found himself settling into college life.

Sedgwick did well academically. He was placed in the top few students in the summer examination of his first year, something which required in-depth knowledge of the classics as well as of mathematics. The examination was a *viva voce*—a spoken test—held in front of the master and senior fellows of the college in Trinity's great hall. Lord Byron, a contemporary of Sedgwick's at Trinity College, remembered the particular terror of this experience:

> High in the midst, surrounded by his peers,
> MAGNUS his ample front sublime uprears;
> Plac'd on his chair of state, he seems a God,
> While Sophs and Freshmen tremble at his nod.
> As all around sit wrapt in speechless gloom,
> His voice, in thunder, shakes the sounding dome;
> Denouncing dire reproach to luckless fools,
> Unskill'd to plod in mathematic rules.[6]

Sedgwick continued to impress his teachers over the next few years—even a bout of typhoid fever in his second year did not derail his progress. He devoted himself wholeheartedly to his studies, ignoring the many distractions open to an undergraduate, as a letter from his dear friend William Ainger shows:

> How possibly can you, deeply immersed as you are in all the sublimities of Mathematical Science, take any interest in the grovelling concerns of one who, since he left you, has merely been scampering about the Fens in order to get rid of time? In truth, Sedgwick, had I anything more important to acquaint you with, I would not presume to inform you that last Tuesday sen'night I was capering at Wisbeach to the sound of a Fiddle, and that the deepest speculation in which I have engaged, has been an attempt to learn the character of an eccentric girl whom you may recollect I once mentioned as the only female likely to make an impression on your iron heart. Positively I think her as great an oddity as yourself; and surely this is saying enough to excite any one's curiosity who is not so much infected with the Mathematical Mania as to scorn everything which is lower than the stars.[7]

Sedgwick's mathematical mania drove him to stay diligently in his rooms, shunning the delights of fiddle music and young women. Instead, he rose each morning at 5 a.m. and worked late into the night, reading the assigned texts until he was overtaken by sleep—sleep that was disturbed by 'the most horrid dreams you have the power of conceiving' as mathematical symbols and philosophical concepts danced through his exhausted mind.[8] But Sedgwick's sleepless nights paid off when he was awarded a scholarship in his third year.

In 1808, Sedgwick took his final examinations—known in Cambridge as the Tripos, after the three-legged stool on which students had traditionally sat while undertaking them—and was placed fifth in the University. Then he faced a decision: to try for one of the small number of fellowships in Cambridge; to train for the bar or the church; or to return to rural Yorkshire to become a schoolmaster. Sedgwick had little real interest in legal matters, struggled to apply himself to theological studies, and, having tasted the delights of university learning, could not bear the idea of returning to the staid life of Dent, and so he determined to become a fellow.[9] After two years sustaining himself in Cambridge by taking on private students, Sedgwick finally won a fellowship at Trinity College in 1810.

The fellowship gave Sedgwick an income, rooms in college, and security, but it also deprived him of much. College fellows at that time could not marry, and so Sedgwick committed himself to a life of celibacy for as long as he held the fellowship. He joked that 'marriage may be all well enough when a man is on his last legs, but you may depend on it that to be linked to a wife is to be linked to misery. From the horrid estate of matrimony I hope long to be delivered.'[10] Yet, Sedgwick did once come close to marrying. He still dwelt upon his former sweetheart as he neared his seventieth birthday, reminiscing to his niece how once in his youth 'I was a dancing-man, and I fell three-quarters in love; but, as you know, did not put my head through love's noose'.[11]

But Sedgwick's fellowship kept him busy, and he had little time to think of love. In-between the daily routines of college life—attendance at chapel, meals in the draughty hall, tea in the combination (i.e. common) room—his duties mainly involved tutoring undergraduates. In Cambridge at this time, mathematics formed the core of the curriculum, so Sedgwick spent his days in cold stone buildings, wrapped in an academic gown, drilling sums into ungrateful students. He found the work burdensome and had to work long hours to keep up with the demands made upon him. His health began to break down under the strain of the work and of his own despondency. But things started to look up from 1815, when Sedgwick was appointed Assistant Tutor at Trinity. The position came with a small pay rise which allowed Sedgwick to spend the University vacation travelling on the continent. For four months, he toured through France, Switzerland, Germany, and Holland. Much of Europe had been closed off to British travellers during the Napoleonic Wars that had raged until 1815, and Sedgwick leapt at the chance to visit these lands about which he had heard so much.

Sedgwick's notes from his first visit to continental Europe are terse, and serve mainly to highlight his dislike of the French: 'the beautiful, gay, and profligate city of Paris is a noble capital,' he admitted, 'but the people are so abominable and detestable that there can be no peace for Europe if they are not chained down as slaves, or exterminated as wild beasts.' Still, Sedgwick's mood improved as he crossed into Switzerland and beheld the majestic Alps for the first time. The mountains left him stunned; their beauty exceeded anything he had previously imagined and their 'exquisite perfection' would remain in his memory long after he had returned from their lofty peaks. He travelled through mountains and forests, saw glaciers for the first time, and picked his way across new lands with only the sun and stars to guide him.[12]

The trip was a revelation for Sedgwick and, after his European adventures, his health improved greatly. He realized that he needed more than the repetitive grind of mathematics tutorials in insular Cambridge. He began to spend more time outdoors, returning to Yorkshire whenever he could; he began to think of studying subjects beyond mathematics. He wished to travel, to see new things, and to stimulate his intellect. And so, when the Woodwardian Chair of Geology became available in 1818, Sedgwick decided to apply for the post.

The Woodwardian Chair had been established in Cambridge in 1728 by John Woodward, a physician who was fascinated by fossils. Woodward had written some of the earliest manuals about how to collect and preserve geological specimens; indeed, many of his instructions are still followed by geologists today. Woodward had used the specimens he gathered as evidence to back his theories of how the earth had formed, trying to link biblical accounts of a flood with physical evidence from the rocks. While many seventeenth- and eighteenth-century naturalists tried to create such theories, their theories were not necessarily rooted in the real world and few engaged in active fieldwork as Woodward did. Upon his death, Woodward left his collections, meticulously arranged in cabinets, to the University of Cambridge[13] (the 'Woodwardian Cabinet'; see Plate 1). He also left enough money to endow a professorship in geology, the first of its kind in Britain. The professor's duties as outlined in Woodward's will would be slight: the professor had to look after the collection and give four lectures each year.

Ever since Adam Sedgwick had arrived in Cambridge, the Woodwardian Chair had been held by the same man: John Hailstone. Hailstone had studied briefly under the famous German geologist Abraham Werner; back in Cambridge, he became an active collector of mineralogical specimens and did much to expand Woodward's original collection. The previous Woodwardian chairs had not been known for their lectures. This, explained Hailstone, was because geology, unlike botany and zoology, 'consisted of a few scattered unconnected facts, incapable of being digested into a system. And what is incapable of being reduced to a system cannot be made the subject of public instruction.'[14] Though he had intentions to rectify this, he seems not to have lectured very widely or regularly during his thirty years in the post, but he did share his geological knowledge through displays of his specimens and tours of the Woodwardian Cabinet. It came as a surprise to many when Hailstone announced his resignation of the post in 1818. The reason? He wished to marry, and the terms of Woodward's bequest demanded that the professor be a bachelor.[15]

Following his wedding, Hailstone moved with his bride Mary to the vicarage of Trumpington, just outside Cambridge, leaving the geology chair unfilled. This was just the opportunity Adam Sedgwick had been waiting for. He was heard to complain that he was 'heartily sick' of the tedious work of tutoring undergraduates in mathematics; he needed a new challenge. When he learned that Hailstone was to step down, he wrote excitedly to a friend of his hopes of attaining the post: 'if I succeed I shall have a motive for *active* exertion in a way which will promote my intellectual improvement, and I hope make me a happy and useful member of society.'[16]

Several candidates put themselves forward for the chair. One, George Cornelius Gorham, a fellow of Queens' College, had even studied geology and was known to have a good working knowledge of the subject. But Sedgwick, hailing from a larger and more powerful college, won the vote—a triumph of 'influence against qualification' according to Gorham who also complained, probably accurately, that some had voted against him because he was a Methodist.[17] The fact that Sedgwick knew little about geology seemed hardly to matter to those who voted for him.

For what it was worth, Sedgwick had sat in on a few lectures by the mineralogy professor, Edward Daniel Clarke, and read a little about geological theories, but really his appointment to the Woodwardian Chair was on the strength of his character, his connections, and his general abilities.[18] This was not unusual at the time, and the University seemed little concerned by its new professor's lack of qualifications. Professors were not necessarily expected to lecture on their subjects. But Sedgwick planned a different approach: he would be a practical geologist; he would lecture weekly; he would lead his students on field trips; and he would collect widely and expand the University's small geological museum.

Sedgwick had had no choice but to devote himself to mathematics in the first years of his career: it was the only way to win scholarships and fellowships at the University. Sedgwick's family was not wealthy enough to support him in an independent career, so he had used mathematics as a means to an end. But the toll it had taken on his health and mental well-being was a heavy price to pay. The many hours cooped up inside, reading late into the night by dim candlelight, had left Sedgwick craving an outdoor pursuit. Accepting the Woodwardian Chair meant a pay cut for Sedgwick, but he didn't mind so long as he could free himself from the drudgery of teaching undergraduates the same mathematical problems over and over, do something that stimulated his mind, and spend more time outside the lecture room.[19]

The Woodwardian Chair had never had such an active incumbent. As soon as he was appointed in 1818, Sedgwick began his research. He spent that summer in the lead mines of Derbyshire, the copper mines of Staffordshire, and the salt mines of Cheshire. Fearlessly, he descended into the pits, lowering himself down the precarious wooden stemples until he was hundreds of feet below the surface of the earth. He followed the mineral veins and examined the rock strata; he began to learn the secrets of the land. And he launched himself into a frenzy of collecting, so that, when he returned to Cambridge in the autumn, every table and chair in his room overflowed with the spoils of the summer.[20] Sedgwick's first lectures began to take shape. It was at this time that he planned that trip to the Isle of Wight with his friend John Stevens Henslow, determined to see the reported vertical beds at Alum Bay for himself.

Sedgwick and Henslow had met some years earlier, through a mutual friend, John George Shaw-Lefevre, who had come up to Cambridge in the same year as Henslow and had studied at Trinity College, where Sedgwick was a tutor. Though ten years apart in age, Sedgwick and Henslow quickly developed a warm friendship. In January 1819, Sedgwick began his first lecture course in Cambridge, a lecture course which Henslow (though no longer a student) eagerly attended. He was greatly impressed by Sedgwick's eloquence and zeal, and much impressed by the science of geology which Sedgwick carefully and engagingly explained to his audience (an audience made up not just of members of the University but also of townspeople, including women, whom Sedgwick enthusiastically welcomed into the lecture room). Henslow, who already had an interest in mineralogy, found the young science of geology thrilling—revealing, as it did, aeons of history in the shape of a landscape.[21] And so, at the end of the lecture course, when Sedgwick proposed to Henslow that he join him on a field trip to the Isle of Wight, Henslow jumped at the chance (see Figure 3).

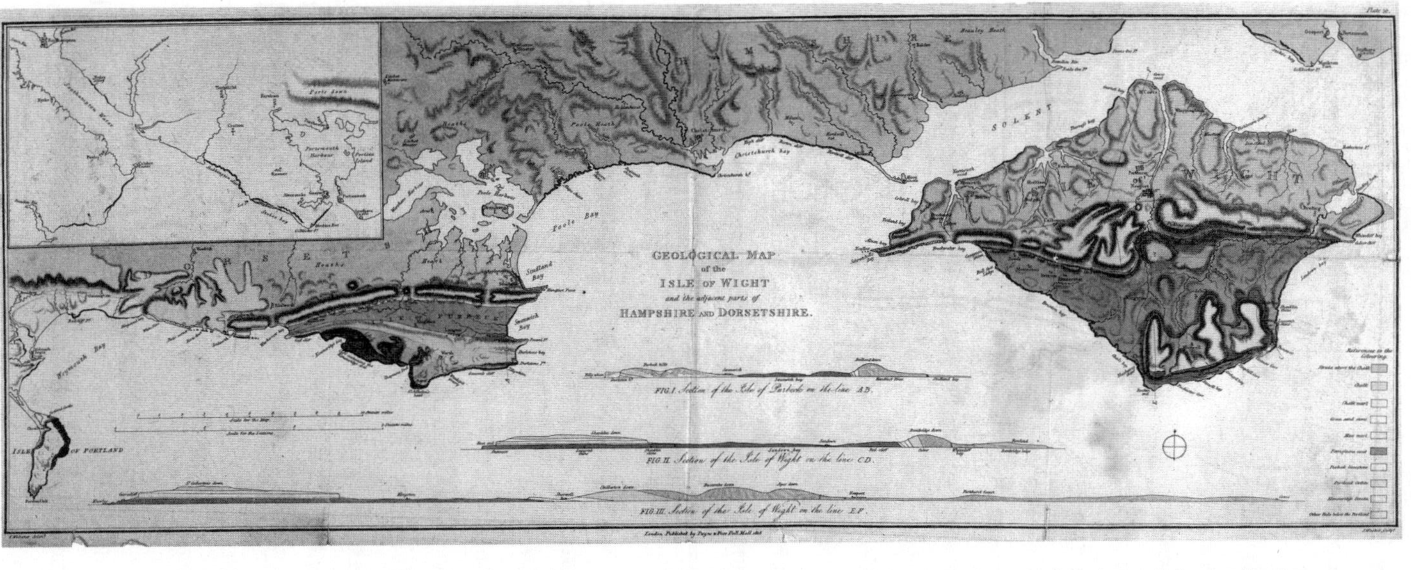


**Figure 3** A geological map of the Isle of Wight belonging to John Stevens Henslow. Henslow and Sedgwick brought this map on their field trip of 1819.

The two friends roved over the whole island, hammers in hand and collecting bags at the ready. Each day they rose early and worked methodically through the strata of the island, examining the fossils of the iron sands, the clays, and the chalks. The fresh bright days seemed to sharpen their senses, and their exertions invigorated them. They travelled many miles on foot, clambering up and down outcrops and cliffs, and almost getting caught by incoming tides as they lost themselves in their work. They uncovered dozens of species of ancient shellfish, zoophytes, corallines, and sea plants. And it was on that trip—stirred perhaps by the potential of the new science of geology, the possibilities open to a new professor or a freshly minted graduate, and the knowledge they were creating and would create—that Henslow and Sedgwick began to realize the inadequacies of their Cambridge world. They were undertaking new research, working at the cutting edge of a young science, and refining the techniques of the discipline—but who in Cambridge would care? Lectures were poorly attended, the geology museum was cramped and little-visited, there was no University journal in which they could publish their research, and a disproportionate amount of energy was expended on out-of-date mathematics.[22] It was on the Isle of Wight—perhaps on that beach at Alum Bay, as the rocks revealed themselves beneath the hammer, or perhaps in an inn that evening, as they were poring over the delicate fossils they had unearthed—that Sedgwick and Henslow dreamed up a new forum where they could show off their research: a scientific society for Cambridge.[23]

Such societies were booming in the late eighteenth and early nineteenth centuries. In London, the venerable Royal Society was joined by a range of new, more specialized societies: the Linnean Society (for natural history) in 1788, the Geological Society in 1807, and the Astronomical Society in 1820. There was also the Royal Institution, which was founded in 1799 and which encouraged new inventions and popularized science for a wide audience. Outside London, the more wide-ranging provincial 'literary and philosophical' societies grew up, most frequently in industrial towns: Manchester's was founded in 1781, followed by the ones in Derby (1783),

Newcastle-upon-Tyne (1793), Birmingham (1800), Glasgow (1802), Liverpool (1812), Plymouth (1812), Leeds (1818), and York, Sheffield, Whitby, and Hull (all in 1822).[24] Many of these societies were in Yorkshire, and perhaps it is significant that Sedgwick was a Yorkshireman and made regular trips to that county—he must have seen the great enthusiasm for the philosophical societies at first hand. There had once been a very short-lived scientific society in Cambridge from 1784 to 1786 but, because of the deaths of some of the instigators and due to its 'not being adequately supported', it was dissolved soon after its foundation.[25]

The London societies were generally reserved for the scientific elite, expensive to join, narrowly focused, and aimed at the higher end of society; on the other hand, the provincial societies were mostly made up of the middle classes, extremely broad in scope, had much more modest membership fees, and required no prior knowledge of any of the sciences. Neither model was right for Cambridge. Henslow and Sedgwick considered several options: at first they thought of a corresponding society, but this idea was soon abandoned. They contemplated a society aimed primarily at students, as a way to introduce a little natural history into their mathematical lives, but that too was rejected.[26]

They left the Isle of Wight at the end of their field trip with the germ of an idea, though without any fixed plans. Easter fell in mid-April that year, so the Easter term was short. Sedgwick was preoccupied with his lecture course, which left him little time to dedicate to setting up a new society. But, later that spring and summer, Henslow and Sedgwick began to write to their friends and colleagues in Cambridge and beyond, outlining their ideas, seeking support, and refining the details.[27] Sedgwick's letters were sent from Suffolk, Somerset, Devon, and Cornwall, where he continued on his mission to learn as much practical geology as possible before the next term began.[28] He was fascinated by the new landscapes he saw, and also by their inhabitants. From the Mendip and Quantock hills of Somerset, he wrote:

> The country I have just been describing wants some of the grander features, but in beauty, luxuriance, and variety, yields to none. The rugged

> cliffs which rise perpendicularly on both sides of the Bristol Channel are in many places exquisitely contrasted with the fine lawns and rich foliage which go sweeping down to the very edge of the water. As for the people of Somersetshire, they seem a mighty stupid good sort of people, who have not wit enough to cheat a stranger. The men get drunk with cider, and the women make clotted cream.[29]

Over the following decades, Sedgwick would tour the whole country in this way—rambling across the counties, observing the contours of the land, collecting fossils, and watching for the strata as they rose and fell below the surface of the earth.

Henslow, meanwhile, had also caught the geology bug. He spent the summer of 1819 in the Isle of Man, where, by happy coincidence, a local brewer had just discovered the fossilized skeleton of a giant Irish elk in a marl pit. The village blacksmith, a man named Thomas Kewish, used his knowledge of horses to reconstruct the skeleton—a feat which Henslow gleefully described to Sedgwick: 'you know I am not much given to the marvellous, but I really think I never saw a more magnificent sight of the kind in my life, and doubt if the Petersburg Mammoth would surpass it...the fellow has really put it together with very great ingenuity.'[30] Henslow was so impressed that he tried to buy the enormous beast for the Woodwardian Cabinet, but the local duke had claimed it for his own and, despite Kewish's attempt to smuggle it into hiding, the duke later gave it to the museum at Edinburgh University.

After a happy summer of geologizing, the autumn closed in and Sedgwick and Henslow had to conclude their tours. In October 1819, the two men returned to Cambridge for the start of a new term. The low, flat fens that surrounded the city made a stark contrast to the dramatic cliffs of the Isle of Wight, but the two friends had not forgotten the vision they had conceived there. As the college and university men reassembled in Cambridge, Sedgwick and Henslow could take the next step in realizing their ambition.

They called upon Edward Daniel Clarke (see Figure 4). A former fellow of Jesus College, Clarke was an eccentric collector of antiquities, an inveterate traveller, and a sometime lecturer in mineralogy. He had once dedicated

**Figure 4** Edward Daniel Clarke (c. 1800).

an entire undergraduate term to the construction of a hot-air balloon—said to be 'magnificent in its size, and splendid in its decorations'.[31] He displayed it in the hall of Jesus College before launching it from a cloistered court in front of a vast crowd of spectators. He spent an almost equal amount of time constructing an orrery (a mechanical model of the solar system) so that he could deliver a lecture course on astronomy to an audience of one: his sister.[32] While travelling in Greece in 1800, he had bribed an official into allowing him to take ownership of a remarkable statue of the goddess Ceres. The statue, which Clarke claimed was unwanted—'lying in a dunghill, buried to her ears'—weighed several tonnes and he had to construct a special machine and enlist the help of sixty peasants to move her from her hillside location to his ship.[33] It can be seen today in the Fitzwilliam Museum.[34] Clarke was a man much given to ambitious schemes, and one who knew how to get things done.

On returning from his travels in Scandinavia, Russia, the Black Sea, Turkey, and the Mediterranean, Clarke was eager to share his new-found knowledge in Cambridge. His time abroad had convinced him that English science, and especially his favourite science of mineralogy, was in a low state compared to its continental neighbours.[35] He suggested to the

University that he might give some lectures on mineralogy, requested a place to conduct the lectures, and dealt diplomatically with concerns that he was stepping on the Woodwardian Professor's toes. The University assented and, in March 1807, Clarke began his series of lectures. He wrote jubilantly to friends of his success:

> Imagine me in a grand room, before all the University, tutors and all! All my minerals around me and models of crystals....
>
> I never came off with such flying colours in my life. I quitted my papers and spoke extempore. There was not room for them all to sit. Above two hundred persons were in the room. I worked myself into a passion with the subject, and so all my terror vanished. I wish you could have seen the table covered with beautiful models for the lecture.[36]

His audience agreed that Clarke's delivery was 'a master-piece of didactic eloquence' and that 'from every stone, as he handled it and described its qualities—from the diamond, through a world of crystals, quartz, limestones, granites, &c. down to the common pebble which the boys pelt with in the streets, would spring some pieces of pleasantry'.[37] Alongside the models of crystals and the samples that Clarke had collected on his many travels sat a cork model of Mount Vesuvius that could spew out lava. Clarke had constructed this while in Naples after multiple ascents of the volcano; it was truly a sight to behold for audiences more used to staid mathematical proofs than dramatic demonstrations. The lectures were so successful and popular that the University made Clarke the first professor of mineralogy in 1808. He was a fine speaker, and his lively lectures drew in unusually large crowds.

It was in the lecture hall that Henslow and Sedgwick had first known Clarke, and thus it was no wonder that they were keen to enlist his support for their new society. They knew him as a dynamic man, full of energy, a lover of the sciences, and passionate about sharing knowledge and circulating new ideas. Their meeting with him in October 1819 transformed a vague notion into a solid plan. Clarke responded warmly to the suggestion of a scientific society for Cambridge and helped set the wheels in motion.[38]

Together, the three men arranged an open meeting which any graduate of the University with an interest in science was welcome to attend. They had notices printed and, on the last Saturday in October, they began to pin them up in the University buildings and circulate them amongst their colleagues. The meeting would take place at twelve o'clock on Tuesday 2 November in the lecture room beneath the University Library. Its purpose was to institute a Society for scientific communications.[39] The invitation was signed by thirty-three people. As well as Sedgwick, Henslow, and Clarke, the signatories included six heads of colleges, six professors, a librarian, and eleven tutors or assistant tutors.[40] Henslow and Sedgwick's summer-time campaign of drumming up support had clearly been successful.

At noon that Tuesday, a large crowd gathered in the lecture room. In 1819, the University's library and lecture rooms were located in what is now called the Old Schools, just to the north of King's College (see Figures 5 and 6). The lecture rooms, or 'schools', sat on the ground floor of a quadrangle, with the library above. The library contained not just

**Figure 5** The central building of this image once housed the University Library and was the site of the preliminary meeting of the Cambridge Philosophical Society. To the left is King's College Chapel, and to the right is Senate House (1845).

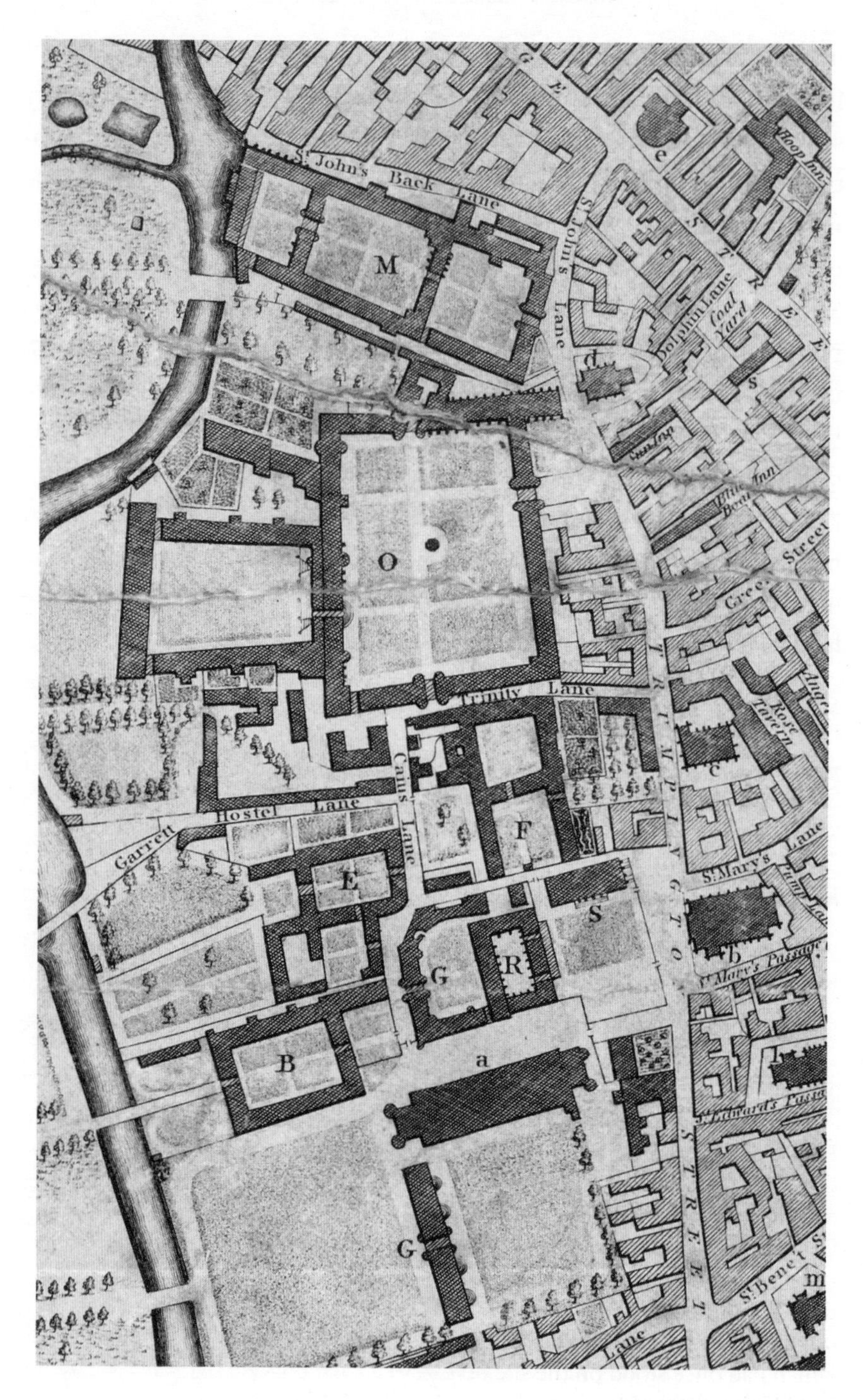
St. John's Back Lane
St. John's Lane
Dolphin Lane
Coal Yard
Hoop Inn
M
O
e
d
s
Green Street
Trinity Lane
Caius Lane
Hostel Lane
Garrett
Rose Tavern
St. Mary's
Lane
St. Mary's Passage
St. Edward's Passage
St. Benet St.
E
F
S
G
R
B
a
G
c
b
m
Lane

books, but curiosities such as mummies, Chinese idols, and death masks. In the court outside stood Clarke's statue of Ceres. The philosophy schools occupied the west side of the court, the divinity schools occupied the north side, and to the south were the schools of law and physic. In the north-west corner was the Woodwardian Museum, with its collection of ores, minerals, and shells.[41] Next to the museum, which was now Sedgwick's responsibility, was the Woodwardian Professor's room—a place that Sedgwick described as being 'small, damp, and ill-lighted, and utterly unfit for a residence or a lecture room'.[42] These ancient halls of learning had seen philosophy, divinity, law, and physic taught for many centuries. On that cold November afternoon of 1819, a newer sort of knowledge was being exalted, and a newer way of sharing it was being created. Those present spoke excitedly of what they might achieve: they would embrace the modern sciences of chemistry, mineralogy, geology, botany, and zoology; they would devise new experiments and hone their observational skills; they would create a new journal in which to publish their results; they would bring like-minded researchers together; and they would excite a spirit of curiosity amongst the students.[43]

Amid all this enthusiastic chatter, John Haviland was called upon to chair the meeting.[44] Haviland had been Professor of Anatomy from 1814 to 1817 and, as soon as he was appointed, he had begun to give regular lectures, something his predecessor had not done in the twenty years he held the chair. In 1817, Haviland became Regius Professor of Physic and taught clinical medicine at Cambridge's Addenbrooke's Hospital, made

**Figure 6** Central Cambridge c. 1800. The University Schools (lecture halls) and the University Library (also called the Public Library) are located at R; this was the site of the preliminary meeting of the Cambridge Philosophical Society. Other buildings of interest are: B. Clare College (then called Clare Hall); E. Trinity Hall; F. Gonville and Caius College; G. King's College; M. St John's College; O. Trinity College; S. Senate House; a. King's College Chapel; d. All Saints' Church (on this map, the lane behind the church is marked as Dolphin Lane after an inn which had once stood on the site here marked as a coal yard, it was also known as All Saints' Passage).

his lectures mandatory, and even tested students on the contents of those lectures.[45] He was the same age as Sedgwick, and clearly shared his passion for reform. He was a good choice to chair that first meeting: well established in the university, respected, yet young and vital enough to push for change. Once Haviland had called the meeting to order, they could proceed to the first order of business: a formal proposal by Adam Sedgwick that an organization be created as a focal point for scientific communication. Those present agreed unanimously: they would form a philosophical society—the Cambridge Philosophical Society—for the advancement of natural philosophy.[46]

Why did Cambridge need such a society? As one of the world's oldest and most famous universities, one might expect that the spirit of inquiry was very much alive there. But the University in the early nineteenth century was an intellectually cautious place. Inspiration for the curriculum came predominantly from the Bible, the classics, and the works of Isaac Newton (which had been published more than a hundred years earlier). The principal aim of the colleges was to fit young men for a life in the Church, not to train them as researchers. Universities had been instituted as places to store knowledge rather than as places to create new knowledge; as one historian has neatly put it, it was 'no more [a professor's] business to add to the existing body of knowledge than a librarian feels an obligation to write new books'.[47] Nor did universities see the knowledge they cultivated as being intended for any particularly practical purpose, as shown by the case of Frederick Thackeray—a surgeon at Addenbrooke's Hospital—who was refused a medical degree by the University because it was not seen as compatible with his hands-on career.[48] One would-be reformer summarized the situation like this:

> A young man passes from our public schools to the universities, ignorant almost of the elements of every branch of useful knowledge; and at these latter establishments, formed originally for instructing those who are intended for the clerical profession, classical and mathematical pursuits are nearly the sole objects proposed to the student's ambition.[49]

The undergraduates were encouraged to study traditional subjects by the fellows who taught them and by the University examiners. These fellows were almost equally conservative in their own studies. The Scottish natural philosopher David Brewster once commented in a letter to a friend: 'I find I have given offence to the *Cantabs* [Cambridge men] for saying that there is no person there carrying on a *train of original research.* Do you know of anybody there who is?'[50]

There were, in theory, a wide variety of subjects on offer in Cambridge. By 1819, there were University professors of divinity, civil law, physic, Hebrew, Greek, Arabic, mathematics, philosophy, music, chemistry, astronomy, experimental philosophy, anatomy, modern history, botany, geology, geometry, natural philosophy, English law, and mineralogy. But, in reality, very few of these professors lectured to undergraduates, and most of their subjects were not included in the degree examinations. Some students took degrees in medicine (though they had to complete most of their clinical training in London or Scotland, as the University only offered limited practical tuition) or law and pursued careers in those fields, but most took a general Bachelor of Arts (BA) degree and hoped for a career in the Church. Almost all college fellows and University professors were members of the clergy: Edward Daniel Clarke, Adam Sedgwick, and John Stevens Henslow were all ordained within a few years of graduating from the University, and the vast majority of early fellows of the Cambridge Philosophical Society were listed with the title 'Reverend'.

Until the mid-eighteenth century, the BA examination (the Tripos) had been reasonably broad in scope and required candidates to discourse on topics in Latin in the Senate House. From about the 1760s, some elements of the Tripos began to change: first, it became more mathematical in character; second, candidates for honours degrees were required to submit written answers; and, finally, it became far more competitive, with increasing emphasis on the ranking of candidates. The students ranked highest in mathematics were known as 'wranglers', and betting on the identity of the highest-ranked student—the senior wrangler—was becoming a popular Cambridge pastime.[51] The emphasis on mathematics was part of Newton's legacy to his *alma mater*. His key works the *Philosophiæ naturalis principia mathematica* (1687) and the *Opticks* (1704) demonstrated how mathematics and experiment could be used to understand the world as it had been created by God. To the minds of the University examiners, they were ideal texts for teaching concepts of divinity at the same time as providing material for written examinations. As more emphasis was placed on the ranking of candidates in strict order of marks, the examinations became more focused on the mathematical elements of Newton's work. By the end of the eighteenth century, one could not gain a degree from Cambridge without an intimate knowledge of Book I of Newton's *Principia* and the related 'mixed' mathematics (subjects like algebra, arithmetic, geometry, fluxional calculus, astronomy, mechanics, hydrostatics, and optics). These mathematical topics pushed almost all other subjects out of the Tripos.[52]

As the examinations set by the University became more mathematical, so did the teaching offered by the colleges. Mathematics tutorials had once consisted of students taking turns to state theorems and proofs, or to solve simple problems orally. Now, tutorials became more intense: long written proofs were drilled into students, and increasingly complex problems were posed to the more able men. The study of mathematics, and particularly applied or 'mixed' mathematics, sounds perhaps like an ideal base from which one could develop an understanding of the sciences, but that was not the point. Mathematics was seen as a tool for cultivating the rational mind; its study was part of a broader spiritual and moral education. The inclusion of mixed mathematics was designed to train the mind to remain calm and ordered when it encountered a problem more convoluted than those of pure mathematics; it showed that clarity was possible in the real world—if one had sufficient mental discipline.[53] Mathematics was not meant to be used in research, nor was it to be seen as a basis for any particular career. If an undergraduate desired to conduct further research after graduation, it was regarded as a failure on the part of their tutors: their explanations must not have been sufficiently convincing.[54]

Few Cambridge men had any interest in pursuing mathematics beyond graduation. There was simply no future in it. Those students who came from wealthy families did not need employment after graduation; those from less wealthy families generally looked to the Church for their living; a few succeeded in legal or medical careers; but students who wished to pursue their mathematical studies further faced a dead end. Apart from fellowships and positions as private mathematics tutors, there were no careers in mathematics. There was almost no state sponsorship of mathematical or scientific jobs.

The mathematics taught at Cambridge was the mathematics of Newton; it was more than a hundred years old. Elsewhere on the continent, particularly in France, Newton's ideas had been taken up, developed, adapted, taken apart, and put back together in new ways. French mathematics—the most sophisticated in the world at this time—was inspired by Newton but it did not require unquestioning devotion to his texts. French mathematicians were encouraged to be creative, playful, and not overly reverent towards the old masters.[55] By contrast, the mathematics studied in Cambridge had ossified. As one commentator put it:

> The mathematical fame of Cambridge is great amongst Cambridge men. The sister University of Oxford limps after Cambridge in the mathematical race, but has long been double-distanced. The examinations at Oxford are, to a Cambridge man, ridiculous; the examinations of Cambridge are, to a French mathematician, beneath notice. Our University science is, on the Continent, in very low repute.[56]

The first stirrings of dissatisfaction about this state of affairs came not from the University lecturers or college tutors, but from a small group of students who revered continental mathematics. It began with Charles Babbage, a bright boy who was from a well-to-do family and had come up to Cambridge in 1810. He had developed a love of algebra as a schoolboy and was already well versed in mathematics, including continental mathematics, when he arrived in Cambridge. He was disappointed by the dull mathematics curriculum he had to study with his college tutors and would often skip classes to go sailing on the river or play sixpenny whist. He also privately kept up his interest in continental mathematics. In 1812, he joked with a friend about translating French mathematics books into English and distributing them to students (the joke was aimed at a Bible Society that was caught up in a controversy about distributing prayer books). Babbage created an elaborate plan for a spoof society that would translate *Traité élémentaire du calcul différentiel et du calcul intégral* by the French mathematician Sylvestre Lacroix. As well as translating this comprehensive work on calculus, Babbage's fictional society would encourage the use of the continental 'd-notation' (developed by Gottfried Wilhelm Leibniz) in place of Newton's 'dot notation', as he thought that the d-notation was more flexible.[57] The notation may have had the advantage of flexibility but, to the minds of the tutors, it had the disadvantage of being popular in France—a breeding ground for all kinds of radical and dangerous ideas.

Very quickly, Babbage realized that these two things—a translation of Lacroix's work on calculus, and wider use of the d-notation—would actually be very useful and so his spoof society became a real society. Babbage met with a few friends on Thursday 7 May 1812 to propose the creation of the Analytical Society and, by the following Monday, they were having their first meeting. The society was small (only sixteen members are known), all the members were undergraduates, and it was really only open to the top students (most members of the society went on to become wranglers). John Herschel, a student at St John's College and son of the astronomer William Herschel, was elected the Analytical Society's first president. Herschel later recalled how he had been drawn to the French mathematics because of the prohibition on it: there was, he wrote, a 'prestige which magnifies the unknown, and [an] attraction inherent in what is forbidden'.[58] Unwittingly, their tutors impelled these inquisitive students towards new knowledge. These young mathematicians met regularly to discuss each other's work, to read the latest papers from Europe, to translate passages from Lacroix, and to despair at the Cambridge curriculum.[59]

In November 1813, Babbage's society made a 'literary assault upon the peace and quietness of the world': it published its first volume of work.[60] In line with the society's aim, all of the papers within the volume were written in the continental d-notation. This posed some problems for the staff of Cambridge University Press as they struggled to find the right characters to print this strange new language. John Smith, the University printer, was heard to complain about the number and size of brackets required to set the text; in the end, he had to send to London for more. Furthermore, the papers (a preface by Babbage explaining the aims of the society, followed by articles on continued products, trigonometrical series, and equations of differences) were almost unintelligible to most readers who had never seen the d-notation before. The general response was one of bewilderment; Babbage wrote of the readers being lost in 'clouds of $\psi$s' while Herschel despaired that, unlike in France, 'the publication of a mathematics work, particularly if it goes one step beyond the comprehension of elementary readers is a dead weight and a loss to its author'.[61]

The society broke up soon afterwards, as the key members graduated. Though most of the men excelled in their final examinations, there was one notable exception: Babbage used a pre-Tripos public disputation to try to prove that God was a material agent—a statement viewed as blasphemous by the fellow who was moderating the session—and so he graduated without honours. But, despite the poor reception of the society's publication and Babbage's growing notoriety, the effects of the society lingered. In 1816, Babbage, Herschel, and George Peacock, another former member of the Analytical Society, published their translation of Lacroix's work: *An elementary treatise on the differential and integral calculus*.[62] By this time, Peacock was a fellow at Trinity College. When he was appointed as a moderator of University examinations in 1817, he used his position to introduce some of the continental notation and ideas into the Tripos. This was highly controversial and many shunned Peacock as a result. He glumly conceded that his efforts had not accomplished anything. But Peacock's actions did have an effect: he had set something in motion and by the 1820s, continental mathematics began to make regular appearances in the University examinations, and the Cambridge curriculum was updated to include more modern developments in mathematics.[63]

Things in Cambridge were beginning to change, but slowly. In 1817, the use of new mathematical notation was still considered shocking to some. In 1818, Sedgwick's decision to lecture regularly from the Woodwardian Chair was surprising to many. And, in 1819, the creation of the Cambridge Philosophical Society caused some bemusement in university circles. Sedgwick wrote to John Herschel, now an astronomer, mathematician, and, occasionally, chemist, to summarize the reactions encountered by the new Philosophical Society:

> Among the senior members of the University some laugh at us; others shrug up their shoulders and think our whole proceedings subversive of good discipline; a much larger number look on us, as they do every other external object, with philosophic indifference; and a small number are among our warm friends.[64]

In November 1819, the Society was voted into existence with much enthusiasm by that group of men gathered together in the University Library, but it was still just an idea: next, it needed structure, rules, and clearly defined purposes. A committee of nine men was appointed to draw up a set of regulations to govern the new society. These nine men were drawn from across the colleges and University, showing what a diverse range of people were keen to involve themselves in this new enterprise. They included Sedgwick, Clarke, and Haviland, as well as John Kaye (Master of Christ's College), William Farish (the Jacksonian Professor of Natural Philosophy), James Cumming (Professor of Chemistry), Fearon Fallows (an astronomer and a fellow of St John's College), Thomas Jephson (a fellow of St John's College and later a candidate for the professorship of mineralogy[65]), and one Mr Bridge (this was probably Bewick Bridge, a mathematician, a fellow of Peterhouse, and a vicar). It was decided that Henslow, who had only graduated with his BA the previous year and had not yet been awarded an MA, was too junior to be appointed to the committee.

The committee worked quickly and, in less than a week, they had drafted the Society's first set of regulations. These formally outlined the Society's objectives, declaring that the Society existed 'for the purpose of promoting Scientific Enquiries, and of facilitating the communication of facts connected with the advancement of Philosophy'.[66] This was a broad remit, not limited to any particular field; it was fitting for a Society that was being shaped by a geologist, a mineralogist, a physician, a natural philosopher, a chemist, an astronomer, and a mathematician. This was to be a truly *philosophical* society.

The following week, on Monday 15 November 1819, the committee presented the full set of regulations to a second meeting of the proto-Society; everyone agreed that their statement of purpose reflected the fine ambition of the group, and the regulations were unanimously accepted. The Society now had a set of rules governing membership, the election of a council, and subscription fees; and they had begun laying the plans for fortnightly meetings, for publishing scientific papers, and for setting up a museum and a reading room. This meeting marked the official foundation of the Society.[67]

It was agreed that membership of the new society would be open to all graduates of the University; potential members needed to be nominated by three others, and then voted in by a two-thirds majority at a Society meeting. Members would pay an admission fee of two guineas, followed by an annual subscription of one guinea—a fee easily within reach of the average Cambridge graduate. The Society would be governed by a council—made up of a patron, a president, a vice-president, a treasurer, two secretaries, and several ordinary members—who would be elected by ballot (with the exception of the patron). There would be presidential elections every two years. Frequent elections, which were also the practice in the newer London societies, including the Geological and Astronomical Societies, were seen as a way to keep the leadership fresh and to avoid the dictatorial tendencies of lifelong presidents.[68] Meanwhile, the requirement that members of the council must have an MA was intended to ensure a high standard amongst council members (though the need for an MA excluded younger men like Henslow from these posts).

Prince William Frederick, the Duke of Gloucester, who had been appointed Chancellor of the University in 1811, was invited to be the Society's patron. The Duke had studied at Trinity College, but was not renowned for his intelligence, and did not take much interest in the Society. The rest of the council was rather more active. William Farish, the Jacksonian Professor of Natural Philosophy, was elected president. Prior to becoming the Jacksonian Professor, Farish had been Professor of Chemistry and was fascinated by the practical applications of chemistry; he was also a clever mechanic who loved taking machines apart to demonstrate their inner workings.[69] John Haviland, Professor of Anatomy, who had chaired the preliminary meeting, was elected vice-president. Adam Sedgwick and Samuel Lee were elected to be the secretaries. Lee had trained as a carpenter as a young man, but his real love was languages and he had spent all of his wages on books so that he could teach himself Latin, Greek, Hebrew, Chaldee, Syriac, Samaritan, Persian, and Hindustani. At the age of 31, under the auspices of the Church Missionary Society, he had entered Cambridge, taken a BA and an MA with incredible speed, and been appointed Professor of Arabic in 1819.[70] The inclusion of Lee in the first committee shows that the Society was welcoming to members from all kinds of scholarly backgrounds. Bewick Bridge was elected treasurer; Bridge divided his time equally between good works in his parish of Cherry Hinton and publishing mathematics books.[71]

As well as these officers, there were also seven ordinary members of council. The first was Edward Daniel Clarke, the professor of mineralogy who had helped set up the preliminary meeting. Next came Thomas Catton. A fellow of St John's College and the chaplain of Horningsea, he was also responsible for the observatory that was housed in a tower of St John's.[72] Thomas Turton was a fellow of St Catharine's College and had been a University examiner in the Mathematical Tripos (he would go on to be appointed Lucasian Professor of Mathematics in 1822, and Bishop of Ely in 1845).[73] Thomas Kerrich had aspired to be an artist but William Hogarth had advised him that the field was overcrowded so Kerrich turned to academia instead. At Cambridge, Kerrich had been

President of Magdalene College in the 1790s before being appointed University Librarian. Kerrich principally devoted himself to art, architecture, and antiquities, and to projects such as the restoration of Cambridge's twelfth-century leper chapel.[74] Robert Woodhouse, a fellow at Caius College, was a mathematician and astronomer who had been agitating for the reform of the mathematics curriculum since the turn of the century.[75] James Cumming had been appointed Professor of Chemistry in 1815; as the University did not possess any scientific apparatus, he constructed a set himself and used it to teach concepts in electricity, heat, magnetism, and many other subjects.[76] The final ordinary member of council was Richard Gwatkin, a fellow of St John's College and a private mathematics tutor who had once been a member of the Analytical Society.[77]

Of these twelve men who made up the first council of the Cambridge Philosophical Society, nine were ordained, and nine had been wranglers in the Mathematical Tripos. Three had won the prestigious Smith's Prize (a set of examinations which was sat two weeks after the Tripos and which allowed the wranglers to compete for further mathematical distinction in the hope of securing fellowships for themselves). These men were very much the products of the Cambridge system: they had excelled in their fields and been rewarded with comfortable fellowships, church livings, and professorships, and yet they were dissatisfied. Their positions did not demand much of them; they could have lived comparatively easy lives of leisure, but they all chose to teach or conduct research to a high standard. They all willingly gave up their time to the Society. They were driven by curiosity, caught up in the fashion for new societies and compelled by a personal desire for advancement or by frustration about the way science was being conducted in the University.

To reform the University system would have been an arduous task, requiring much political manoeuvring in the Regent House (the University's governing body) and the passing of elaborate 'Graces', as the legislative bills published by the University were called. The creation of a scientific society, open only to members of the University but officially independent of it, was a neat way around the problem of reform.

It was certainly easier to start a new society than to change the English universities: Oxford and Cambridge were both conservative institutions that looked upon reform with suspicion. There was talk of setting up a third English university—a non-denominational university in London (which was the only major European capital city not to boast a university of its own), but that did not happen until the foundation of University College London in 1826.[78] Scottish higher education was in a better state: the five universities there were open to non-Anglicans. Furthermore, the Scottish universities were far more receptive to new ideas from the continent. The political turmoil that had rocked France in the later years of the eighteenth century and the Napoleonic wars of the early nineteenth century terrified the English establishment. The English elite could not accept new scientific and mathematical ideas coming out of France because they were so associated with radical politics. The Scots were far more relaxed about this, and many Scottish university professors had spent time studying in France.

Because the primary function of the English universities was to supply clergy for the Church of England, and because they shunned so many foreign ideas, they could not (and did not necessarily wish to) keep up with the most modern developments in science and mathematics. Classics and theology had always been central to the curriculum, and remained so in this period. Many within the University of Cambridge were perfectly happy with their role and with their achievements and did not see any benefits to changing the system. In fact, many railed against the rise of the natural sciences in Cambridge, as they saw them as a distraction from the colleges' principal duty of giving a religious and moral education to the future leaders of church and state. Hugh James Rose, a clergyman who had been the fourteenth wrangler in 1817, campaigned against the teaching of science in Cambridge as he feared that the cult of 'reason' was a threat to religious orthodoxy. He wrote that

> the man of science may scoff at the names of Henry More and of [Joseph] Mede, and at their gross ignorance of all he knows; and doubtless they are

> as much below his contempt as they are above it. They could not arrange all the products of this material world in their scientific order, they could not use the tools of the laboratory, nor the engines of the mechanist; but who would lessen the dignity of man and of his intellect by comparing their elevated views, their thoughtful hearts, their exquisite conceptions, their gentle desires, and their Christian peace, with the million facts, the hurry, the fever, and the impatience of the experimentalist and the discoverer?[79]

In place of these hurried facts, Rose advocated a return to the slow, patient study of theology and literature, and quite dismissed the idea that studying the sciences 'could render the mind more susceptible of elevated knowledge'. Though some within Cambridge defended the sciences and argued that there was no incompatibility between scientific investigations and religious feelings, many supported Rose's views.[80]

Outside the universities, English science was also falling behind its European and Scottish neighbours. The Royal Society of London, so vibrant in the seventeenth century, had declined by the early nineteenth century. Some complained that it had become little more than a fashionable club; it had hundreds of fellows but few conducted scientific research, published anything, or showed any apparent interest in the sciences.[81] The fees at the Royal Society were so high as to exclude most aspiring men of science—like the men from the industrial towns of the north who were making genuine advances in science and technology.

The growing numbers of scientific societies outside London, and the more specialized societies within London, grew up to counter this malaise in the older institutions. It was scientific societies, rather than universities, that led the way in scientific research in the pre-Victorian period.[82] Many of the key figures of eighteenth- and nineteenth-century science are more closely associated with a society than with a university: the chemist John Dalton and the physicist James Joule with the Manchester Literary and Philosophical Society; the astronomer William Herschel with the Bath Literary and Philosophical Society; and the natural philosophers Humphrey Davy and Michael Faraday with London's Royal Institution. The societies were revitalizing the sciences, because the establishment would not.

In a small market town in the flat fenland of East Anglia, the Cambridge Philosophical Society found itself in a unique position: it was an independent society like those in industrial Manchester or fashionable Bath, but it was located in a university town. All of its fellows were also members of the University and so it could exert subtle (and quite unofficial) influence upon the University. The Society was only open to graduates and so did not have a direct effect on undergraduate learning, but it encouraged new research amongst the senior members of the University. This research would eventually trickle down to the students—perhaps via the classes of their college tutors or, occasionally, through the written papers of the University examiners. Slowly, the work of the Society would become more visible in the University. A new generation would grow up familiar with its aims and methods, and the University would be changed. Science itself would be changed.

# 2

# THE HOUSE ON ALL SAINTS' PASSAGE

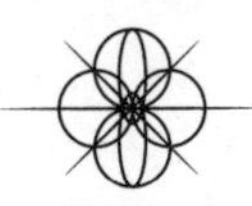

From the Great Gate of Trinity College, past the church that people still called All Saints in the Jewry (though Cambridge had not had a Jewish quarter for many centuries), past the old Sun Inn and alongside the remains of what had once been the Dolphin Inn, there ran a narrow and gloomy laneway (see Figure 7). Here, in the jumbled medieval alleys of Cambridge, one might have been surprised to see a pale, pristine new edifice nestling itself in amongst the older soot-stained buildings. A visitor to Cambridge in the 1830s would, upon entering this grand new house, leave behind the familiar streets of the city and find himself transported to a strange and wondrous place. For hidden away upstairs was Cambridge's most modern and impressive museum. Polished mahogany and glass cases lined the walls, filled almost to bursting with treasures from around the world: stuffed animals, the riotous colours of birds' plumage, glittering insects, and ancient bones perfectly preserved. There were riches from the sea too—pale and shining shells that told of another realm beneath the waves—and precious stones dug from the ground. There were dried plants and seeds, the life within suspended in time. And then there were the jars with strange-looking beasts preserved in spirits; the colour drained from them, only the form remained. All these things competed to draw the visitor's attention. The collected objects represented the totality of life on the earth—each continent was represented, with thousands of taxa—and they had been gathered together here in Cambridge in the name of cutting-edge science.

**Figure 7** A map showing the location of the Society's house on All Saints' Passage (marked).

The creation of a museum had been planned almost from the Society's first meeting. Prior to the foundation of the Society, there had been no natural history museum in Cambridge. The Woodwardian Cabinet had many fine geological and mineralogical specimens, but little space to display them. There was a small anatomical collection too, but access to it was limited. It was John Stevens Henslow who took responsibility for the Philosophical Society's new museum, labouring over it painstakingly for many years.

Henslow had long been a student of nature. He had been born into a middle-class family in Kent in 1796; his mother was said to have been an admirer and collector of curiosities, while his father, a solicitor, kept an aviary and devoted much time to his garden. Their house was filled with books, including many on natural history. So, young Henslow grew up with a strong interest in the natural world and loved to collect in the fields near his home, once bringing home a fungus almost as big as himself which his parents dried and hung up in their hall. At boarding school, Henslow's love of nature was further encouraged by his drawing master, George Samuel, who was a keen entomologist, and Henslow spent many happy hours chasing insects through the school's orchard with his butterfly net. It was from Samuel that Henslow learned how to preserve and display the insects he caught, proudly showing them off to his family.

In 1814, Henslow entered St John's College in Cambridge. Henslow dedicated most of his time to following a curriculum of mathematics, classics, and theology, but he did manage to continue his natural history expeditions, and he learned a little more of the sciences by attending Professor Cumming's chemistry lectures and Professor Clarke's spectacular mineralogy ones. Henslow took the Tripos examinations in 1818 and was placed sixteenth wrangler (his friend Shaw-Lefevre was senior wrangler that year). This was good, but not good enough to win a fellowship. Yet, Henslow decided to stay in Cambridge, continuing his scholarly work, collecting, and accompanying his friends on field trips.[1]

The museum of the Philosophical Society began with Henslow. Its first major collection was Henslow's entire set of British insects and shells. These were objects he had been gathering since boyhood; from those early excursions near his parents' home and around his school, he progressed to day trips around Cambridge and the Cambridgeshire fens, and on to longer, more elaborate trips like the one he took with Adam Sedgwick to the Isle of Wight in 1819. Everywhere he went, Henslow was on the lookout for interesting new specimens. Back home, he would meticulously sort his finds into their taxonomic groups and arrange them in special cabinets and boxes. The idea of a safe repository for his cherished collections must have appealed greatly to Henslow as his cramped rooms began to overflow with objects.

The Society did not yet have its elegant house in All Saints' Passage when Henslow began his curacy of the museum. The Society's early meetings were generally held in rooms at the original site of the University Botanic Garden—a few acres of land in the centre of Cambridge, bordered by Free School Lane and Pembroke Lane (the area now called the New Museums Site) (see Figure 8). They took place every second Monday during term time, beginning at seven o'clock and ending by nine o'clock—a tradition barely altered to this day—and fellows were permitted to bring guests (including women). It was the same room in which Edward Clarke had once lectured so vigorously on mineralogy. There were few spaces for science in Cambridge at the time: only two scientific lecture halls existed—one for anatomy and one for chemistry—and, though some of the other professors had rooms, they were not really suitable for teaching anything more than a handful of students.

So, the earliest meetings of the Society were held in borrowed rooms. Sedgwick, Henslow, and the new committee did not know what to expect at their first gatherings; they did not know how many people would show up to hear the first papers or how many would pay their two guineas to

**Figure 8** The lecture room in the Botanic Gardens, site of the Society's first formal meetings.

join. They had carefully crafted their preliminary meeting to create a buzz of excitement about the new Society; this had taken place at the end of 1819, and consisted of a welcome address by the president, William Farish; an impassioned speech by Edward Clarke about the purpose of the Society; and a letter of support from the patron, Prince William Frederick, with the letter being read aloud. Though this meeting had not featured any actual scientific papers, it had attracted much attention around the town and piqued people's interest: the idea of a modern philosophical society was an exhilarating one. The spectacle that an audience might expect from a gathering of a philosophical society is captured in these lines written by a member of London's City Philosophical Society in 1816:

> But hark! A voice arises near the chair!
> Its liquid sounds glide smoothly through the air;
> The listening Muse with rapture bends to view
> The place of speaking and the speaker too.
> Neat was the youth in dress, in person plain;
> His eye read thus, *Philosopher-in-grain*;
> Of understanding clear, reflection deep;
> Expert to apprehend, and strong to keep.
> His watchful mind no subject can elude,
> Nor specious arts of sophist e'er delude;
> His powers, unshackled, range from pole to pole;
> His mind from error free, from guilt his soul.
> Warmth in his heart, good humour in his face,
> A friend to mirth, but foe to vile grimace;
> A temper candid, manner unassuming,
> Always correct, yet always unpresuming.[2]

The poet, a man called Dryden, about whom little is known except that he attended meetings of the City Philosophical Society, wrote these lines about Michael Faraday, then an assistant at the prestigious Royal Institution. But they epitomized an idealized natural philosopher—a lover of truth who could turn his mind to any one of a number of subjects.

The ideal of this natural philosopher and his wide-ranging study of the sciences was what had appealed so much to Sedgwick and Henslow. Outside Cambridge, in the scientific societies of London, the industrial towns of England, and the universities of Europe, new scientific discoveries were being announced, technical innovations were being revealed, and new ways of understanding the world were being created. In the early nineteenth century, John Dalton's atomic theory was still a novelty, George Stephenson invented his first steam-powered locomotive, and Humphry Davy invented the safety lamp for the nation's miners and refined his theory of electrolysis, which allowed for the discovery of several new elements. Abroad, the French naturalists Étienne Geoffroy Saint-Hilaire, Jean-Baptiste Lamarck, and Georges Cuvier were rethinking evolutionary theory and the fossil record, while their colleagues in the physical sciences tried to unravel the mysteries of light, heat, and gravity. And now that knowledge would come to this little market town in the flat fens of Cambridgeshire. Experiments in electricity, new chemical elements, and steam-powered engines—there was a whole new world of science to be explored.

The words of Farish, Clarke, and the prince at the Society's preliminary meeting had had the desired effect. For, when the new term began in 1820, and the date of the Society's first proper meeting rolled around, the room in the Botanic Garden was packed full with aspiring men of science. The first paper was to be given by William Farish, who had been chosen not only because he was the president of the Society but also because his work spanned the worlds of science and technology, linking the rarefied realm of Cambridge to the marvels of the industrial age. As Jacksonian Professor of Natural Philosophy, Farish had a small room—dark, with only tiny windows, and cold, without a fireplace—on the corner of Slaughterhouse Lane and Bird Bolt Lane, just next to the old Beast Market.[3] In this uninspiring little chamber, often overpowered by the stench of offal from the adjoining laneway, the professor worked with his assistants to create a sort of magic: around them hummed and whirred a hundred different devices. Using just simple apparatus—'loose brass wheels, the teeth of which, all fit into one another: axes, of various lengths, on any part of which the wheel required may be fixed: bars, clamps, and frames'—they brought to life the wonders of the modern age. In their workshop, they recreated every single important machine that existed in industrial Britain, machines the professor had seen first-hand on his many tours of the industrial areas of England.

Farish was fulfilling the duties laid out to him by Richard Jackson, who had established the Natural Philosophy Chair in 1782. Jackson had specified that the holder of the post must promote 'real and useful knowledge'; lectures could be on any topic within the broad bounds of natural philosophy, just as long as they were 'of an experimental character'.[4] When Farish had accepted the Jacksonian Chair in 1813, he already had form as a practical lecturer: since 1794, he had been Professor of Chemistry and had taught his students about everything from mining and manufacturing to the properties of metals, ceramics, glass, and dyes (though, of course, none of these topics were included in the University examinations, and attendance was optional for undergraduates).[5]

With the creation of the Cambridge Philosophical Society in 1819, Farish's horizons expanded even further. And so it was that Farish took the floor at that first meeting and, in the flickering candlelight, began to speak of the incredible mechanical feats that took place in his workshop. He could conjure everything from cotton mills and steam engines to simple wheelbarrows, and then collapse them back again into their component parts—those loose brass wheels, bars, clamps, and frames. The focus of Farish's talk that evening was the ingenious geometrical method he had developed for drawing the plans for these constructions and deconstructions, allowing him to translate between the practical world of machinery and the mathematical *lingua franca* of Cambridge (see Figure 9).

His paper was followed by two more on that February evening. One was by Edward Clarke on the discovery of cadmium in some English ores. The new element cadmium had been discovered a few years earlier in Germany but this was the first proof of its existence in English rocks.[6] There were still fewer than fifty elements known at this time and the discovery of a new one was cause for excitement. The other paper was given by one Captain Fairfax on a series of experiments to determine soundings

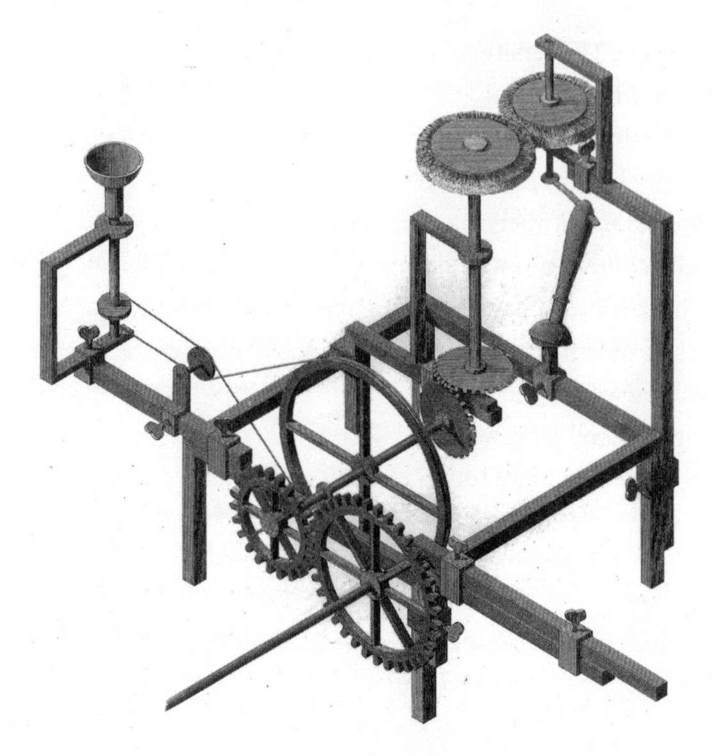

**Figure 9** One of the machines displayed by William Farish, Jacksonian Professor of Natural Philosophy, at the first ever talk presented to the Society.

at sea by observing the times of descent of different waves. After the talks came the exhibition—something that was to be a feature of the Society for many decades to come—in which a small number of new or curious objects were shown to the assembled fellows. The objects that were used to illustrate talks could be examined more closely at the end of the meetings: magnified drawings of plants, animals, and machinery; natural history specimens; or models of apparatus and natural objects.[7] The meetings were a treat for the senses: colourful drawings and maps adorned the stage; wonderful specimens covered the tables; and the sounds of machines and apparatus whizzing could be heard throughout the building. On that first night, Farish showed off a new kind of banknote designed in such a way as to prevent the possibility of forgery, and Frederick Thackeray (the Addenbrooke's surgeon who had been denied a medical degree by the University) displayed a fossil that had been found on the coast of Scarborough.[8]

The meeting was deliberately different from what one might expect at a university lecture. Though two of the presenters—Farish and Clarke—were professors, neither Thackeray nor Fairfax was a traditional university man. The content too was far removed from what a student might learn in the course of his degree studies: industrial machinery, new elements, practical problems of navigation, a display of the latest technologies from the banking world, and fossils from the fashionable field of geology—the meeting had it all. This was a forward-looking society, not afraid to break away from the standard topics of Cambridge academia.

Sedgwick wrote to John Herschel a few days later, reporting news of the Society and assuring him that

> [n]ow that we have launched I have little fear: we shall, I doubt not, go on and prosper...We may count on the zeal of our members for a sufficient number of communications; we may also venture to found some hopes on an active spirit infused by a new system.[9]

The meeting—its speakers, and the objects displayed—was reported in the journals of London and Edinburgh.[10] Momentum began to build and membership of the Society began to rise. With each meeting that term, attendance increased. The topics, and the speakers, pulled in large crowds of Cambridge men: there were papers on mathematics, palaeontology, geology, volcanology, optics, and mechanics. New ideas flowed in from France, Italy, Germany, and London. Speakers included established Cambridge names like Edward Clarke, but there was also space for younger scholars, like a mathematics tutor from Trinity College, called William Whewell, who presented a new way of calculating planetary orbits.[11]

Whewell was the man who had filled Adam Sedgwick's Trinity tutorship when the latter had given it up to become Woodwardian Professor. The son of a carpenter, he hailed from Lancaster. Like Sedgwick, he had travelled to Cambridge from the north and had found himself in an alien world of gowns, high tables, tutorials, and college politics. Even poorer than Sedgwick, he entered Trinity College as a lowly subsizar in 1812. His humble beginnings marked him out from his peers; once, shortly after his arrival in Cambridge, as he watched a herd of pigs being driven past the gates of Trinity College, he was heard to mutter, 'they're a hard thing to drive—very—when there's many of them—is a pig'—a statement which caused much amusement amongst his fellow students. But Whewell was bright and hardworking and his academic achievements quickly earned him respect. Moreover, he was cheery and fun-loving and devoted many hours to 'shooting swallows, bathing by half dozens, sailing to Chesterton, dancing at country fairs, playing billiards, turning beakers into musical glasses, making rockets, riding out in bodies' and winning himself many friends along the way. In 1817, after being placed second wrangler in 1816, he was awarded a fellowship at Trinity College, rising to Assistant Tutor in 1818 after Sedgwick gave up the role. He joined the Philosophical Society as soon as it was begun, throwing himself wholeheartedly into this new venture and giving more than twenty papers in its first decade alone.[12]

After just a few meetings, it became clear that the rooms at the Botanic Garden were not sufficient to contain the burgeoning new Society, and so they began to look for a home of their own. It was March 1820 when Adam Sedgwick, along with his co-secretary Samuel Lee and council member Thomas Turton, began their search for new accommodation.[13] They soon found what they were looking for: a large house which was located in the centre of Cambridge and which could host their meetings and also allow them to create other spaces for science. The house was on Sidney Street, opposite Jesus Lane. It belonged to Elliot Smith—a local businessman who was involved in the auction of land, rare books, and fine paintings and who went on to become Mayor of Cambridge three times in the 1850s and 1860s.[14] Smith had been using the house for his auctions but, seeing a good business opportunity, he signed a ten-year agreement, at a price of ninety guineas per annum, allowing the Society use of the whole premises. By the end of April, it was theirs.[15]

If the Society had a house, they also needed a housekeeper. They employed a man named John Crouch, paying him £25 each year for attending on the Society, and charging him £5 rent for his accommodation in the house. He was also given three guineas a year for fires, and an extra guinea for the upkeep of the fireplaces. Crouch had use of the whole house except for a few function rooms, and was only required to work five hours each day. He was also given regular pay increases—his annual income had risen to £40 per year by 1824.[16] The Society would later come to regret the generous terms of his employment.

The house was ready to host its first gathering on 1 May 1820. That spring evening, in a meeting room overlooking the gardens of Sidney Sussex College, a large audience assembled to hear William Farish speak of the difficulties of polar navigation, to hear John Herschel talk of light bending as it journeyed through a crystal, and to hear Charles Babbage (the former *enfant terrible* of Cambridge, who by 1820 was busy setting up London's Astronomical Society and publishing extensively on mathematics) speak about the intricacies of calculus.[17] But the most momentous part of the evening's meeting was the exhibition that followed. John Stevens Henslow had carefully packed up boxes and boxes of his specimens—butterflies with shimmering wings, beetles neatly arranged by class and order, and delicately whorled shells gathered from the seashore—and had carried them from his nearby rooms to the Society's house. Now he revealed them to the hushed gathering, handling them with great delicacy, describing each creature's history and habitat, and pointing out those features that made it unique, and those that linked it to other species.

These were the objects that would form the basis of Cambridge's first natural history museum—Henslow donated them to the Society after the meeting and they formed the start of the Society's collection. Earlier that same day, the Society's council had voted to set aside some funds to support a museum, earmarking £30 for the glass-fronted cabinets needed to show off Henslow's specimens.[18]

Henslow's gift to the Society was special. Previously, the town had lacked such a collection and a scholar could study natural history only by collecting a cabinet of his own, which could be a costly and time-consuming process. Museums in the early nineteenth century were important sites of research where multiple specimens could be gathered together, scientifically studied, described, compared, named, and categorized according to the principles of taxonomy.[19] This search for an underlying order in the natural world was one of the most central aims of the sciences, and museums were an ideal location for this kind of research. Though we now think of museums as public spaces for education or entertainment, many nineteenth-century museums were open only to scholars (though some, particularly in London, functioned primarily as public spectacles). The Society's museum was intended as a scholarly venture, though members of the public could apply for permission to visit. The creation of a museum, as well as the promotion of natural history, was a key aim of the Society; now that it had its own premises, Henslow and his colleagues could begin to build one up.[20]

This they did with great relish. Henslow's main collaborator on the museum was a man named Leonard Jenyns. Jenyns, like Henslow, had studied at St John's College, beginning his education there in 1818, just as Henslow graduated. Jenyns had had an interest in natural history since boyhood. As a child in London, he used to receive parcels of birds from the keeper of the family's country estate in Cambridgeshire: 'these birds,' recalled Jenyns, 'after the species had been determined and their characters noted down, were mostly skinned for preservation, the bodies being subjected to a rough dissection'.[21] As the years went on, Jenyns's love for collecting grew until he found himself in possession of countless specimens of birds, mammals, fishes, insects, shells, and dried plants. When Henslow and Jenyns were introduced by a tutor at St John's, they found that they had a natural affinity, and thus began an enduring friendship, as well as a series of collaborations that would last throughout their lives.

The museum of the Philosophical Society was their most striking joint endeavour. Both men gave generously from their own collections. Henslow donated a collection of Chinese and Brazilian insects, which had been carried back to Britain on merchant ships from those far-away lands to be pored over, named, described, and compared to their European counterparts. He gave minerals that he had gathered himself, beginning with a collection dug from the exposed outcrops of North Wales in the summer of 1820. Both men gave birds. Sometimes these were familiar ones: preserved sparrows; goldfinches, with their bright splashes of yellow and red; rosy-hued chaffinches; owls and pheasants; rooks and jays; and buzzards, bitterns, and goatsuckers. Sometimes they were more exotic: a black-billed auk that stood sentinel in its glass cage; guans from Honduras and Demerara, with their scarlet hats; and a peregrine falcon. The two also gave little mammals caught around Cambridgeshire—moles, shrews, and even an otter—and less common creatures, like polecats.[22]

Other fellows of the Society followed suit and gave generously to the nascent museum. As well as ready-made collections of shells, insects, birds, and dried plants, there were many curiosities: an enormous elephant's tooth; an even larger fossilized antler from an ancient elk; wasps' nests and edible birds' nests; and petrified wood. There were globes that showed the distant countries from which many of these specimens had come, and globes that showed the heavens. There was a clever new cosmosphere that showed the earth and sky together. There were some rather more alarming gifts, such as the head of a New Zealander sent by Edmund Storr Haswell (who later lived in New Zealand and held the position of 'protector of aborigines').[23] And there were celebrity objects, like fossils from the famous Kirkdale Cave in Yorkshire, where the preserved bones of creatures, including elephants, hyenas, rhinoceroses, hippopotamuses, and bison, had been discovered in 1821. At first, many assumed that these remains had been deposited by some kind of massive flood, perhaps even Noah's flood, with the carcasses of the unusual animals carried long distances to their final resting place in the dales—it was the furthest north such animals had ever been found. But the Oxford geologist William Buckland soon realized that the animals had lived locally and found evidence that a family of hyenas living in the cave had brought the bones in, leaving their gnawed remains. Buckland's work reshaped contemporary ideas of history and geology: he proved that animals like elephants had once lived in Yorkshire, and he showed the power of fossils in verifying such seemingly outlandish theories.[24]

These were the kinds of stories that could be told through museum objects. By studying the collections held in the Society's museum, the boundaries of knowledge could be pushed outwards. Connections could be made between apparently unconnected objects: foreign specimens could be compared to local ones, and older objects could sit alongside newer ones, allowing changes to be tracked over time. The museum, visually arresting though it must have been—a cavern of wonders filled with exotic animals, strange fossils, and curiosities from around the world—was not just a place for display; it was a place for discovery.

But, though it was intended as a space for its fellows to conduct research, outsiders could be brought as guests. Joseph Romilly, a fellow of the Society and indefatigable diarist, recalled bringing a young relative to the museum: 'In the evening [we] took the little girl to the Philos. Society to see the stuffed birds.—Began reading loud Miss Austin's [sic] "Persuasion".'[25] A visit to the museum could enlighten a man of science about the order of nature, and at the same time it could offer the same kind of entertainment as a Jane Austen novel.

Henslow and Jenyns curated this eclectic collection with much enthusiasm, bringing order to the great multitude of things that began to fill the Society's rented house. They conceived the aim of making a comprehensive collection of all the fauna of Cambridgeshire—including all 281 known vertebrates and over 9,000 invertebrates—as a step towards building up a definitive picture of all British wildlife, and towards working out the distribution of different animals. This ambitious collection policy meant that the museum grew even more rapidly and it was constantly shuffled around the house into ever larger rooms, and ever more expensive cabinets.[26]

The growing museum was just one reason that the Society began to look for a more permanent home. Membership of the Society rose rapidly in its first decade: in 1820 there were 171 fellows; in 1830 there were almost

300.[27] And, when the lease on the house in Sidney Street expired, Elliot Smith did not renew it. The council began to search for new premises. The fellows decided that, rather than simply renting another house, it would be better for them to build one of their own. Initially, they sought to buy a piece of land behind the Blue Boar Inn on Trinity Street but, when that deal fell through, they turned their attention to a plot of land belonging to St John's College. This was the plot located on All Saints' Passage, in the crook of the laneway behind All Saints' Church (see Figure 10). In May 1832, Sedgwick, who had been elected President of the Society in 1831, approached Charles Black, the bursar of St John's, to negotiate the lease of the land and they quickly agreed a forty-year lease at a price of £20 per year.[28] The terms of the lease paint a vivid picture of town life in Cambridge, for Sedgwick solemnly signed a document agreeing not to allow the premises to be used as

> a soap manufactory, tallow chandlery, distillery, sugar bakery, slaughter-house, common brewhouse, malthouse, doghouse, furnace, forge, cooks or oilshops, butchers shop, farriers shop, blacksmiths shop, carriers shop, or pewterers shop, braziers shop, beer shop, or tobacco manufactory.[29]

**Figure 10** A drawing of All Saints in the Jewry by Rev William Cole (c. 1743). The windows of the Society's reading room overlooked this church.

Having agreed to all this, the lease was placed in the Society's iron chest for safekeeping and the business of building the house could begin.

The Society engaged the services of Charles Humfrey as the architect and started to design their new house. Humfrey was responsible for many of the new buildings and features that shaped nineteenth-century Cambridge: he designed the tower of St Clement's Church on Bridge Street, a colonnade at Addenbrooke's Hospital, Trinity College's Lecture Room Court (now called Angel Court), and Ely's session house. He would later be put in charge of draining and levelling the rough pastureland of Parker's Piece in time for a coronation feast held to honour Queen Victoria in 1838. Though he was a town man at a time when the Cambridge worlds of town and gown were mostly separate, he and his family must have mingled socially with the University men, for his daughter later married Robert Willis, who was the Jacksonian Professor from 1837 to 1875 and a prominent fellow of the Society.[30]

For the plot in All Saints' Passage, Humfrey designed a four-storey house in gault brick (see Figure 11). Its tall windows were elegantly arched; its front door, topped with an imposing neoclassical pediment, was recessed into the house to make space for it on its narrow lane. Inside were high ceilings, delicate stucco cornices, and modern gaslights. It had a meeting room, a council room, a library, reading room, and space for the museum, as well as space for Mr Crouch, the housekeeper. By June 1832, the Society had contracted one Mr Bell to build the house.[31] They agreed his fee of £1,716 on the condition that the house be completed by 1 May 1833. The Society raised the necessary funds by selling £50 shares to its fellows.[32] In the autumn of 1833, only a few months later than planned, the Society moved into its new house.

Henslow and Jenyns, with help from a few others, eagerly set about arranging the museum in the new house. The thousands of specimens were carefully packed up and moved the short distance—only a hundred yards or so—to their new home. Henslow's initial donation of shells and insects had been housed in just two cases. That number had grown over the years and now more than thirty cases were needed to display the

**Figure 11** The Society's house on All Saints' Passage as seen from the churchyard. The tall arched windows above the railings were those of the reading room. Mr Crouch's rooms were in the basement (their windows just visible above ground level).

museum's treasures. Henslow had added a collection of marine invertebrates gathered during summers spent on the Dorset coast, and Jenyns had contributed fishes caught along the southern shores of Britain. During the 1820s, the Society had begun to buy preassembled collections: a magnificent set of British birds collected by the London surgeon John Morgan, and an arrangement of over 2,000 insects, which had been named and ordered by James Stephens of the British Museum.[33] The new house allowed all these objects to be displayed properly for the first time, telling the story of the natural world in all its complexity.

By the time the museum moved to the house on All Saints' Passage, Henslow had become a well-known figure in Cambridge science. He had been elected Secretary of the Philosophical Society as soon as he was awarded his MA. In 1822, following the death of Edward Daniel Clarke, Henslow had been appointed Professor of Mineralogy. Clarke's death was much lamented by his friends in the Society—Sedgwick and Henslow always insisted that Clarke be recognized as an equal co-founder of the Society—and, following his funeral in Jesus College Chapel, they began to collect subscriptions to have a bust carved.[34] In 1824, Henslow was ordained, allowing him to supplement his meagre University salary by acting as the curate of Little Saint Mary's Church. In 1825, Henslow's career progressed again when he became Professor of Botany, the role that would come to define him. He gave up the mineralogy chair shortly after being appointed to his botanical post, and the vacant professorship was taken by another rising star of the Philosophical Society—William Whewell.

These were happy years for Henslow. At around the same time as he became Professor of Mineralogy, he began to fall in love. He had spent much time at the family home of his friend Leonard Jenyns in Bottisham, just seven miles east of Cambridge, getting to know Leonard's four sisters and two brothers. It was Harriet Jenyns, Leonard's youngest sister, who caught his eye. He wooed her in the only way a university man knew how: with books. 'Tell Miss Harriet,' he implored her brother, 'that Schiller's works are in 12 volumes & that I have sent 3 & that if this is not enough I will forward an additional supply—& also say—That Miss Benger's Queen of Scots is *ordered* but has not yet arrived.'[35] The mineralogy professor, unlike the Woodwardian Professor, the post then held by Henslow's dear friend Sedgwick, was allowed to marry and, following a successful courtship, John and Harriet wed in December 1823.

Harriet, who was clearly fond of reading Schiller in the original German, was an intelligent woman and a good match for Henslow. Together they set up home in Cambridge, in a house overlooking the open green space of Parker's Piece. There, their Friday evening soirées became legendary: everyone was invited, from lowly undergraduates to college masters, and guests were encouraged to mix freely, discussing a range of topics across the arts and sciences, or displaying drawings, objects from natural history, or new books. Mrs Henslow was just as much a presence at these meetings as her husband, as a letter from the young Charles Darwin—Henslow's most famous student—reveals: 'I have not seen Prof. Henslow, but am going to a party there tonight... [Mrs Henslow] is a devilish odd woman. I am always frightened whenever I speak to her, & yet I cannot help liking her.'[36]

Leonard Jenyns did not pursue a University career. He was drawn towards the religious life and, after being ordained deacon in 1823 and priest in 1824, he was given the curacy of Swaffham Bulbeck—a parish just a mile from his family home in Bottisham. There he combined his religious duties and charitable works with his natural historical researches. He continued to travel into Cambridge frequently, speaking at the Society's meetings, tending to the museum, and stopping by his sister and brother-in-law's soirées.

While Henslow and Jenyns busied themselves with the museum, the other fellows set about arranging the rest of the house. To begin with, there was the reading room. This was the first room one came to upon entering the house and climbing the three steps in the vestibule. It was light and airy with a high ceiling and three large windows looking south-west over the churchyard. Round the room were ranged chairs and desks, and in the centre were tables neatly displaying the latest newspapers and journals. The reading room was an immensely popular feature of the Society. In the 1820s and 1830s, though the colleges had combination rooms, they were often only open to senior fellows, and did not necessarily provide any reading material.[37] The University had a library, but its focus was not scientific.[38] The Society's reading room was open to all of its members, and took in a huge range of publications—current affairs and general interest as well as scientific. The first reading room had been set up by William Whewell in the rented house on Sidney Street. He diligently devoted himself to its management: deciding which periodicals were worth taking, discontinuing the less popular ones, and instituting rules to prevent ungentlemanly hogging of the newspapers.[39]

In the new house, the reading room and the selection of periodicals could be greatly expanded. George Peacock (the one-time member of the Analytical Society and mathematical reformer) and Robert Willis (who would go on to become the Jacksonian Professor of Natural Philosophy) were assigned to set it up. Carefully they oversaw the minutest details, proudly placing the Society's astronomical clock at the north-west end of the room, and hanging the Society's copy of George Bellas Greenough's geological map of England and Wales in a prominent position.[40]

Alongside the reading room, the library also began to grow. Initially, it had been built up through donations of books from fellows. Sedgwick, Henslow, and Clarke had all given generously, as had other council members like Whewell, and prominent Cambridge men of science like Herschel and Babbage. They crammed the library's shelves full with the most advanced volumes on every branch of natural philosophy: there were journals from around the world, and books on all kinds of subjects—from mathematics and mineralogy to physics and physiology, and even a book on calculating machines, courtesy of Babbage. There were maps too, their bright colours showing the contours of lands known and unknown, and geological maps, which were fast becoming indispensable scientific tools used to unravel the mysteries of the strata: Henslow gave maps of England, Europe, and the world, while the astronomer Richard Sheepshanks sent the copy of Greenough's geological map of England which now adorned the reading room. And there were curiosities, like a copy of the Rosetta Stone's inscription, whose hieroglyphics were not decoded until the 1820s.

Gifts also came to the library from outside Cambridge, mostly from the honorary fellows. These were distinguished men of science who had not studied at the University. The condition that members had to hold a Cambridge degree excluded many from the Society; the honorary fellowships were a way of celebrating their achievements, while also connecting the Society to the wider scientific world. Though the honorary fellows might never visit the Society in person, it became customary for them to send copies of their books or articles for the library. So it was that William

Buckland, the Oxford professor who had decoded the Kirkdale fossils, gave works on geology; the Scottish natural philosopher David Brewster kept his Cambridge colleagues abreast of developments in Edinburgh; and luminaries like the geologist William Conybeare, the polymath Mary Somerville (who, despite Adam Sedgwick's best efforts, was denied an honorary fellowship on account of her gender), and the star of the Royal Institution, Michael Faraday, sent their latest papers. Books rolled in from all over Europe too: from France, André-Marie Ampère delighted the Cambridge men with his tales of electrical experiments; Giorgio Santi, Professor of Natural History at the University of Pisa, transported the fellows from the chilly fens to the rolling hills of Italy with his *Viaggio per la Toscana*; and there were works sent from Leiden about Roman and Phoenician archaeology, from Lithuania about astronomy, from Switzerland about celestial mechanics, and from France about the sciences of nature.[41] As the years went on, these many gifts were supplemented with carefully selected purchases to create the most extensive scientific library in Cambridge.

Upstairs in the house, next door to the museum, the lecture room was made ready to host its first meetings. The speaker would address his audience from a sunken platform furnished with chairs for him and the president; they were faced by rows of raked benches, as if in a theatre. Unlike a regular theatre, though, each seat would have a little wooden desk, allowing the audience to take notes. The whole would be overlooked by galleries on each side. Above, the roof would open to a magnificent skylight, seven feet square; its clever design was copied from one recently installed in the operating room of Addenbrooke's Hospital and its moveable parts allowed for ventilation[42] (see Figure 12). To light the evening meetings, a modern system of gaslights was installed; the house was one of the first in Cambridge to boast of such a thing. The lecture room was decorated with specimens borrowed from the museum, changed at irregular intervals: a big cat might watch impassively as scientific findings were revealed to enthralled audiences; brightly coloured birds might flock round the benches as a professor covered the blackboard with mysterious

**Figure 12** The meeting room of the Philosophical Society. This is the only known image; it was drawn during the 1845 meeting of the British Association for the Advancement of Science.

symbols; and, always, rows and rows of glass spirit jars would glint from the upper galleries, their inhabitants bobbing unawares as the greatest scientific minds of the day took to the stage.

This new house was not just a home for the Society: it was also a site of research. The museum, of course, is the earliest evidence that the Society was conducting original research of its own. While the meetings provided a platform for Cambridge's men of science to present research they had conducted independently, the museum actually facilitated new discoveries within the Society. And discoveries were central to the Society: since the earliest meetings, the fellows had followed the rule that the announcement of new discoveries must take precedence over all other papers at general meetings, regardless of the speaker's seniority or rank within the University.[43] There was other research too: William Whewell was given permission to install a wind anemometer of his own design above the portico of the house, and there, in All Saints' Passage, it whirred away contentedly for many years, disturbed only occasionally as Whewell or Mr Crouch took a reading.[44]

The house was a great success and a profitable home for the Society, and also for the University more widely: in the 1830s, William Smyth, the Regius Professor of History, used the house's meeting room for his lectures as the University was unable to provide a suitable lecture room.[45] Many other Cambridge groups used the meeting room too, turning the house into a hub of Cambridge life, with constant comings and goings.[46]

In the spring of 1832, just as they were poring over Humfrey's plans for the new house, and considering the best way to raise funds for its completion, the fellows began a campaign to be awarded a royal charter. A charter—a document issued by a monarch granting rights or privileges to an organization—was the only means for creating a legally recognized corporate body in the 1830s. The Royal Society of London had held a royal charter since 1662, and many of the newer scientific societies of London had also been awarded charters. Sedgwick and Peacock were deputized to prepare the petition for the charter.[47]

A recently discovered letter from Adam Sedgwick to Prince William Frederick, Chancellor of the University and Patron of the Cambridge Philosophical Society, explains just why the Society wanted a charter and reveals their ambitions for the future. Sedgwick wrote of how a charter would increase the 'honor', 'respectability', and 'influence' of the Society but, most importantly in Sedgwick's eyes, it would give the Society permanency. The ever-changing population of their university town was a concern for Sedgwick, as he explained to the Prince:

> [W]ith our present constitution there is a constant risk of our dissolution—from the fleeting nature, as your Royal Highness well knows, of our academic population—from the indifference of some, and perhaps the hostility of others; who might perhaps wish that the funds should be employed on works of fancy or amusement, rather than on science and dry investigations.

A royal charter would protect the future interests, and funds, of the Society, and would help to ensure its longevity. On a more practical level, being an incorporated body would ease the building of the Society's new house, allowing the Society to 'build a meeting room, museum and all other accommodations necessary for our present and future well being', and it would make it simpler for fellows to give donations or bequests to the Society.[48] Their petition was successful and, on 6 August that year, a royal charter was officially granted (see Figure 13).

The charter was awarded by King William IV (Prince William Frederick's first cousin), who had taken the throne in 1830 following the death of his older brother, George IV. It acknowledged the achievements of the Society, touching on its library, museum, meetings, and journal:

> Adam Sedgwick [and other graduates of the University] have collected and become possessed of a valuable library and various collections in natural history...and have also been and continue to be actively employed in

William the Fourth by the Grace of God

By Writ of Privy Seal

**Figure 13** The Royal Charter of the Cambridge Philosophical Society, granted by King William IV in 1832.

> the promotion of philosophical and natural knowledge...by offering encouragements for original research and especially by the publications of volumes of transactions composed of original memoirs read before the Society.[49]

The charter had been granted solely in Sedgwick's name to lessen the fees charged by the Stamp Office. As both the founder of the Society and its president at the time of incorporation, this seemed fitting; but this practical step tickled Sedgwick: 'I was the Society', he joked to his friends.[50] When the University term began in October, a general meeting of the Society officially accepted the charter.[51] The Society delighted in their new status. They celebrated with a special meeting followed by a dinner at the Eagle public house on Bene't Street, and the merriment continued long into the evening. Joseph Romilly recorded the events in his diary: 'Meeting at the Philosoph to accept our Charter.—Dinner at Eagle to commem. the same. Sedgw (as Pres) in the Chair: good speeches from him, Airy and Whewell.—Thence to Sedgwick's where WW[hewell], Lodge, Thirlwall and self sat till past 2.'[52] Adam Sedgwick, nursing a hangover, also wrote about the celebrations in a letter to a friend:

> Yesterday after lecture I presided at a public meeting held by our Society for the purpose of accepting a Charter. We afterwards adjourned to an inn and had a blow-out. Finally, three or four of my friends came to my rooms and kept me up till two this morning—for which I do not now much thank them.[53]

Once granted their new charter, the Society required an official seal. George Peacock was put in charge of this; he commissioned William Wyon, the chief engraver of the Royal Mint, to create a seal bearing an impression of Isaac Newton[54] (see Figure 14). The image of Newton was based on the statue of him located in the chapel of Trinity College. That statue, sculpted by Louis-François Roubiliac in 1755 at a cost of £3,000, was much admired: Newton's parted lips and upturned eyes were said to show the loftiness of his mind, while his flowing gown and carefully clasped prism situated him in the academic world of Cambridge science.

**Figure 14** The seal of the Cambridge Philosophical Society, designed in 1832 by William Wyon, chief engraver of the Royal Mint. The original seal was lost in the 1870s; this is its replacement.

By choosing Newton—Cambridge's (and England's) greatest ever natural philosopher—as their mascot, the Society celebrated the history of science in their University while declaring their intention to emulate his successes in the future.

The Society came to specialize in providing services that were not generally available in Cambridge; and, just as they saw the need for a museum, a reading room, and a library, they saw the need for Cambridge to have a scientific journal of its own. Though the University had a press, it had traditionally focused on publishing books rather than periodicals. Periodicals like the *Philosophical Transactions*, which had been published by the Royal Society since the late seventeenth century, were considered an expedient way of publishing up-to-the-minute scientific research.[55] The creation of a journal was central to the aims of the Cambridge Philosophical Society and, by May 1820, arrangements were being made for publication. The Society looked not to the venerable (but increasingly stagnant) Royal Society for their publishing model; instead, they turned to the much newer and more vibrant Geological Society of London. The Geological Society had been founded in 1807 and had begun publishing the *Transactions of the Geological Society* in 1811. The Geological Society was widely admired for possessing 'the freshness, the vigour, and the ardour of youth in the pursuit of a youthful science'. Sedgwick, Henslow, and several other members of the Cambridge Society were also fellows of the Geological Society and had read and admired its *Transactions*.[56]

With the rise in steam-printing in the early nineteenth century, publishing was becoming easier and books were becoming cheaper. Improvements in the techniques of lithography meant that colour printing was becoming more affordable too. Even so, the young Cambridge Philosophical Society had limited financial means, and Adam Sedgwick wrote to the Cambridge University Press's 'Syndics'—the senior members of the University who governed the Press—asking them to help cover the costs of producing a periodical. The Press generously agreed to defray all printing costs.[57] It had only just begun to venture into periodical printing with the annual volumes recording observations from the University's new observatory; the journals it printed for the Society were the Press's second foray into periodical publishing.[58] Other practical matters also had to be arranged: the price of the journal was set (the first volume cost a guinea for fellows of the Society, and nine shillings more (£1.10s) for non-members) and letters were written to an engraver, Mr Lowry, asking him to prepare the plates; to John Smith at the Press, asking him to print the volumes; to a bookbinder, Mr Bowtell, asking him to stitch the pages once printed; and to John Murray and others in London, asking them to act as official booksellers.[59]

The practicalities were in hand, but what would the journal contain? The intention was to publish the papers read at the Society's meetings;

however, simply presenting a decent paper would not be enough to guarantee publication. The Society's council set about creating a system for rigorously reviewing all potential papers. A formal system of peer review did not exist in the early nineteenth century. For many scientific journals, the editor acted as an informal reviewer; others created committees of expertise to offer opinions on particular papers. Most journals aspired to publishing high-quality papers, but it was not a given that a paper was subjected to any particular scrutiny before appearing in print.[60] At the Cambridge Philosophical Society, it was decided that no paper should be printed until it had been examined by two fellows of the Society. These fellows would be nominated by the council, and would be required to submit a written report on the paper's merits.[61]

It was 1821 when the first part of the first volume of the *Transactions of the Cambridge Philosophical Society* appeared in print; the second part followed the next year, in May 1822. It was available to all members at a discount, and to the general public through bookshops in Cambridge and other towns. The twenty-seven papers contained in this first volume paint a perfect picture of the scientific concerns of the day, and its list of authors reads like a roll call of key Cambridge scholars of the early nineteenth century. There were Sedgwick's first papers on geology, in which he led his reader across Devonshire and Cornwall, hammer in hand, uncovering the structures hidden beneath the landscape. There were papers by Herschel and Babbage pushing their beloved analytical mathematics. There was a paper by Whewell on calculating planetary orbits. There was Henslow's paper on his geological field work in Anglesey, describing how he traversed the 'low and undulating' island, plotting out the rocks he came across and sketching out cross sections in pale pastels (see Plate 2). There was also a lengthy preface, written by Adam Sedgwick, George Peacock, and William Whewell and which told the story of the Society and outlined its hopes for the future: in it, the authors hoped that, by bringing together 'men of cultivated understandings', they could 'keep alive the spirit of inquiry', discovering truths of nature and creating a home for science in Cambridge.[62]

On publication, Sedgwick wrote to his old friend William Ainger: 'I have myself turned author since we met last. I will send you a copy of my paper in the *Cambridge Transactions* when I have an opportunity. Not that I wish you to read it. I will therefor give it to Mrs A. to tie up sugar-plumbs [sic] for my god-daughter.'[63] Though Sedgwick joked, his paper was rather well received; indeed, reviews of the first volume were universally warm, with reviewers welcoming not just the *Transactions*, but the Society itself. *The British critic* commended the 'zeal and ability' of the Society's fellows. The reviewer saw the creation of the Society as an important step for the sciences; he hailed the rise of local centres of communication that would bring together many individuals and so ensure the advancement of science.[64] *The Cambridge quarterly review* thought that the *Transactions* could be 'ranked among the most scientific [volumes] of the day; dreading no comparison with the Transactions of the National Societies themselves'. The reviewer congratulated the writers on their 'able and spirited productions'. He praised Farish for his 'valuable' paper on drawing his mechanical marvels; he thought the works of Babbage and Herschel's papers on algebraic analysis were 'highly ingenious' (though perhaps not of a nature to interest anyone but a mathematician); and Whewell's 'masterly' application of differential calculus to the higher branches of astronomy aroused feelings of admiration. Significantly, the reviewer saw the Society as part of a more modern Cambridge, writing that 'Cambridge is in an improving state. In our young days of academic pilgrimage we had neither Museum for the imprisonment of pictures; nor Philosophical Society for the propagation of knowledge.' The Society, according to this reviewer, could do more than just propagate knowledge; it could shake up the University—acting as a catalyst for the modernization of that ancient institution.[65]

The *Transactions* were a key record of the talks given at the Society's meetings, and they quickly became the primary means of publicizing the Society's work. They also performed another useful function: they became objects of exchange. Copies of the first volume of the *Transactions* were sent as gifts to the Institute of France, to London's new Astronomical

Society, to the Royal Irish Academy in Dublin, to the Royal Societies of Edinburgh and London, to the recently established Fitzwilliam Museum in Cambridge, to the Linnean and Geological Societies in London, and to the Asiatic Society of Calcutta, amongst others.[66] The societies which received a copy of the *Transactions* were thus exposed to the new discoveries of the Cambridge fellows. Moreover, they almost universally felt obliged to reciprocate, meaning that the Cambridge Society received dozens of volumes of scientific works for its new library.

With each number of the *Transactions* that appeared, the Society circulated it to more provincial, national, and international societies; and, in return, they were sent more and more journals. New volumes of the *Transactions* were published in 1827, 1830, 1833, 1835, 1838, and so on fairly regularly throughout the rest of the century. By the 1830s, copies of the *Transactions* were systematically being sent to Paris, London, Edinburgh, Dublin, Leiden, Brussels, Amsterdam, Geneva, Lisbon, Swansea, Calcutta, and Philadelphia, amongst others; and learned societies in all the great European and world cities were sending their journals to Cambridge.[67] What began as an informal giving of gifts was being cemented into a formal periodical exchange. Thanks to these exchanges, the library of the Cambridge Philosophical Society became the most modern and well-stocked library in the city. It had, by far, the widest range of journals in Cambridge, and it was to the Society that people went when they wanted to read the most up-to-date research. Many joined the Society because of the lure of the library.[68] The colleges and University could not compete.

Though the holdings of the museum and library varied widely and represented all aspects of the sciences, the papers published in the *Transactions* began to become more focused on the mathematical sciences. Since the beginning, it had been a stated aim of the Society to apply mathematical principles to chemistry, mineralogy, geology, botany, zoology, and other fields. This was because the fellows believed that mathematizing a subject would confer a greater degree of precision and certainty upon it.[69] This was another of Newton's legacies. Ever since he had described the *mathematical principles* of natural philosophy—bringing together the previously separate fields of mathematics and philosophy—natural philosophers had been trying to apply numbers and equations to the natural world. Because all the fellows of the Cambridge Philosophical Society had been trained in the Cambridge system, they were all fluent in the language of mathematics, and it seemed natural to them to use that language to talk about sciences that might, at first glance, seem more qualitative than quantitative.

Sedgwick's geological style, for example, was considered mathematical. Trying to convince Sedgwick to accompany him on a geological trip to Europe, Roderick Impey Murchison, a prominent fellow of the Geological Society, wrote: 'Let me...invoke the spirit of inquiry which prevails at Cambridge, and urge you, who are really almost our only *mathematical* champion, not to let another year elapse without endeavouring to add to the stock of your British Geology some of the continental materials.'[70] While many of Sedgwick's fellow geologists—such as Murchison, William Conybeare, and William Buckland—were primarily interested in fossils and palaeontology, Sedgwick's main interest was in geological structures. He thought about rocks in geometrical terms. He wanted to know about dips, strikes, and synclines. Fossils could be a useful tool, but they were not the central feature of Sedgwick's geology.[71] Sedgwick's first paper to the Cambridge Philosophical Society, which was delivered on 20 March 1820 and in which he described the landscapes and underlying rocks of Devonshire and Cornwall, perfectly illustrated his interest in applying geometrical ideas to the earth. The mining districts of Cornwall—bleak as they were, with their 'wild and dreary aspect', sparse vegetation, and towering mounds of boulders and waste—allowed Sedgwick to see the internal geometry of the earth.[72] Seeing the shapes of the bands of rock that ran beneath the surface meant that Sedgwick could begin to understand their three-dimensional interrelations, their histories, and their origins. For him, this mathematical view was a key to unlocking the earth's past.

Mathematizing phenomena became ever more important in Cambridge science as the century progressed.[73] Other fellows of the Society were also determined to apply mathematical reasoning to new fields. In the spring of 1831, William Whewell (who was by now the head tutor at Trinity College as well as being Professor of Mineralogy) read a paper on the mathematics of political economy and taxation. This was typical of Whewell's broad approach; often, the papers he brought to the Society delved into topics unrelated to the natural sciences, and yet his methods of exploring those topics fitted perfectly with the general aims of the Society. Whewell believed that explaining economics in mathematical terms would make it 'more clear, compendious, and manageable', and that his mathematical formulae would be both 'more exact and much more general' than the examples used by other authors.[74] A few years later, Whewell applied the same idea to the tides: he believed that, if he could mathematize the tides, he could understand them better and use them to develop a more profound understanding of universal gravitation.[75] In March 1837, Henry Bond, a physician at Addenbrooke's Hospital, presented a statistical analysis of the patients who had used the hospital in the previous year, breaking them down by age, gender, occupation, and illness. His results were later published in the *Transactions*, in an enormous and elaborate fold-out.[76] This application of statistics to medicine had been developed in France in the late eighteenth century and was just beginning to be applied in England.[77]

There were very few papers in the *Transactions* that did not include formulae or mathematical reasoning, though occasionally Henslow, Jenyns, or another natural historian would publish an article on botany or zoology and would follow the conventions of those fields rather than the mathematical style that was growing in Cambridge. Sedgwick once jokingly complained that he could not understand most of the papers presented to the Society:

> I rejoice in the progress of mathematical science; I measure it in this way; I am a stationary kind of being with regard to mathematics; the progress of the science may be measured by the small amount of that which I am able to understand; and I give you my word of honour that I have not been able to understand a single paper that has been read before this Society during the last twenty years.[78]

In reality, of course, Sedgwick's beloved geology was becoming increasingly mathematical in this period, and he was a driving force behind that mathematization.

Sedgwick explained this mathematical methodology in a series of letters to his close friend William Wordsworth. Sedgwick had first met the poet in the early 1820s through Christopher Wordsworth—William's younger brother and the master of Trinity College from 1820 to 1841—and the two had quickly formed a bond. Twenty years later, Sedgwick still recalled how 'some of the happiest summers of my life were passed among the Cumbrian mountains, and some of the brightest days of those summers were spent in your society and guidance'.[79] Since the turn of the century, the Lake District had exercised a fascination over English society. The works of the Lake Poets—figures like Wordsworth, Samuel Taylor Coleridge, and Robert Southey—seeped into the public consciousness as the Romantic Movement swept the nation. Sedgwick spent much time in the Lakes, hammer always in hand, seeking to unravel the mysteries of its lofty peaks. Sedgwick was not alone; his letters, and those of his friends, reveal a district almost overrun with the brightest lights of English arts and sciences. 'It was near the summit of Helvellyn that I first met Dalton,' writes Sedgwick casually of his encounter with the chemist who had proposed the atomic theory; 'it was, also, during my geological rambles in Cumberland that I first became acquainted with Southey,' he continued.[80] William Whewell meanwhile, another eager visitor to the Lakes, gives a sense of the sociable nature of the place: 'I got here on Thursday last, and next day saw Wordsworth at Rydal, and Southey at Keswick, by whom I was informed where to look for Sedgwick. I found him on Saturday at the base of Skiddaw, in company with [Richard] Gwatkin, as I had expected.'[81] Sedgwick found himself elevated by the irresistible combination of place and companion:

> [Wordsworth] joined me in many a lusty excursion, and delighted me (amidst the dry and sometimes almost sterile details of my own study) with the outpourings of his manly sense, and with the beauteous and healthy

images which were ever starting up within his mind during his communion with nature, and were embodied, at the moment, in his own majestic and glowing language.[82]

The friendship of the poet and the geologist must have seemed odd to many, for it was only a few years before they met that Wordsworth had written these scathing lines about geology:

Nor is that Fellow-wanderer, so deem I,
Less to be envied, (you may trace him oft
By scars which his activity has left
Beside our roads and pathways, though, thank Heaven!
This covert nook reports not of his hand)
He who with pocket-hammer smites the edge
Of luckless rock or prominent stone, disguised
In weather-stains or crusted o'er by Nature
With her first growths, detaching by the stroke
A chip or splinter—to resolve his doubts;
And, with that ready answer satisfied,
The substance classes by some barbarous name,
And hurries on; or from the fragments picks
His specimen, if but haply interveined
With sparkling mineral, or should crystal cube
Lurk in its cells—and thinks himself enriched,
Wealthier, and doubtless wiser, than before![83]

This view of Wordsworth's, that geologists were scarring the countryside for little purpose, was changed by seeing how Sedgwick's fieldwork allowed him to attempt to answer bigger, more philosophical questions about how the earth was put together, and how it all worked. The two must have talked through these questions as they worked their way across the Cumbrian countryside, and Sedgwick later formalized his thoughts into a series of letters to Wordsworth explaining his methodology:

When physical phenomena are well defined, and their laws made out by long and patient observations, or proved by adequate experiments: they then, by an act of thought, may be made to pass into the form of mere abstractions, and so come within the reach of exact mathematical analysis:

> and many new physical truths, unapproachable in any other way, and far removed from direct observation, may thus be brought to light, and fixed as firmly as are the truths of pure geometry.[84]

This was the motivation behind Sedgwick's methodology; mathematization was a useful tool for him, as it was for many researchers in Cambridge, and so mathematical thinking had begun to dominate the pages of the Philosophical Society's *Transactions*.

The Society gave confidence to Cambridge's men of science. It provided more than simply a place to meet and books to read; it bestowed upon them a sense of fellowship and common purpose. Along with the many other philosophical societies springing up around the country, it gave a public face to natural philosophy and demonstrated the growing interest in the sciences. Within a few years of its foundation, the Society had accomplished much, but some of the fellows thought that there was still more to achieve. Charles Babbage felt this particularly. He had looked to Europe and seen visions of professional men of science who were funded by governments and pursuing research undreamed of in England. He lamented that 'the pursuit of science does not, in England, constitute a distinct profession, as it does in many other countries'.[85]

The Society aimed to promote scientific enquiries and facilitate communication but it was not a campaigning body and kept itself out of politics—both within the University and nationally. Though Babbage had joined the Cambridge Philosophical Society in its first year of existence, he spent most of his time in London, forging a career for himself in science. He lectured on astronomy at the Royal Institution, became a fellow of the Royal Society in 1816, helped to establish the new Astronomical Society in London in 1820, published widely on mathematics and mathematical tables, and wrote about a diverse range of subjects: chess, geology, solar eclipses, and lighthouses, amongst others. In 1821, while working with his old friend John Herschel on some newly calculated astronomical tables, Babbage had first conceived the idea of a computing engine which would be powered by steam and would produce error-free tables. By 1823, he had secured a government grant to begin developing such a machine. But the haphazard nature of his funding and lack of support for the sciences more generally worried him.

In Cambridge, there was a growing appetite for the sciences amongst the undergraduates; Adam Sedgwick realized this when the attendance at his lectures began to rise dramatically. He described to a friend 'a kind of philosophical mania which broke out among the Cambridge Blues. Unfortunately for me their madness took a geological turn, so that I was obliged, out of pure compassion, to administer to them a sedative dose in the form of a three hours lecture.'[86] But, despite increasing attendance at lectures and, indeed, increasing numbers of professors who gave lectures, there were still few paid positions for an aspiring natural philosopher, and subjects like geology were not included in the University examinations. Herschel, who had tried Cambridge life after graduation, was frustrated by the gulf between what he wished to teach and what the University curriculum demanded he teach. This gifted mathematician found teaching the prescribed curriculum to uninterested students exasperating; he wrote to Babbage that he had to 'set from 8 to 10 or 12 hours a day examining 60 or 70 blockheads, not one in ten of whom knows his right hand from his left, and not one in ten of whom knows anything but what is in the book...In a word, I am grown fat, full and stupid. Pupillizing has done this.'[87] Having missed out on election to the chemistry chair by one vote, Herschel decided to leave the Cambridge system and embark on a new scientific adventure: he became, like his father William, an astronomer.[88]

However, the University was, slowly, beginning to allocate more space and money to the sciences. The University's new observatory building was completed in 1823 (see Plate 3). It had first been proposed in 1790, but no practical steps had been taken towards building it. In 1816, George Peacock, who had recently co-published on calculus with Babbage and Herschel and who had been elected a fellow of Trinity College, began seeking support for a new observatory amongst his colleagues in Trinity and the University more widely. 'Trust me,' he wrote to Herschel, describing his hopes for the new observatory, 'the golden age of the University is approaching.'[89]

Peacock's desire for a new observatory was driven by his attraction to analytical mathematics: the continental methods of analysis were particularly useful in astronomy, and Peacock hoped to increase interest in calculus by increasing interest in astronomy. The heads of the colleges, led by James Wood, the master of St John's College, vigorously opposed the building of an observatory. Perhaps the expense put them off, perhaps it was seen as a distraction from the real work of the University, or perhaps Peacock's desire for reform alarmed them. At last, in 1820, after much campaigning, the University agreed to the formation of a new observatory. The Reverend Wood, warming to the new scientific scene in Cambridge, even joined the Philosophical Society. 'This was more than we expected,' wrote Sedgwick to Herschel, 'and certainly more than Dr Wood intended last year. It seems as if we had risen in his good opinion.'[90] Wood must have found the Society a congenial place, for he went on to become its president in 1821.

The observatory would be overseen by the Plumian Professor of Astronomy and Experimental Philosophy, it would employ two assistants to act as professional observers, and it would have two sets of telescopes and observational instruments: a more sophisticated set for the assistants' research, and a simpler set for teaching students. The observatory was ready for use in 1824 but the Plumian Professor, Robert Woodhouse, who had been a member of the Philosophical Society's first council, was in poor health and conducted little research in the observatory in the years before his death in 1827. In 1828, when George Biddell Airy was elected to the Plumian Chair, he took over the observatory, giving new vigour to Cambridge astronomy.[91]

Airy had come to Cambridge as a student in 1819 and graduated as senior wrangler in 1823. He had studied under Peacock, and presumably it was Peacock who had encouraged Airy—while still a student—to send several papers to the Philosophical Society. Being considered too young to present his work himself, his first paper—on silvered telescope mirrors—was read to the Philosophical Society on his behalf by Peacock.[92] Airy joined the Philosophical Society as soon as he graduated, and continued to read his papers at their meetings and publish in their *Transactions*. His topics were incredibly varied: eye defects, fluid mechanics, cogwheels, Laplace's theory of the attraction of spheroids, spherical aberration in lenses, pendulums, longitude, and perpetual motion. Airy took part in all kinds of scientific investigations, from plunging himself, along with his friends William Whewell and Richard Sheepshanks, hundreds of feet down a Cornish mineshaft in a bid to measure the increasing effects of gravity below the earth's surface, to tramping around the Lake District with Wordsworth—who nicknamed him 'Orson the hairy'. In 1826, Airy published his *Mathematical tracts on physical astronomy*, which quickly became a university textbook; in the same year, he became Lucasian Professor of Mathematics. Two years later, he became Plumian Professor of Astronomy and Experimental Philosophy, and thus took control of the observatory.[93]

His new post not only allowed Airy to pursue his scientific work; it also meant that he could marry. The Lucasian Chair came with only a small salary—not enough to keep a family—but the Plumian Chair came with £500 each year, as well as a house. Airy married Richarda Smith in March 1830, six years after he had first proposed, and they moved together into the apartments in the east wing of the observatory, looked over by the building's fine copper dome. Richarda Airy entered enthusiastically into the social life of Cambridge: by May 1830, she could be heard singing at one of the Henslows' soirées and, by November that year, she was hosting her own lively parties at the observatory, at which, recorded Joseph Romilly, 'Mrs Airey [sic] sang the Marseillaise, Mrs Henslow Beethoven, Miss Smith a singularly coarse milkmaid (Haydin [sic])'.[94]

While the east wing of the observatory rang with music, the central domed portion was given over to serious investigation. One of the stated reasons that the University had built an observatory was because of the utility of astronomy. The star charts the observatory produced could be used to refine the tables of the *Nautical almanac*—an important tool for the powerful Royal Navy.[95] The interests of the Navy had long been a spur to scientific research; in 1714, the government had founded the Board of Longitude to encourage ingenious solutions to the problem of determining longitude at sea. The potential monetary rewards of the Board had driven much scientific research in the eighteenth and early nineteenth centuries. John Harrison's famous chronometers were created as a result of the Longitude fund, as were a host of other, less well-known inventions and innovations that improved navigation: lunar telescopes, more detailed lunar distance tables, improved sextants, and several different kinds of timepiece.[96] The Board of Longitude was one of the primary means by which the government distributed funding to men of science. There were some disputes about how the Board was run, and some of its members, including Herschel and Airy, hoped to reform it from within. Before they could succeed, the Board of Longitude was dissolved by an Act of Parliament in 1828, cutting off a valuable resource to aspiring inventors and men of science.[97] This caused further despair to Charles Babbage and he began to pen a book on what he saw as the depressing state of science in England.[98]

By the time his book was published in 1830, Babbage had succeeded Airy as the Lucasian Professor of Mathematics at Cambridge (thanks, in part, to the support of Adam Sedgwick[99]): he was no longer the scrappy undergraduate who had set up the fringe Analytical Society; he was part of the establishment. He showed a draft of the book to John Herschel, who hated it, writing that he wished he could give Babbage 'a good slap in the face' and advising him to burn the book, or rewrite it. David Brewster, to whom Babbage also showed the proofs, loved it, and wrote an essay in support of Babbage and the book.[100]

*Decline of science* was a sustained, bitter, sarcastic, and personal attack on the Royal Society and its members. Babbage believed that the Royal Society had inexcusably let its standards slip and that most of its fellows did not conduct any research; indeed, that many were ignorant of the basic tenets of natural philosophy. He believed that the Royal Society had become a cliquish and expensive private club, rather than the upholder of scientific values and promoter of scientific careers that it should be. Babbage also touched on the need to reform the universities and improve the teaching of science in schools. He envisioned a curriculum that embraced history, law, political economy, applied science, chemistry, mineralogy, geology, botany, and zoology. He also suggested compulsory lectures at the universities (though he never once lectured from the Lucasian Chair in the eleven years he held it). In contrast to the venerable but staid Royal Society, Babbage praised the newer societies, particularly the Geological Society.[101]

*Decline of science* was, as Babbage had intended, highly controversial. A few, like Brewster, wholeheartedly supported him. Most fellows of the Royal Society, particularly those who had been specifically named or alluded to in the book, took great offence. The Cambridge men too, led by William Whewell, were generally hostile to the book.[102] But *Decline of science* did have the desired effect: it made people talk about the state of science in the country, about the role of scientific societies, and about the reforms needed in the universities.

The response to *Decline of science*, along with the growth of provincial societies, gave David Brewster an idea: if the old guard of natural philosophy was no longer promoting the cause of science, a new kind of scientific body must be formed that would. Brewster proposed holding a meeting for all British men of science, away from the metropolis, that would bring together researchers to meet and exchange ideas, thus enlivening all branches of science. He proposed a meeting at York—halfway between the English and Scottish capitals—to be held in the autumn of 1831.[103]

In Yorkshire, the discovery of the Kirkdale Cave fossils in 1821 had fuelled desire for a local repository to keep the treasures and, in response to this desire, the Yorkshire Philosophical Society was set up the following year.[104] This society, like its sister societies around the country, became a thriving centre of popular science. The provincial societies were on friendly terms with each other, often engaging in specimen exchanges and even discussing plans for cooperative research and for establishing a network of provincial lecturers. The Yorkshire Society in particular, under the leadership of its president, William Vernon Harcourt, sought out cooperation with other societies and looked for new ways to promote its work. It collaborated with Brewster in Edinburgh on a series of joint meteorological observations, and kept in touch with Oxford's Professor William Buckland, who had meticulously studied the Kirkdale fossils. The Yorkshire Society thus maintained important links to metropolitan and university science.[105]

The members of the Yorkshire Philosophical Society, led by Harcourt, threw themselves into the organization of Brewster's proposed meeting, inviting as many men of science as they could think of. And so, on 26 September 1831, the British Association for the Advancement of Science was born in York. It was a large gathering—353 men attended to give papers on their research and hear about the latest developments in science. Many of the attendees were drawn from the provincial philosophical societies. They bustled excitedly into the meeting rooms of the grand new Yorkshire Museum to meet their peers and catch up on the latest scientific news from across the country. An atmosphere of hope and expectation abounded, a feeling that this was the beginning of something significant. Sadly, the London and university elites largely snubbed the meeting, deeming it too distant, too peripheral, and too unimportant. Sedgwick did not attend because he did not want to give up a geological field trip. Whewell did not attend because he was annoyed with Brewster's support for Babbage's *Decline of science*. In fact, only three Cambridge men attended.[106]

In 1832, with support from William Buckland, who presided over the meeting, the British Association met in Oxford. This time, a large number of Cambridge men attended. Buckland tempted a reluctant Sedgwick to Oxford by appealing to his sense of comradery, and a promise of a warm welcome: 'I exhort you by all your love for Professorial Unity and the eternal fitness of things, to locate yourself in a fraternal habitat within my domicile during the orgies of the week beginning on the 3rd of June.'[107] Sedgwick was persuaded, and found himself delighted by the Oxford meeting, playing an active role in the geological section. He was greatly impressed by Buckland's presidency, recalling his friend 'diffusing sunshine' throughout the proceedings.[108] In fact, Sedgwick and the entire Cambridge contingent were so impressed that they became determined to hold the next meeting at Cambridge. And Sedgwick, still President of the Cambridge Philosophical Society, was chosen to preside over the 1833 meeting of the British Association in Cambridge.

The meeting was held under the auspices of the Cambridge Philosophical Society. Sedgwick as Society President and the three secretaries—Whewell, Henslow, and the University librarian, John Lodge—effectively organized all practical aspects of the meeting: setting the dates, negotiating the use of University rooms, and taking care of fundraising. The house of the Philosophical Society played a central role: it was there that the meeting's participants could collect their tickets, arrange lodgings, find lists of the Association's regulations, obtain a map of Cambridge, and sign the guest book. The Society threw open its reading room to visitors and ordered in extra copies of all the major newspapers to cope with demand. Almost 900 people flocked to the city, throwing themselves into the genteel chaos of the Association. They moved between meetings in six specialist sections (mathematical and general physics; chemistry and mineralogy; geology and geography; natural history; anatomy and medicine; and statistics); they attended committee meetings, dined at the Hoop Inn, or took tea in Senate House; and they queued patiently for admittance to the popular evening lectures, listened to concerts, worshipped in the college chapels, marvelled at a fireworks display on the banks of the River Cam, followed Professor Henslow as he led a botanical barge trip downriver, and looked forward to the cold collation for six hundred guests that was to be held in the hall of Trinity College and would be the grand finale.

The meeting was a great success: at the start of the Cambridge meeting, there were just under 700 members of the Association; by the end, there were almost 1,400. The town had buzzed like never before as the members circulated between talks in the University's Senate House, the Schools, and in the hall of Caius College, hearing about cutting-edge science across a range of topics. The report produced after the event filled over 500 pages. But perhaps the surest sign of the Association's success, and confirmation of its growing visibility in British society, was Charles Dickens's parody of it as the Mudfog Association for the Advancement of Everything.[109]

The programme of the 1833 Cambridge meeting overflowed with talks and events, but the meeting is best remembered as the occasion of the coining of a new word: 'scientist'. Though many now laud the creation of the word as the beginning of a brave new world for disinterested studies of the natural realm, it was originally intended as something of a put-down. As natural philosophy moved away from dealing with larger questions about God and His creation, and towards precise investigations of very specific phenomena, many despaired. The poet and philosopher Samuel Taylor Coleridge admitted that he was 'half angry with [the chemist Humphry Davy] for prostituting the name of Philosopher...to every Fellow who has made a lucky experiment'.[110] A new word was needed to differentiate between the lofty knowledge of the philosopher, and the grubby workings of a scientific laboratory. Coleridge discussed this problem with William Whewell, who outlined what he saw as the problem:

> Formerly, the 'learned' embraced in their wide grasp all the branches of the tree of knowledge...but these days are past...[T]he disintegration goes on, like that of a great empire falling to pieces; physical science itself is endlessly subdivided, and the subdivisions insulated....The mathematician turns away from the chemist; the chemist from the naturalist; the mathematician, left to himself, divides himself into a pure mathematician and a mixed mathematician, who soon part company; the chemist is perhaps a chemist of electro-chemistry; if so, he leaves common chemical analysis to others... And thus science, even mere physical science, loses all traces of unity. A curious illustration of this result may be observed in the want of any name by which we can designate the students of the knowledge of the material world collectively....[T]his difficulty was felt very oppressively by the members of the British Association for the Advancement of Science, at their meetings at York, Oxford, and Cambridge, in the last three summers. There was no general term by which these gentlemen could describe themselves with reference to their pursuits.

The term 'philosopher' was 'too wide and too lofty'; the German *naturforscher* was good but did not lend itself to any elegant English translation, while the French word *savans* was 'rather assuming' and too, well, French. Instead, in front of the British Association gathered in Cambridge's Senate House, Whewell proposed the word 'scientist' as an analogy with 'artist'. At the time, Whewell reported, his idea was not 'generally palatable'. In Britain, it was rejected as being ugly and utilitarian. But, over the course of the century, both the word and the concept of distinct scientific disciplines began to become more visible.[111]

Though Whewell's new word did not find any supporters immediately, this meeting of the British Association played an important role in pushing the agenda of science. It brought the latest science and the most innovative thinkers to Cambridge and it showcased the work of Cambridge men, and the Cambridge Philosophical Society, for a wider audience. The rise of the British Association for the Advancement of Science showed the increasing importance of societies, and particularly those philosophical societies located outside London, in the development of the sciences. In the eyes of the nation—including the working and middle classes—it was the societies, and especially the public-facing British Association, that represented modern science. For the Association, which perhaps more than any other body campaigned for the creation of what we now think of as 'science', it was important that their activities had a practical bent. They did not focus exclusively on the esoteric knowledge of natural philosophy but looked towards industry and threw themselves into the modern world. Though the British Association held some meetings in university towns, the organizers also made a point of touring Britain's industrial and trade cities: Bristol in 1836, Liverpool in 1837, Newcastle in 1838, and Birmingham in 1839. In the first decades of the nineteenth century, the societies comprised the most vital force in the scientific life of the country.

The Cambridge Philosophical Society was in a unique position: it had both the freshness of the rising societies, and access to the resources of a long-established university. Its members allied themselves closely with the British Association and with the elite London societies, while maintaining their fellowships and professorships. Though its activities were designed to be self-contained, it did not operate in a vacuum: the Society was perfectly positioned to influence the University and, consequently, to shape the future relationships between the sciences and the academy.[112] Its ability to operate in so many different contexts made the Society one of the most important organizations in the scientific life of Britain in this period.

Some years after his initial geological field trip with Henslow, Sedgwick wrote that he would 'never forget the glowing beauty of the shores of the Isle of Wight'.[113] The Society that had first been conceived of on that island had grown and prospered. Each year, on the anniversary of the founding of the Society, even many decades after Sedgwick and Henslow had stood beneath the cliffs of Alum Bay wondering what those ancient fossils meant, the Society marked the occasion with a dinner at the Eagle. The fellows would crowd into the inn, eager to hear one of Sedgwick's legendary speeches. An anniversary speech by Sedgwick was said to be

> [r]emarkable—a wild exuberance that nearly touched upon the region of nonsense, and then, apparently without effort…rose to the solemn and almost to the sublime; the combination, without incongruity, of lofty morality with almost boyish fun was quite wonderful, and almost Shakespearean.[114]

For Sedgwick and the other fellows, the Society was all of these things: solemn, sublime, lofty, and yet enormous fun. It was their playground: a place of intellectual freedom beyond the strict confines of the University curriculum, and a place of cooperation and friendships. It was a place where things got done: if Cambridge lacked a decent scientific library, the Society would assemble one; if the town didn't have a natural history museum, the Society would create one; if the Press failed to produce a natural philosophical journal, the Society would write one themselves; and, if there was no dedicated space for science in Cambridge, the Society would build one—their house in All Saints' Passage.

# 3

# LETTERS FROM THE SOUTH

The stars in the southern skies shine differently. As John Herschel sailed south from England in the winter of 1833, he saw the familiar stars above him slip over the horizon and out of view. Into their place rolled new constellations: he would have seen the Southern Cross for the first time; the Chamaeleon; Tucana—the toucan; Dorado—the dolphin-fish; and, perhaps most strikingly, Centaurus—the centaur. Centaurus had been included in ancient Greek celestial maps; Ptolemy had catalogued its stars. But, as the earth's axis underwent precession, Centaurus had dropped out of view in the northern latitudes, becoming just a dim memory for European astronomers. As Centaurus reappeared in Herschel's vision, it reminded him of his mission: to become the first astronomer to survey both the northern and the southern skies with a powerful telescope. More specifically, Herschel was travelling to the southern hemisphere to complete the work that he and his father William and aunt Caroline had begun in the north: mapping nebulae. These cloudy patches of light, dotted across the heavens, fascinated the Herschels. Were they clusters of stars, or was William right in thinking that they were they made up of a mysterious shining fluid? By the time he left England in the early 1830s, John Herschel had spent countless hours staring through his telescope, documenting the thousands of nebulae visible there. Now he would continue his work from the Cape of Good Hope.

After more than two months at sea, Herschel and his wife Margaret disembarked at the British-controlled Cape Colony in January 1834 (see Figures 15 and 16). Five years earlier, on the occasion of John and Margaret's marriage, Adam Sedgwick had used his presidential address to London's

**Figure 15** Margaret Herschel (c. 1829).

**Figure 16** John Herschel (c. 1829).

Geological Society to wish his old friend well, provoking peals of laughter from his audience when he launched into an ornate speech: 'May the house of Herschel be perpetuated, and, like the Cassinis, be illustrious astronomers for three generations. May all the constellations wait upon him; may Virgo go before, and Gemini follow after.'[1] Sedgwick's benediction must have worked, for the Herschels already had three children before setting off for the Cape, and Margaret had another three in South Africa, plus a further six after the couple returned home; many of the children had an interest in astronomy and one—Alexander—would go on to become a professor of the subject.

Within a month of the arrival of the young family in South Africa, Herschel had found them a house a few miles outside Cape Town. This pretty gabled house, Feldhausen, surrounded by luscious vegetation and overlooked by the imposing Devil's Peak, would be their home for the next four years (see Figure 17). Though they had come south to see the stars, the rich flora of the countryside around them inspired Margaret to pursue a different kind of scientific pastime—she began an extensive botanical collection. John too was much taken by the exotic plants of South Africa, and so the couple began a joint venture: John would use a *camera lucida* to draw the outlines of the plants Margaret collected, and she would use her artistic skill to fill in the details and add colour. The result

**Figure 17** Herschel's telescope at Feldhausen in the 1830s.

was a striking, and scientifically important, collection of the flora of South Africa (see Plate 4).

But John could not forget the real reason he had come here. Work on setting up his enormous twenty-foot reflecting telescope began as soon as he found Feldhausen. Herschel's mechanic, a man named John Stone, whom Herschel described as 'my Astronomical blacksmith, a most excellent and useful man', had travelled with him to Africa and oversaw the assembly of the telescope; Herschel and Stone were helped by four local labourers whose names Herschel recorded only as Abdul, January, Thorn, and Jacob.[2] Once the telescope was ready, Herschel could begin his systematic sweeps of the night sky—searching for smudges of light that might be new nebulae.

Herschel was a man of means, and his work was undertaken independently, but he benefited enormously from the recent completion of a British Royal Observatory at the Cape. Under the direction of the astronomer Thomas Maclear (who had arrived in Cape Town just a few days before Herschel), the Observatory began to remeasure the locations of the southern stars and update the star charts. With Maclear's charts next to him, Herschel spent his nights scanning the clear skies of South Africa and, in his four years there, he observed and logged 1,708 nebulae, over 1,200 of them previously unrecorded. Herschel saw hundreds of stars within each nebula. In the Large Magellanic Cloud, he counted 919 stars, and painstakingly drew them in his notebook; in the Small Magellanic Cloud, he catalogued 244 individual stars. Herschel also saw nebulous structures within these two great Clouds: nebulae within nebulae. This puzzled him. Herschel began to question the prevalent theories about nebulae, and about the way the universe was put together.[3]

The nebulae filled Herschel's nights and days. He wrote to a friend back in England that he had time for nothing but 'stars stars stars'.[4] When he wasn't searching for nebulae, Herschel would look for double stars. The prize here was to find pairs that were gravitationally linked: stars that did not just appear to be close when viewed from the earth, but that really were close enough to exert a gravitational force on each other. Herschel catalogued over 2,000 double stars, carefully looking for the slight wobble that indicated the stars were pulling themselves around each other. In 1835, Herschel's gaze was distracted from the stars by the long-expected reappearance of Halley's Comet. Anxious to view and record the comet from his southern vantage point, Herschel began looking out for the comet as early as February 1835, but it was October before it became visible. Despite his best efforts to view the comet from his house, the thick cloud that rolled off the top of nearby Table Mountain and the avenue of oak trees that led to the property blocked his view. Cutting a gap in the oak trees did little to improve the situation, so Herschel knocked up a temporary stand for his seven-foot equatorial telescope, dismantled it, carried it out to the Sand Hills on the Flats a few miles distant, then reassembled the stand, mounted the telescope and, just as the sun was setting, he was, as he wrote in his diary, 'rewarded with the first glorious sight of Halley's Comet!!!'[5]

Though Herschel claimed that he did nothing but look at the stars, he found time to attend meetings of the South African Literary and Philosophical Institute, of which he was elected President almost as soon as he arrived. He and Margaret embarked on their joint botanical project, and they cultivated the flora of South Africa in their garden. Herschel also expounded on new geological theories, and took careful meteorological readings. Herschel wrote regularly to his colleagues in England about his scientific labours. His letters were frequently addressed to his old Cambridge associates, and their contents often found their way into meetings of the Cambridge Philosophical Society.

On the evening of 16 November 1835, William Whewell took the floor of the Society's meeting room to read aloud just such a letter. In it, Herschel described how, while sailing across the equator, he had noticed that the barometer unexpectedly dipped. Determined to find out why this happened, Herschel had commissioned various ships' captains and gentlemen travellers to make observations for him as they sailed north and south along the coasts of Africa and India. Analysing the results from these different sources, Herschel proposed that the depression of the barometer might be caused by the trade winds.[6] Herschel's letter represented more than just some interesting meteorological data; it shows how scientific information from across the globe found its way into meetings of the Cambridge Philosophical Society. By staying in communication with fellows who had left Cambridge, the Society broadened its horizons; by engaging with the most up-to-date data from a variety of sciences, the Society kept itself relevant. Through this breadth and this relevance, the Society ensured that it had become, and would remain, desirable: to have a letter read at a meeting of the Cambridge Philosophical Society was considered a great honour.

After Whewell had read Herschel's letter aloud, the meeting's next item of business was a letter from another Cambridge man who was far from home. The meeting was being chaired by William Clark, Professor of Anatomy and President of the Society (nicknamed 'Bone' Clark to distinguish him from the old mineralogy professor, Edward Daniel 'Stone' Clarke, and the professor of music, John 'Tone' Clarke). Clark called John Stevens Henslow to the floor to read the letter, for it had been written to Henslow from a former student of his, a young man named Charles Darwin[7] (see Figure 18). Darwin had studied at Cambridge between 1828 and 1831, and had been a favourite student of Henslow's. Through Henslow's lectures, field trips, and soirées, the young Darwin had deepened his understanding of the natural sciences. Upon graduating, Darwin had vague hopes of pursuing a career in natural history but no clear plan until, in August 1831, he was offered a place on HMS *Beagle* as it mapped the coastlines of South America.

The offer came about through the machinations of several fellows of the Cambridge Philosophical Society. George Peacock, who was still a mathematics tutor at Trinity and an active member of the Society, had been asked by his friend Francis Beaufort, the hydrographer of the Navy, to recommend a young gentleman who could travel on the *Beagle*, acting as both naturalist and companion to the captain. Peacock first thought of Leonard Jenyns, who had worked so hard to build up and curate the Museum of the Philosophical Society. Peacock wrote to Henslow to ask

**Figure 18** Charles Darwin (1830s).

his opinion of Jenyns as a candidate, and also to highlight how the voyage could contribute to Cambridge's natural history collections, most of which, at that time, were held by the Philosophical Society. 'If Leonard Jenyns could go, what treasures he might bring home with him,' mused Peacock; and, a few days later, 'what a glorious opportunity this would be for forming collections for our museums.'[8] But Jenyns declined to travel on the *Beagle*, citing his duty to his parishioners in Swaffham Bulbeck. For a fleeting moment, Henslow considered going along on the voyage himself but, though Harriet Henslow agreed to let him go, it was said that she looked so miserable that Henslow could not bring himself to leave her. And so it was that Henslow suggested Darwin as a candidate.[9]

The *Beagle*, with Darwin on board, had set sail in December 1831. It sailed south, past the Canary Islands, before crossing the Atlantic and landing in Brazil. The ship then spent many years painstakingly tracing the outlines of South America. Darwin would take advantage of longer stops to venture ashore, often spending months inland exploring the local terrain. Whenever he could, Darwin wrote to his friends at home, especially to Henslow. Through Henslow, Darwin would send messages to his many Cambridge friends. He was particularly keen to reassure Jenyns that he had made the right decision: 'I think L Jenyns did very wisely in not coming... Remember me most kindly to him, & tell him if ever he dreams in the night of Palm trees he may in the morning comfort himself with the assurance that the voyage would not have suited him.'[10] Despite its many material discomforts, the voyage suited the young Darwin perfectly, and the five years he spent aboard the little vessel would come to define his life and work.

By November 1835, the survey of South America was complete and the *Beagle* had sailed on, passing though the Galapagos Islands before arriving at the South Sea island of Tahiti on 15 November. The following day, Darwin began his exploration of the island in earnest:

> [I] ascended the slope of the nearest part of the mountain to a height of between two and three thousand feet.... From the point which I attained, there was a good view of the distant island of Eimeo... On the lofty and broken pinnacles, white massive clouds were piled up, which formed an island in the blue sky, as Eimeo itself did in the blue ocean. The island, with the exception of one small gateway is completely encircled by a reef. At this distance, a narrow but well-defined line of brilliant white was alone visible, where the waves first encountered the wall of coral. The glassy water of the lagoon was included within this line; and out of it the mountains rose abruptly. The effect was very pleasing.[11]

While Darwin was exploring the dramatic landscape of Tahiti, his mentor and friend Henslow, back in England, was stepping up to the stage of the Cambridge Philosophical Society. In the soft gaslight, Henslow read to the assembled philosophers.

The letters Henslow read aloud on the evening of 16 November 1835 told tales from Darwin's travels in South America. In the sombre hall of the Philosophical Society, the spellbound audience heard Darwin's accounts of storms that lashed the *Beagle* in the southern seas; of the wretched inhabitants of Tierra del Fuego who were struggling to keep warm in a bitter winter; and of wars being waged against the native populations around Buenos Aires. Henslow's voice, reading Darwin's words, summoned the endless plains of Patagonia and the immense peaks of the Andes. Darwin's words made the listeners understand the force of fierce volcanoes, and of earthquakes that could level entire towns. And he told of many marvels too: of fossil bones belonging to giant sloths and armadillos; seashells found 13,000 feet up a mountain; and a forest made of petrified trees. Darwin's letters brought the wonders of South America to life in that hushed chamber.[12]

The letters excited much curiosity about the wildlife and geology of South America, and they introduced Darwin to a scientific audience, showing him to be a man capable of making sound observations. It was a momentous day in the career of a young naturalist, but one that, strictly speaking, didn't exist for Darwin: the *Beagle* reached Tahiti on 15 November but the following day was recorded in the ship's log as 17 November. This was, as Darwin put it, 'owing to our so far successful chase of the sun'—what we would now refer to as crossing the International Date Line.[13] Unaware of the stir that his letters were causing back in Cambridge—indeed, completely unaware that Henslow had read them to the Philosophical Society—Darwin continued with his exploration of Tahiti and then sailed on with the *Beagle* to New Zealand and Australia, before crossing the Indian Ocean and landing at the Cape Colony of South Africa.

At the Cape, Darwin planned a short expedition into the African desert to explore the landscape, but first he hoped to meet the celebrated John Herschel. Darwin wrote to his younger sister Catherine that he and the *Beagle*'s captain, Robert Fitzroy, planned to visit Herschel, explaining: 'I have heard so much about his eccentric but very amiable manners, that I have a high curiosity to see the great man.'[14] Darwin and Fitzroy were welcomed by Herschel into his home, and a few weeks later Darwin wrote to Henslow about how they had

> enjoyed a memorable piece of good fortune in meeting Sir J. Herschel. We dined at his house and saw him a few times besides. He was exceedingly good natured but his manners, at first, appeared to me, rather awful.... He appears to find time for every thing; he shewed us a pretty garden full of Cape bulbs of his own collecting; and I afterwards understood, that every thing was the work of his own hands.[15]

It had been chance that letters from Herschel at the Cape and Darwin in South America had been read to the Cambridge Philosophical Society on the same evening, but their letters, and their later meeting in Cape Town, show how the geographical spread of science was expanding in the nineteenth century. It was necessary for groups like the Cambridge Philosophical Society to engage with this increasingly global science if they were to maintain their place at the cutting edge of knowledge.

It was while at the Cape that Darwin learned in a letter from his older sister Caroline that his correspondence had been read to the Cambridge Philosophical Society. Following the first reading of the letters on 16 November, the fellows' curiosity had been piqued. News of the letters and their contents quickly spread beyond Cambridge; periodicals in Edinburgh and London reported some of Darwin's findings. It was while reading the respected journal *The Athenaeum* that Darwin's own father learned that his son's letters had been read at the Geological Society of London.[16] This too was Henslow's doing: he had persuaded Adam Sedgwick to read parts of the letters to a meeting of the Geological Society and was most pleased to hear that '[Darwin's] remarks have excited so much interest'.[17] In Cambridge, there was demand for a further reading. So a few weeks later, on 14 December, Henslow again stood before a general meeting of the Cambridge Philosophical Society and told them of the astonishing things Darwin had seen in South America: lizards that could bear live young; snow that fell red like blood; and how, when climbing through the Andes at an elevation of 12,000 feet, 'there is a transparency in the air, and a confusion of distances, and a sort of stillness, which give the sensation of being in another world'.[18]

The lizards were particularly interesting because they belonged to a species that normally lays eggs. Darwin and other naturalists observed that in certain regions around Mendoza, rather than laying eggs, the species became viviparous. Naturalists debated whether different climates might cause animals to respond in different ways.[19]

The local variations in geology were just as fascinating to the Cambridge philosophers as the variations in flora and fauna. In the 1830s, geology was a rapidly growing science and new theories of how rocks, mountains, and landscapes formed were being hotly debated by geologists like Adam Sedgwick and the London-based geologist Charles Lyell. Darwin had known Sedgwick in Cambridge, and it was Sedgwick who taught Darwin the basics of geological fieldwork during a trip to Wales just a few months before Darwin embarked on his *Beagle* adventure. From Rio de Janeiro, Darwin wrote to Henslow, asking him to pass on his regards to Sedgwick:

> Tell Prof Sedgwick he does not know how much I am indebted to him for the Welch [sic] expedition.—It has given me an interest in geology, which I would not give up for any consideration.—I do not think I ever spent a more delightful three weeks, than in pounding the N[orth] W[ales] mountains.[20]

And, a year later, having explored many new landscapes, Darwin wrote again to Henslow with some information about the origins of a fossilized giant sloth whose skeleton was causing much excitement back in England, 'Professor Sedgwick might like to know this: & tell him I have never ceased being thankful for that short tour in Wales.'[21]

Darwin was also deeply affected by the writings of Lyell whose work *Principles of geology* argued for a theory called 'uniformitarianism'. The central tenet of uniformitarianism was that the geological processes that occur in the present are the same as those that occurred in the past, and that present-day geological features have been formed by slow, incremental change over very long time spans. The word had been coined by William Whewell to distinguish it from 'catastrophism'—a geological theory that attributed present-day landscapes to sudden, catastrophic events in the history of the world, such as a major flood. As Darwin crossed the Andes, he made careful observations of geological phenomena. He noted specimens of minerals and rocks, recorded their positions and their relationships to one another, and tried to understand how the Andes had been created from 'the breaking up of the crust of the globe'.[22]

Henslow considered Darwin's observations important enough to be printed. On 30 November 1835, he suggested this to the council of the Philosophical Society, and Henslow, Peacock, and Whewell were appointed to act as peer reviewers.[23] The letters passed the review process and it was agreed to gather the most important passages into a booklet, which would be called simply *Extracts from letters addressed to Professor Henslow*, to be published and distributed amongst fellows of the Society and other interested readers. Apart from their *Transactions*, this was the first volume published by the Society. By Christmas, the book had been printed. A preface to the booklet, almost certainly written by Henslow, enthused about 'the interest which has been excited by some of the geological notices which [the letters] contain'. It also included the proviso that 'the opinions here expressed must be viewed in no other light than as the first thoughts which occur to a traveller respecting what he sees, before he has had time to collate his notes, and examine his collections, with the attention necessary for scientific accuracy'.[24]

Despite this wise proviso, Darwin was utterly alarmed to learn that Henslow had published extracts from his letters. He had not yet seen the booklet and he wrote to his sister: 'I can only suppose they refer to a few geological details. But I have always written to Henslow in the same careless manner as to you; and to print what has been written without care and accuracy, is indeed playing with edge tools.'[25] By the time Darwin learned of the booklet's existence in June 1836, it had been in circulation for more than six months, so he had little choice but to philosophically accept Henslow's actions. Darwin, however, was mollified when he learned from his sister that the booklet had been well received and that Henslow had told the Darwin family that young Charles would surely be placed 'among the first naturalist[s] of the day'. He was particularly gratified to learn that his father, who had had doubts about Charles's decision to pursue a career in natural history, 'did not move from his seat till he had read every word of your book...he liked so much the simple clear way you gave information'.[26]

A few months after visiting the Cape Colony, the *Beagle* returned home to England in October 1836. Darwin planned to spend several months in Cambridge before migrating to London. He was drawn to Cambridge because of the presence of his mentor Henslow and also because of the welcoming embrace of the Philosophical Society and the University men; he wrote to Henslow:

> I am very sure the assistance I shall get in Cambridge will be infinitely more than I ever should receive in London...I am out of patience with the zoologists, not because they are overworked but for their mean quarrelsome spirit. I went the other evening to the Zoological Soc[iety of London] where the speakers were snarling at each other, in a manner anything but like that of gentlemen. Thank Heaven, as long as I remain in Cambridge there will not be any danger of falling into any such contemptible quarrels....Mr. Lyell owned that second to London, there was no place in England, so good for a naturalist as Cambridge.[27]

Another reason that Darwin wished to be in Cambridge was because most of his collections were there. Initially, Darwin had hoped to send the specimens he gathered on the *Beagle* voyage to London to be part of the 'largest and most central collection' at the British Museum. He worried that sending them to a collection outside London, even a respected one like that at the Cambridge Philosophical Society, would displease the Admiralty, which was funding the expedition.[28] But, in the end, all of the specimens he gathered during his five years in South America, the Galapagos Islands, Tahiti, and countless other places were carefully packed up and sent back to Cambridge, care of Henslow.

When Darwin arrived back in Cambridge in the autumn of 1836, he was faced with the task of sorting the specimens and finding appropriate homes for them. Sedgwick had referred to the objects Darwin had sent home from South America as 'a collection above all price', and yet Darwin struggled to get the London naturalists interested in them, writing to Henslow:

> I have scarcely met anyone who seems to wish to possess any of my specimens...It is clear that collectors so much outnumber the real naturalists, that the latter have no time to spare. I do not even find that the collections care for receiving the unnamed specimens. The Zoological Museum is nearly full and upward of a thousand specimens remain unmounted. I daresay the British Museum would receive them, but I cannot feel, from all I hear, any great respect even for the present state of that establishment. Your plan will be not only the best, but the only one, namely to come down to Cambridge, arrange & group together the different families and then wait till people, who are already working in different branches may want specimens.[29]

Thus, Darwin spent several months in Cambridge, cataloguing specimens and communicating with naturalists, hoping that he could persuade them to help him describe and house the objects. Many animals, plants, and minerals from his collection eventually found their way to London, but a significant number (especially the fishes) remained in Cambridge, adopted by the ever growing museum of the Cambridge Philosophical Society.[30]

The Society took advantage of Darwin's presence in Cambridge to invite him to speak at a meeting that winter. On 27 February 1837, he gathered up some specimens from his geological collection and arrived at the Society's house on All Saints' Passage. The specimens he showed transported his listeners from the lava flows of the Cape Verde Islands, off the west coast of Africa, to the shores of the Rio Plata in Argentina, where lightning bolts had fused sand into fantastical shapes, and then back across the Atlantic to Ascension Island, hundreds of miles from any other speck of land, where the rocks had long been battered by heavy seas and changing tides.[31] Elated with the success of his paper, Darwin wrote to his sister Caroline that night:

> It is nearly twelve o'clock...I have been reading a short paper to the Philosoph. Socy of this place, and exhibiting some specimens and giving a verbal account of them. It went off very prosperously and we had a good

> discussion in which Whewell and Sedgwick took an active part. Sedgwick has just come from Norwich and we have been drinking tea with him.[32]

Shortly after this talk, Darwin moved to London. Though Darwin did not speak at the Cambridge Philosophical Society again, Henslow and others continued to present talks on Darwin's collections and findings. Darwin praised Henslow's presentations as 'magnificent' and thanked him for continuing to publicize his work in Cambridge.[33] Because Darwin had spent such a short time in Cambridge after graduating, he never got round to formally joining the Philosophical Society. Despite that, the Society had been the publisher of Darwin's first scientific writings, had welcomed his collections when no one else would, and had given him an important platform from which to address a scientific audience.

In the 1830s, Darwin was just at the beginning of his career. No one yet had any inkling of the ways in which his work would unfold. But already the battle lines were being drawn in the most significant scientific controversy of the century. Disputes about the idea of 'transmutation' (or evolution) of species long predated the 1859 publication of Darwin's *On the origin of species*. Since the latter half of the eighteenth century, many theories had been proposed by respected naturalists postulating that animals and plants could fundamentally alter their nature, thereby becoming different species.[34] These theories were often written in a technical style, intended to be read primarily by other serious naturalists; they held limited appeal for the general reader. But, in 1844, a new book was published—one that combined details from astronomy, geology, botany, zoology, chemistry, psychology, and anthropology to argue for the evolution of species. It was beautifully written, filled with compelling detail, and it was immensely popular. The book was called *Vestiges of the natural history of creation* and at least 100,000 people read it.[35]

It was a sensation. Newspaper reviews called it 'extraordinary', 'ingenious', and 'powerful'.[36] It was a hit with everyone from Queen Victoria, who had Prince Albert read it aloud to her in the afternoons, to political radicals and from aristocrats to shopkeepers.[37] The book's anonymous author described his theory of how the world and its inhabitants had been created. It began, said the author, with the creation of the sun and planets as the force of gravity pulled particles of nebulous matter into nuclei. Then he told the story of the development of the earth and its rocky surface before describing the earth's first life forms as preserved in the fossil record:

> the unpretending forms of various zoophytes and polypes, together with a few single and double-valved shell-fish...it is probable that there were sea plants, and also some simpler forms of animal life, before this period, although of too slight a substance to leave any fossil trace of their existence.

Next to appear in the fossil record were fishes, the first vertebrates. Then, as the seas receded and dry land appeared, came land plants, reptiles, birds, and mammals.[38] The author used results from recent studies in chemistry and electricity to explain how these life forms might have first been sparked into creation: he cited the synthesis of organic chemicals in laboratories and recounted the controversial experiments of Andrew Crosse, who claimed to have created insects when he applied electricity to various chemical solutions.[39] The author of *Vestiges* argued that

> the first step in the creation of life upon this planet was a chemico-electric operation, by which simple germinal vesicles were produced...simple forms are produced at first, but afterwards they become more complicated...the first step was an advance under favour of peculiar conditions, from the simplest form of being, to the next more complicated, and this through the medium of the ordinary process of generation.[40]

The author needed a mechanism to explain how simpler species could develop into more complicated species, and he looked to the theory of embryology to furnish this mechanism. He speculated that the embryos of all animals passed through similar developmental stages: in the earliest stage, the embryo resembled a fish, and then it developed to seem more like a reptile, then a bird, and finally a mammal. Fish embryos usually only developed as much as necessary before hatching and becoming mature fish *but*, if a fish embryo should happen to develop for longer, it could turn into a reptile—thus changing the species.[41]

These bold speculations, if true, would mean that man was nothing more than a highly developed animal, little different from the beasts. They would also mean that there was little to separate living things from inanimate matter. Such wild, provocative theories drove the success of the book. The printers churned out increasingly large print runs to try to keep up with the enthusiastic demand from the public.[42]

In Cambridge, however, *Vestiges* received a rather frostier reception. There were concerns on multiple fronts: the book contained several inaccuracies, and there were seen to be gaps in the author's logic. Furthermore, the central premise of the book—that one could write a natural history of creation—was deemed to be fundamentally wrong-headed, a contradiction in terms: creation was an event quite apart from natural history.[43] Finally, *Vestiges* challenged several central tenets of Christian belief. Cambridge was still a devoutly Anglican centre of learning at this time, and it was felt that the young men who studied there had to be protected from some of the dangerous and anti-religious ideas contained within the book. Adam Sedgwick called it 'mischievous, and sometimes anti-social, nonsense' and decried the author's theory of embryonic development as 'a monstrous scheme'.[44]

Within a year of the first appearance of *Vestiges*, it went through several editions; sales continued to rise. When it became clear that the book wasn't going to go away, the Cambridge men realized that they needed to tackle it head-on. Whewell responded to the success of *Vestiges* with a book of his own titled *Indications of the Creator*. By this time, Whewell had gone from being Mineralogy Professor to being Professor of Moral Philosophy, and had been elected as Master of Trinity College in 1841. As Master, Whewell took on an air of authority. Indeed, many colleagues commented on his new-found imperiousness, as when he told off his old friend Sedgwick for bringing his pet dog into the College: 'your frequent appearance in the College courts accompanied by a dog is inconsistent... [with the] Rules [of the College]'.[45] But Whewell was just as likely to turn this new sense of authority to scientific debates as he was to wayward pets.

*Indications of the Creator* was short, easy to read, and 'daintily dressed for dainty people'—in other words, it was aimed at the same people who were buying *Vestiges*.[46] Whewell compiled almost 200 pages demolishing the theory put forward in *Vestiges*, but his most damning assessment of *Vestiges* could be summed up in these words: 'no really philosophical book could have had such success'.[47] To Whewell's mind, the idea that major philosophical concepts or the full spectrum of modern science could be learned and understood from a little light afternoon reading was ridiculous. Cutting-edge science was best discussed and interpreted by those who dedicated their working lives to carrying it out; a properly scientific stance was necessary for destroying the theories put forth in *Vestiges*.

Of course, the Cambridge Philosophical Society was an ideal forum for discussing the latest developments in the sciences, and so it was there that Sedgwick launched his assault on what he saw as an inaccurate work that was receiving undue public attention. On 28 April 1845, Sedgwick chaired a general meeting at which William Clark, Professor of Anatomy, took to the stage to 'offer remarks on the theory of developments as that theory is propounded in a work entitled *Vestiges of the natural history of creation*'.[48] Sedgwick, who, though well versed in the minutiae of the fossil record, knew little about generation and developmental theory of organisms, had asked his friend Clark to deliver this paper.

Clark had previously published a paper in the Society's *Transactions*—a study of conjoined twins, with a lengthy discussion of embryology—which was centred around the idea 'that no organ in an individual of a lower class, however it may deviate from its perfect type, ever represents the type of a higher class'.[49] In other words, there was a hierarchy of species, and each creature was assigned a set place in it: neither species nor parts of species ever underwent true transmutation. Sedgwick saw in Clark an ally against the tide of transmutationist thought that was sweeping through popular culture. Like that earlier paper, Clark's 1845 paper used arguments from embryology to undermine the theory of transformation put forth in *Vestiges*. Clark strongly refuted the *Vestiges* author's claim that embryos from all species are essentially the same and pass through 'lower' forms (fish or reptiles) before becoming 'higher' forms like mammals. Clark described how

> when frogs and fishes are beginning to breathe by bronchial tufts and gills, other amphibia and birds are breathing by allantöid, and never, for an instant, breathe by gills. At the same period of foetal development, hot-blooded quadrupeds are breathing by allantöid and placenta jointly, while man is breathing by placenta alone. These are essential foetal differences, connected with the last perfection of animal structure, and they form a wide anatomical separation so as to bar all interchange or confusion of organic type.[50]

Clark also cited evidence about the different kind of eggs particular to different species, the different ways in which the hearts of particular species develop, and evidence about how sexual characteristics develop in embryos.[51]

By the 1840s, we know that discussion was allowed following papers presented at the Philosophical Society but, sadly, no record of what was said after Clark's paper has survived.[52] As Clark's work was championed by Sedgwick, we can assume that the paper would have been well received. Sedgwick was certainly pleased with it: Clark's words were used by Sedgwick in a lengthy review of *Vestiges*, in the *Edinburgh review* a few months later. The review was an almost paragraph-by-paragraph dissection of *Vestiges*, with each point painstakingly refuted. William Whewell also quoted Clark's 1845 paper in *Indications of the Creator*.[53]

To the modern reader, the reaction of Sedgwick and Clark against this theory of evolution may seem oddly backward-looking, but it must be remembered that the theory put forward in *Vestiges* is not the same as the evolutionary theory of today. Sedgwick, Clark, and many other fellows of the Society who opposed *Vestiges* were open to new ideas in science, provided that they were properly evidenced and argued according to the fellows' standards, standards that were conceived in a Christian context: faith and method went hand in hand. It was Clark who had chaired the meetings of the Cambridge Philosophical Society at which Darwin's letters from South America had been read—he was known for being welcoming to new voices in science. Clark had also done much to advance the study of anatomy in Cambridge: he bought huge numbers of specimens for the University's anatomical museum, he worked with his wife Mary Willis Clark to produce detailed embryological models for the museum, he did his own dissections, he gave hands-on lectures, and he oversaw the move of the museum from a small building opposite Queen's College to a larger building on Downing Street—a move which allowed Clark to build the museum into the best anatomy museum outside of London.[54]

Sedgwick too was a great defender of innovation in the sciences. At the 1844 York meeting of the British Association for the Advancement of Science, he had achieved national notoriety for defending modern geology against the accusation by the Reverend William Cockburn that it was anti-scriptural. Cockburn believed that geological findings should be interpreted according to the texts of the Old Testament and in 1844 gave a talk to the British Association in which he attacked a recent book by Oxford's Professor of Geology, William Buckland. As Buckland was not present at the meeting, Sedgwick defended Buckland's work as a fair interpretation of geological knowledge. Riled, Cockburn replied shortly afterwards with a fierce sermon and a book called *The Bible defended against the British Association*. The press, naturally, jumped on this supposed clash between science and the church, and the incident was reported in newspapers across the country. But, of course, it was possible to be both a devout Christian and a man of science: Sedgwick, ordained in the Church of England, was deeply religious, and his beliefs were entirely compatible with his geological work.[55]

In 1845, the British Association for the Advancement of Science met again at Cambridge. As before, the arrival of the British Association in

Cambridge was intimately connected with the Cambridge Philosophical Society. The meeting was largely arranged by the Society's men (Peacock was a trustee, Sedgwick and Airy were vice-presidents, and there was much overlap between the councils of the two bodies); much of the planning took place in the Society's house; and the reading room, lecture room, council room, and museum of the Society were made available to the Association's members.[56] As preparations advanced, an air of anticipation built in the city. The 1845 meeting was presided over by another man long associated with the Philosophical Society—John Herschel. On his return to England in 1838, Herschel had been welcomed with a dinner attended by 400 gentlemen of science and other prominent figures, and had been elevated to the baronetcy at the coronation of Queen Victoria—he had become one of the most celebrated scientific characters of the day.[57]

Herschel had a long-standing interest in what he called the 'mystery of mysteries'—the origins of life on the earth.[58] Herschel was open to new theories but he objected to the evolutionary theory of *Vestiges* and to the anonymity of the author. The book, which cited Herschel's own work in support of the author's wild speculations, freely mixed established science with out-of-date theories like spontaneous generation and quinarian circles (a theory which stated that all taxa can be divided into five subgroups), while the author's anonymity meant that his or her credentials, and therefore legitimacy, could not be established. As a leader in the sciences, Herschel saw it as his role to defend scientific authority against the anonymous assault of *Vestiges*, and he saw his presidential address to the British Association as the perfect place to do this.[59]

Attendees at the British Association meeting that summer gossiped excitedly about the contents of *Vestiges*, and about the true identity of its anonymous author. Sedgwick was convinced that it was a woman, 'from the hasty jumping to conclusions'.[60] Many suspected Richard Vyvyan (a politician and natural philosopher), Ada Lovelace (a mathematician, and daughter of Lord Byron), George Combe (a phrenologist), or Harriet Martineau (a writer with a particular interest in economic and social issues). Few suspected that the real author of *Vestiges* walked among them that summer in Cambridge. As the audience filed into Senate House to hear John Herschel deliver his damning verdict against *Vestiges*, they did not suspect that its author sat quietly in the front row, listening to his work being attacked by the most revered scientific mind of the generation. Of course, he could not reply.

As far as Herschel was concerned, *Vestiges* did not give any explanation or mechanism for the series of events proposed by the author. As he explained in his presidential address, it did not deal with *causes*:

> as when we are told, for example, that the successive appearance of races of organised beings on earth, and their disappearance, to give place to others, which Geology teaches us, is a result of some certain law of development, in virtue of which an unbroken chain of gradually exalted organization from the crystal to the globule, and thence, through the successive stages of the polypus, the mollusc, the insect, the fish, the reptile, the bird, and the beast, up to the monkey and the man (nay, for aught we know, even to the angel), has been (or remains to be) evolved. Surely when we hear such a theory, the natural human craving after *causes*, capable in some conceivable way of giving rise to such changes and transformations of organs and intellect,— *causes why* the development at different parts of its progress should divaricate into different lines . . . And when nothing is offered to satisfy this craving, but loose and vague reference to *favourable circumstance* of climate, food, and general situation, which no experience has ever shown to convert one species into another; who is there who does not perceive that such a theory is in no respect more *explanatory*, than that would be which simply asserted a miraculous intervention at every step of that unknown series of events by which the earth has been alternately peopled and dispeopled of its denizens?[61]

Against the sloppy and sensationalist arguments of *Vestiges*, Herschel contrasted the sober scientific work of the Cambridge Philosophical Society, which he saw as the embodiment of cool rationality. The intellectual rigour of the Society and its fellows' 'steady concentration of thought' and 'stern mathematical discipline' were what the scientific world needed if it was to combat *Vestiges*.[62] It was not just the words spoken by Sedgwick or Clark at meetings of the Society that would counter the feeble logic of the book; the very existence of the Society was, to Herschel, an emblem of rationality. It stood for those who had spent years training their minds to understand the smallest nuances of natural philosophy, in contrast to those with a 'propensity to crude and over-hasty generalizations'.[63] Herschel's talk was widely reported in the press and, despite his best efforts to dampen controversy (e.g. he did not mention *Vestiges* by name), naturally added to the general sensation around the book.

For Cambridge men, the popularity of *Vestiges* was an affront to their core beliefs. Not just about whether the universe and living things had evolved according to materialistic principles, but about how knowledge should be created, and about who should be considered a scientific authority. Prior to the publication of *Vestiges*, Sedgwick, Whewell, and Herschel had each written about their views on the relationship between science and religion, methodology in the sciences, and the purpose of a university. John Herschel's *Preliminary discourse on the study of natural philosophy* was published in 1831 as part of a 133-volume series called the *Cabinet Cyclopædia*.

The *Cyclopædia*—the brainchild of the professor of natural philosophy at London University, Dionysius Lardner—covered topics ranging from history and biography to arts, manufacturing, natural history, and natural philosophy. The series was part of a trend towards publishing high-quality educational books for a general audience at reasonably affordable prices; a trend made possible by both rising literacy rates and the invention of cost-effective steam-printing. If the books were to have wide appeal, thought Lardner, it was not sufficient that they contain up-to-date information; they must also be essentially *moral*, as the prospectus for the series clarified:

> [N]othing will be admitted into the pages of the 'Cabinet cyclopaedia' which can have the most remote tendency to offend public or private morals. To enforce the cultivation of religion and the practice of virtue should be a principal object with all who undertake to inform the public mind.[64]

This idea of a wholesome cultivation of knowledge was growing in this period and would blossom into a more formalized system of education in the Victorian period. There was a long-standing Christian tradition of studying nature as a route to understanding the glory of God's creation and, at the same time, understanding one's place in the world, thus enabling an individual to lead a more moral life. For English men of science in the nineteenth century, this sort of 'natural theology' usually meant studying nature alongside scripture to learn something about God's creation—their religious faith and their scientific works were mutually reinforcing.[65]

Herschel certainly believed firmly in a mutual relationship between religion and natural philosophy. He wrote in his *Discourse* that natural philosophy 'places the existence and principle attributes of a Deity on such grounds as to render doubt absurd and atheism ridiculous'.[66] William Whewell too saw religion and science as being intimately linked: he thought that religion played a part in setting the goals, but also the limits, of science. In 1833, he published *Astronomy and general physics considered with reference to natural theology* as part of a series of books called the Bridgewater Treatises. These books, funded by a bequest by the Earl of Bridgewater, were designed to explore 'the power, wisdom, and goodness of God, as manifested in the Creation'. Whewell wrote explicitly about the relationship between science and religion:

> We are not to expect that physical investigation can enable us to conceive the manner in which God acts upon the members of the universe.... Indeed, science shows us, far more clearly than the conceptions of every day reason, at what an immeasurable distance we are from any faculty of conceiving *how* the universe, material and moral, is the work of the Deity. But with regard to the material world, we can at least go so far as this;—we can perceive that events are brought about, not by insulated interposition of divine power exerted in each particular case, but by the establishment of general laws. This...is the view of the universe proper to science, whose office it is to search out these laws.[67]

This passage shows how immensely useful natural theology was to natural philosophers in the early nineteenth century. This was because one of its central tenets was that God had created the universe according to regular laws, and so philosophers could generalize from a small set of observations or experiments to a more general theory.

Whewell and Herschel were both influential thinkers in this period. Herschel especially, already a famous figure by 1831, was perhaps the best-known man of science in the country, and his *Discourse* became an influential book.[68] He wrote explicitly about the methods of science, and particularly the importance of observation and experiment:

> A clever man, shut up alone and allowed unlimited time, might reason out for himself all the truths of mathematics, by proceeding from those simple notions of space and number of which he cannot divest himself without ceasing to think. But he could never tell, by any effort of reasoning, what would become of a lump of sugar if immersed in water, or what impression would be produced on his eye by mixing the colours yellow and blue.[69]

For Herschel, observation and experiment were the 'fountains of all natural science'. Once one had gathered some facts through the art of observation or experiment, a philosopher could analyse the facts, group like facts together, perfect the nomenclature, and begin a series of inductive steps to discover the laws that governed phenomena, or the proximate causes that lay behind them. (A proximate (or secondary) cause was the event or action immediately responsible for a phenomenon, while the ultimate (or final) cause was the real reason something happened. Most natural philosophers of this time would have seen God as the ultimate cause behind all events in nature.) Adam Sedgwick elegantly explained the essence of natural theology when he wrote: 'Tis the crown & glory of organic science that it *does* thro' *final cause,* link material to moral; & yet *does not* allow us to mingle them in our first conception of laws'.[70]

Herschel believed that philosophers needed to be cautious about causes. He was keen to differentiate between 'real' and 'imaginary' causes; he explained the difference like this:

> The phenomenon of shells found in rocks, at a great height above the sea, has been attributed to several causes. By some it has been ascribed to a

> plastic virtue in the soil; by some, to fermentation; by some, to the influence of celestial bodies; by some, to the passage of pilgrims with their scallops; by some, to birds feeding on shell-fish; and by all modern geologists, with one consent, to the life and death of real mollusca at the bottom of the sea, and a subsequent alteration of the relative level of the land and sea. Of these, the plastic virtue and celestial influence belong to the class of figments of fancy. Casual transport by pilgrims is a real cause, and might account for a few shells here and there dropped on frequented passes, but it is not extensive enough for the purpose of explanation. Fermentation, generally, is a real cause, so far as there *is such a thing*; but it is not a real cause of the production of a shell in a rock, since no such thing was ever witnessed as one of its effects, and rocks and stones do not ferment. On the other hand, for a shell-fish dying at the bottom of the sea to leave his shell in the mud, where it becomes silted over and imbedded, happens daily; and the elevation of the bottom of the sea to become dry land has really been witnessed so often, and on such a scale, as to qualify it for a *vera causa* available in sound philosophy.[71]

To Herschel's mind, the problem with *Vestiges* was that the author had not been sufficiently careful when talking about causes. That is why, in his address to the British Association, Herschel complained that *Vestiges* barely mentioned causes and, where it did, the causes given fell into Herschel's category of 'figments of fancy'.

Sedgwick's book *A discourse on the studies of the University of Cambridge* gives some further clues about how the Cambridge elite viewed knowledge, and why they were so concerned by the popularity of *Vestiges*. The book had begun life as a sermon in Trinity College Chapel in December 1832. At Whewell's suggestion, Sedgwick expanded and published it in 1833. The book illustrated the relationship between mathematics, natural philosophy, and the intellectual and moral goals of the University. As Sedgwick wrote:

> A study of the laws of nature for many years has been, and I hope ever will be, held up to honour in this venerable seat of the discoveries of Newton.... Before [a student] can reach that elevation from whence he may look down upon and comprehend the mysteries of the natural world, his way is steep and toilsome, and he must read the records of creation in a strange, and to many minds, a repulsive language, which rejecting both the

> senses and the imagination, speaks only to the understanding. But when this language is once learnt, it becomes a mighty instrument of thought, teaching us to link together the phenomena of past and future times; and gives the mind a domination over many parts of the material world, by teaching it to comprehend the laws by which the actions of material things are governed.... The laws by which God has thought good to govern the universe are surely subjects of lofty contemplation; and the study of that symbolical language by which alone these laws can be fully decyphered, is well deserving of your noblest efforts....
>
> A study of the Newtonian philosophy... teaches us to see the finger of God in all things animate and inanimate.... We find that no parts of the visible universe are insulated from the rest; but that all are knit together by the operation of a common law. We follow this law into its remotest consequences, and we find it terminating in beauty, and harmony, and order.[72]

Sedgwick believed that knowledge was hard won; a student had to spend years mastering the basics of mathematics and its endless symbols before he could begin to understand how the world worked and appreciate the true beauty of the universe. Because the author of *Vestiges* was anonymous, there was no evidence that he or she (and Sedgwick, as we have seen, thought the author might very well be a woman) was qualified to make pronouncements on natural philosophy. This undermined everything he had worked so hard for in his Cambridge career.

The first to fourth editions of Sedgwick's *Discourse* were published before the appearance of *Vestiges*. In 1850, when *Vestiges* had been on the market for six years, Sedgwick brought out a fifth edition of his *Discourse* with a mammoth preface and appendix—many times the length of the discourse itself—dedicated in large part to attacking the ideas of *Vestiges*. In the preface, Sedgwick discussed how he objected to the removal of God as the ultimate cause behind the events of the natural world. He wrote with disgust about how materialists (especially French ones) approached the question of causation in nature: 'they deny the indications of a God, they deify dead matter'.[73] Though (unlike the French materialists who were so disliked by Sedgwick) the author of *Vestiges* did speak with apparent reverence of a Creator, his proposed scheme for the creation of the world and its inhabitants was most closely allied to the materialistic model. Sedgwick devoted more than 300 pages to unpicking the materialism of *Vestiges*. This, too, was done out of concern for the proper education of young men:

> The reader who has gone through the preceding pages may perhaps have complained, that the criticism of *The vestiges of creation* is too long, and out of proportion to the other discussions in the Preface. But that work has had a very wide circulation; partly from the good language and positive form in which it is written, and perhaps still more from the novelty of its conclusions. For the Author (apparently without seeing the end of his philosophy) has been the first to put material Pantheism, decked in a good English dress, before the readers in this country. He has also written *A sequel to the vestiges*, to which no formal reply has yet been given. I was therefore constrained to notice this Author's two works in some detail; and through them I have been led to my main object: viz. to expose to our Undergraduates the mischief of modern pantheistic doctrines when applied to physical, moral, and religious questions.[74]

For Sedgwick, a key part of a University education was to demonstrate how intellectual beliefs and religious beliefs could sit comfortably together, so long as each was properly interpreted.

Though these works of Herschel, Whewell, and Sedgwick were not necessarily aimed at the fellows of the Cambridge Philosophical Society (Herschel's and Whewell's were aimed at general readers, while Sedgwick's was aimed at undergraduates), they explain some of the underlying thinking about the roles and methods of natural philosophy in early nineteenth-century Cambridge, and in the British establishment more widely. The use of the Society's meetings to attack *Vestiges* shows how these beliefs percolated into the Society, and how the leading members of the Society took on a role as defender of scientific authority and orthodoxy.

The Society's reaction to *Vestiges* was hostile not just because the fellows disagreed with the theory of transmutation, but also because they saw the book's brand of logic, its materialism, and its anonymous authorship as contrary to the fundamental ideals of the University and the

Society. Few English men of science in the first half of the nineteenth century had much time for transmutationist theories; theories that were, to Sedgwick's mind, 'no better than a phrensied dream'.[75] Throughout the 1840s and 1850s, there were occasional talks at meetings of the Society on topics such as spontaneous generation or homologies of vertebrate skeletons but the *Vestiges* was never mentioned again—clearly, the fellows felt that they had done enough to suppress its influence in Cambridge.[76] Few in Cambridge suspected that the next significant reappearance of evolutionary theory would be brought about by one of their own.

In 1859, Charles Darwin wrote to his old teacher and friend John Stevens Henslow, and to his former mentor Adam Sedgwick, who had trained him in the ways of the field geologist thirty years earlier. He wrote to tell them of a new book he had just authored, and to tell them to expect a copy direct from the publisher. To Henslow, whom he called his 'dear old master in natural history', Darwin wrote that he feared 'that you will not approve of your pupil in this case'.[77] To Sedgwick, he admitted that 'the conclusion at which I have arrived... [is] diametrically opposed to that which you have often advocated with much force'.[78] The book, of course, was *On the origin of species* and in it Darwin detailed his theory of evolution by natural selection. Unlike *Vestiges*, Darwin's theory had a carefully thought-out mechanism behind it, and Darwin was a well-known and highly respected naturalist; but the book was still controversial.

Sedgwick raced through the book in just a few days. Once he had finished, he wrote at once to his former student, admitting to Darwin that 'I have read your book with more pain than pleasure. Parts of it I admired greatly; parts I laughed at till my sides were almost sore; other parts I read with absolute sorrow; because I think them utterly false & grievously mischievous.'[79] Sedgwick explained to Darwin that he believed his theory of evolution deserted the principles of inductive reasoning, and that Darwin had confused secondary causes with final ones. Sedgwick also, more damningly, compared Darwin's tone to that of the author of *Vestiges*. But despite his criticisms, Sedgwick avowed that he was still Darwin's 'true-hearted old friend' and jokingly referred to himself as 'a son of a monkey'.[80]

Henslow, knowing the many attacks Darwin had been subjected to, was far more measured in his response, replying to Darwin not directly about the content of the book, but writing: 'I don't think it is at all *becoming* in one Naturalist to be bitter against another—any more than for one sect to burn the members of another.'[81]

Like *Vestiges* before it, *On the origin of species* caused a furore in Victorian Britain. It sold so quickly that a second edition was hurried out just two months after the first had been published. It too was discussed at all levels of society but, unlike *Vestiges*, with its anonymous author, it was much harder for the scientific establishment to dismiss outright. Seeing its popularity rising, Sedgwick felt that he had to say something publicly about the book and, once again, he chose the Cambridge Philosophical Society as the place to say it. Henslow, who had remained close to Darwin throughout the years, wrote to tell him that 'Sedgwick is to illuminate us on Monday at the Philosophical Society in regard to your supposed errors!'[82] The talk took place on the evening of Monday, 7 May 1860. It is recorded, drily, in the Society's minute book as being about 'the succession of organic forms during long geological periods, and on certain theories which profess to account for the origin of new species'.[83] That evening, Sedgwick took to the floor, as brilliant an orator as ever. He led his audience down through the geological strata, speaking of the wondrous forms preserved in rocks and what they told of the history of life on the earth—a history that did not, he believed, tally with Darwin's account. He asserted firmly that Darwin—his own former student—had given up on logical thought, while he himself adhered firmly to 'the stern inductive truths of the Newtonian Philosophy'. He finished on an exhortation to the young men of Cambridge to deal in facts rather than speculative theories.[84] William Clark, Sedgwick's old ally, followed him that evening, attacking both Darwin's theory and Darwin himself.

Henslow, sitting in the audience, could not listen to his friend Darwin being so roundly condemned and rose in his defence. He recorded the events of the evening in a letter to the botanist Joseph Dalton Hooker at Kew Gardens:

> Sedgwick's address last Monday was temperate enough for his usual mode of attack, but strong enough to cast a *slur* upon all who substitute hypotheses for strict induction, & as he expressed himself in regard to some of [Darwin's] suggestions as *revolting* to his own sense of wrong and right & as Dr Clark who followed him, spoke so unnecessarily severely against Darwin's views; I got up, as Sedgwick had alluded to me, and stuck up for Darwin as well as I could, refusing to allow that he was guided by any but truthful motives …
>
> I believe I succeeded in diminishing, if not entirely removing, the chances of Darwin's being prejudged by many who take their cue in such cases according to views of those they suppose may know something of the matter …
>
> When I had had my say, Sedgwick got up to explain, in a very few words, his good opinion of Darwin.[85]

In the same letter, Henslow admitted that he thought Darwin's theory went too far, but he was still willing to publicly defend his former student. Perhaps more surprisingly, Henslow told Hooker that he was discussing material from *On the origin of species* in his Cambridge lectures. When Darwin learned how Henslow had defended him against Sedgwick and Clark, he was touched, writing: 'I must thank you from my heart for so generously defending me as far as you could against my powerful attackers.'[86] To other friends, Darwin blithely dismissed Sedgwick's attack: 'as for the old fogies in Cambridge it really signifies nothing', he wrote to Hooker, aware that Sedgwick, now aged 75, was no longer at the forefront of geological thinking.[87]

But Sedgwick was still playing an important role in shaping young minds, and shaping the University curriculum. The examination questions he set in the 1850s regularly asked students to discuss the evidence for and against evolution; but, even though he conscientiously told the students to state their own opinions even if they should differ from his, his views on evolution were too well known for any but the most strong-willed student to dare to disagree with him in an examination. In 1859, for example, he set the following question: 'is there a shadow of proof from the ethnographical, and physical history of man that any one of his oldest varieties was derived from a quadrumanous [i.e. primate] progenitor?'[88]

Though the students might struggle to reconcile what they heard in Sedgwick's lectures with what they read in Darwin's book, at least one person did well out of Sedgwick's evolutionary examinations—as Darwin reported to Henslow: 'Murray, the Publisher, thought it splendid for selling copies to the unfortunate Students.'[89]

Through the 1860s, as Darwin's work came to be more accepted, his theories found their way into many talks at the Cambridge Philosophical Society. John William Salter, a geologist who had once worked alongside Sedgwick, presented a talk illustrated by a striking diagram showing the succession of plant life through the fossil record[90] (see Figure 19). He also published several articles in the Society's *Transactions* on fossils and shellfish. In one of these papers, he described how he had once opposed evolutionary theory 'on geological grounds alone' but now he had come to accept it and he concluded that shellfish display 'an almost infinite gradation of species, which may well be explained on the Darwinian theory'[91] (see Figure 20). Alfred Newton, a zoologist who was an early believer in Darwin's theory and later became Professor of Zoology and Comparative Anatomy and President of the Society, gave papers on the ancient zoology of Europe and on the many species he had collected on his travels throughout the continent—all informed by his Darwinian views.[92]

As the years passed and Darwin's theory gained more followers, he came to be seen as one of the great sons of Cambridge. When Darwin died in 1882, the Society sent their condolences to his widow Emma, and praised the 'extraordinary value of [his] scientific labours'.[93] By the 1890s, there was talk of commissioning a statue of Darwin for the Society's library.[94] When Darwin's centenary and the fiftieth anniversary of the publication of *On the origin of species* rolled around in 1909, the Society decided to commemorate him with a book celebrating his life and works. By 1909, the Society was seen as a natural home for Darwinian theory, and most people were unaware that Darwin's ideas had not always been so welcome there. The Society produced a lavish book with chapters on almost every aspect of the modern life sciences: selection, variation, heredity, cell theory, embryology, palaeontology, and ecology, as well as a

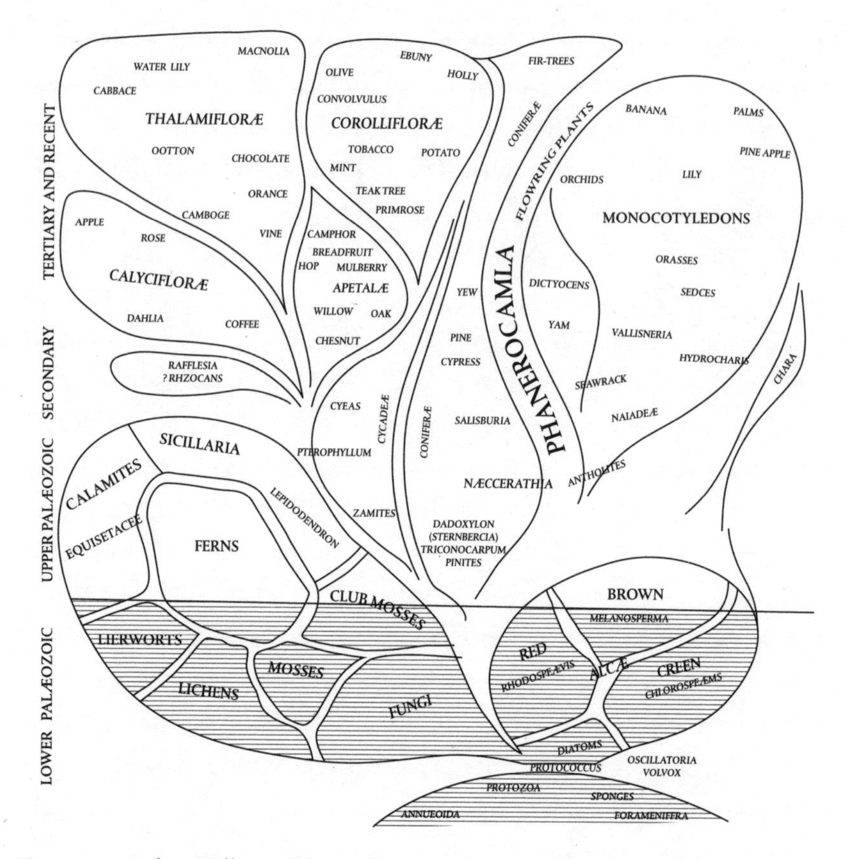


**Figure 19** John William Salter's diagram showing the succession of plant life upon the Earth, from a talk he presented to the Society in February 1869.

curious essay on double stars by Charles Darwin's son, George Darwin, Plumian Professor and President-elect of the Cambridge Philosophical Society.[95] Just as interesting as what was included in the book is what was not: strangely, despite listing Darwin's publications, the 1835 booklet of his letters printed by the Society was not included, nor was there a single mention of the controversy Darwin's ideas had once aroused in the Society's very own meeting room.

Had the Society forgotten the debates of the mid-nineteenth century, or did they simply not wish to dwell on them? The willingness of the Society's early fellows to engage head-on with the controversies of the

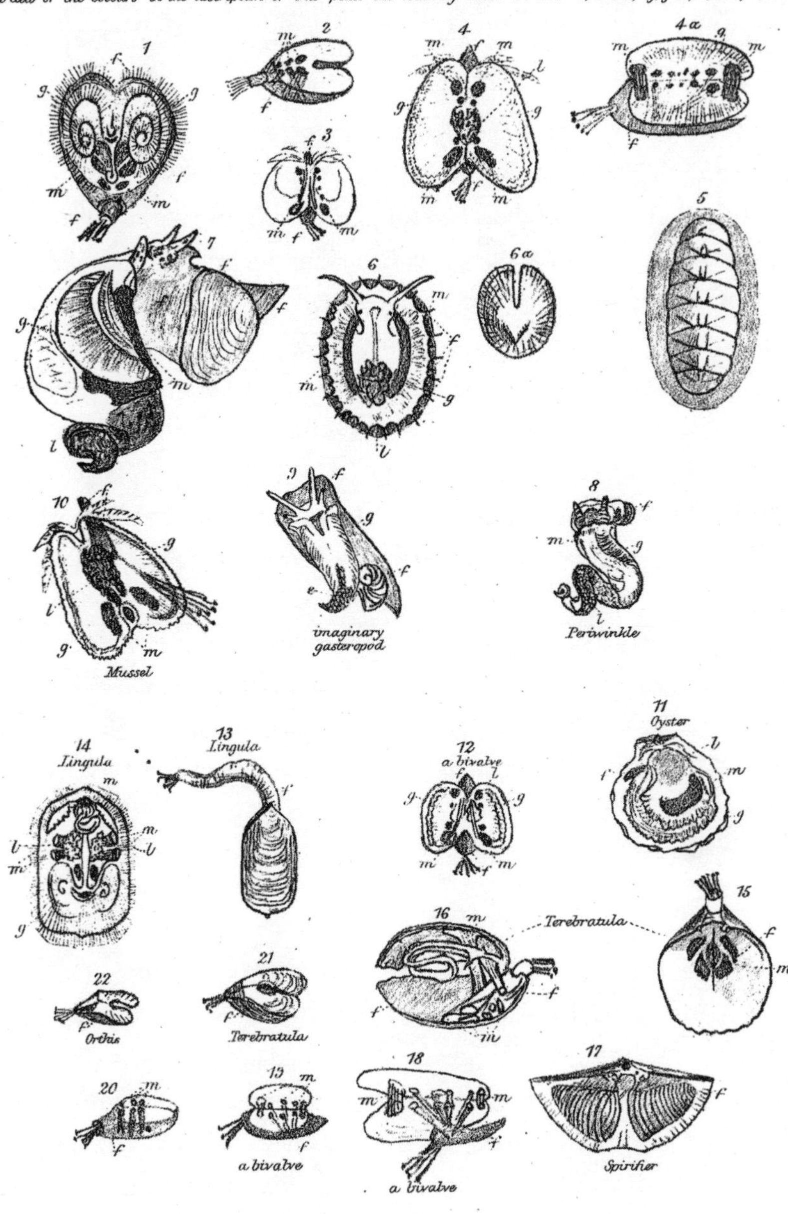


**Figure 20** John William Salter's diagram showing the relations between various shellfish, from a talk he presented to the Society in February 1869.

day had been one of their hallmarks. Though the Society was always polite and orderly (unlike the quarrelsome Zoological Society of London, about which young Darwin had complained), they defended their beliefs, and their theories of knowledge, unflinchingly against attack. They maintained their high standards for scientific training, their high standards for scientific evidence, and their carefully worked-out scientific rationale in the face of a new popular science which increasingly catered to audiences who demanded simple answers in attractive packaging. The Society was open to new ideas and new knowledge—whether that was Herschel's meteorological data from Africa, or Darwin's geological observations from South America—but they carefully vetted that knowledge (and the people behind it) before admission into their hallowed meeting room or journal. The Scottish publisher Robert Chambers, who was unmasked as the author of *Vestiges* after his death in 1871, had never really stood a chance against establishment figures like Sedgwick. Because he was unwilling to reveal his identity, Chambers could not be considered a true man of science according to the standards set by bodies like the Cambridge Philosophical Society, where proper training in morals, religion, and the methods of science all mattered.

# 4

# 'A NEW PROSPERITY'

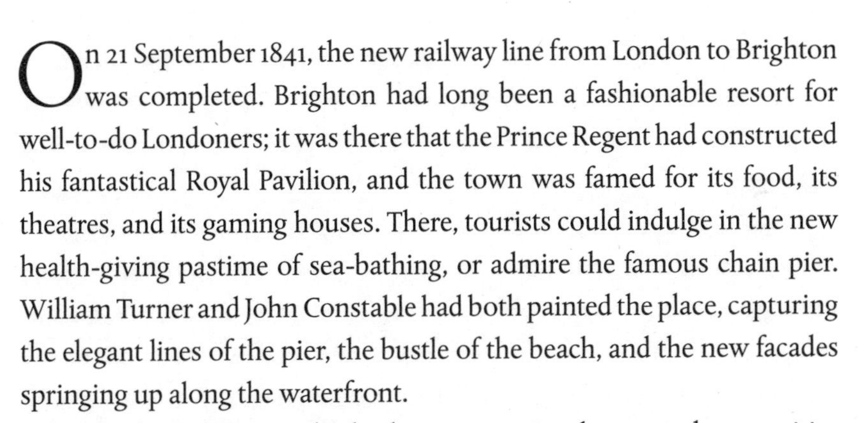

On 21 September 1841, the new railway line from London to Brighton was completed. Brighton had long been a fashionable resort for well-to-do Londoners; it was there that the Prince Regent had constructed his fantastical Royal Pavilion, and the town was famed for its food, its theatres, and its gaming houses. There, tourists could indulge in the new health-giving pastime of sea-bathing, or admire the famous chain pier. William Turner and John Constable had both painted the place, capturing the elegant lines of the pier, the bustle of the beach, and the new facades springing up along the waterfront.

Before the railway was built, the journey time between the two cities was about six hours by stage coach; now, it took a mere two and a half hours to reach the Brighton seaside from the centre of London. The railways were fast, they were modern, and they caught the public imagination. New tracks were unrolling themselves across the countryside at an astounding rate: in 1835 Britain had just over 300 miles of railway; in 1840 there were 1,500 miles; by 1845 that had jumped to 2,500 miles. It wasn't just the passengers who loved the railways; share prices rocketed as speculators rushed to buy into the network—the so-called railway mania.

There was a sense of optimism in the air on a fine Saturday morning, less than two weeks after the line opened, as a train rumbled along the track towards the delights of Brighton (see Plate 5). The train of eleven carriages was pulled by two engines: a four-wheeled Bury engine in front, followed by a six-wheeled Fairburn. It steamed across the magnificent viaduct high above the Ouse Valley at a pace of more than twenty miles an hour and then increased its speed to a cracking thirty miles an hour as it entered the cutting in Copyhold Hill. In the cutting, the front engine began to 'joggle' on the line, oscillating with increasing and alarming force. Within half a minute, the front engine derailed to the left of the track. Its driver, Charles Goldsmith, was thrown clear of the engine just seconds before the boiler exploded; the engine's stoker, Robert Marshall, was killed by the explosion. The driver of the second engine, James Jackson, wrenched up the reversing lever, hoping to save the rest of the train. Despite Jackson's best efforts, his engine was also derailed, flying off to the right of the track. Crawling out of the wreckage, Jackson saw the mangled body of his engine's stoker, Robert Field, a young man who had celebrated his wedding only three weeks earlier.

Hot coals were flying through the cutting, landing on the terrified passengers, most of whom were travelling in open second- and third-class carriages. The shrieks of the passengers filled the air. Through the smoke, Jackson saw that the front three carriages had followed the engines off the track. These included the enclosed first-class carriage. Its roof trapped its passengers inside, where they were almost suffocated by smoke and steam until one Mr Loud of Dover succeeded in kicking a door open. Two servants travelling in another part of the train were less fortunate. Henry Palmer, a footman, and Jane Watson, a housemaid, both twenty-three years old and employed in the house of the surgeon Joseph Carpue, were killed instantly. Henry's skull was split open by the force of the impact; Jane was crushed beneath a wheel.[1]

This was one of the first fatal train crashes in Britain. The public were horrified to learn that the new engines which they had come to love and depend upon could bring such devastation. As one commentator wrote,

> [such a crash] would have been deemed by our forefathers too extravagant even to be allowed a place in the wildest fictions. Colossal vehicles, weighing several tons, shivered to pieces; rods of iron, thick and strong enough to sustain a vast building, bent, twisted, and doubled up as though they were rods of wax; massive bars of metal snapped and broken like glass;

> bodies of the killed dispersed here and there, amongst the wrecks of vehicles and machinery, so mangled as to render identification impossible—limbs, and even heads, severed from the trunks, and scattered right and left, so as to render it impossible to re-combine the *disjecta membra* of the same body—the countenances of the dead, where countenances remain at all, having a ghastly expression of the mingled astonishment and horror with which the sufferer was filled in the brief instants which elapsed between catastrophe and death.[2]

The railway authorities, and the public, demanded to know what had caused this terrible crash. Was it the design of the front engine, with only four wheels instead of six? Did the engine's water tank make it too top-heavy? Had there been clay on the tracks? Was the cutting structurally unsound? All of these possibilities were gone over at the inquest, which took place the following Monday in a local beer shop. They were also discussed at length in a report by the railway company, and debated in considerable detail by the public and the press.[3] In Cambridge, they were the subject of some discussion at the Philosophical Society.

Less than two months after the accident, the Reverend Joseph Power, who was a mathematics tutor at Trinity Hall and would later go on to become University Librarian, stood up before the Society to deliver his verdict on the causes of the accident. Trained, of course, in the ways of Cambridge mathematics, Power delivered his solution to the problem of railway accidents in the form of an equation: minimize $F^2h^2/[M^2g(a^2 + k^2)]$, he said, and lives would be saved (see Figure 21). Power dismissed all previous theories about the cause of the accident and focused instead on the way the two engines were coupled together. His main contention was that the train had derailed because of the 'joggle' in the front engine, and that this 'joggle' had been caused by the chain connecting it to the second engine. The driver of the first engine had felt that it 'wavered backwards and forwards'; the driver of the second engine said that he had seen the engine ahead of him display a 'rocking motion'; meanwhile, a labourer who saw the train from the road described 'a jumping up and down of the fore wheels' of the front engine.[4]

**I will now give the mathematical details to which allusion has been made.**

**It will be as well to begin with the very simple case of a body whose centre of gravity is $G$, supported on a perfectly smooth horizontal plane $PQ$ by two props $BD$, $CE$, and urged by a horizontal impulsive force $F$ at the point $A$. Draw $GHKL$ vertical.**

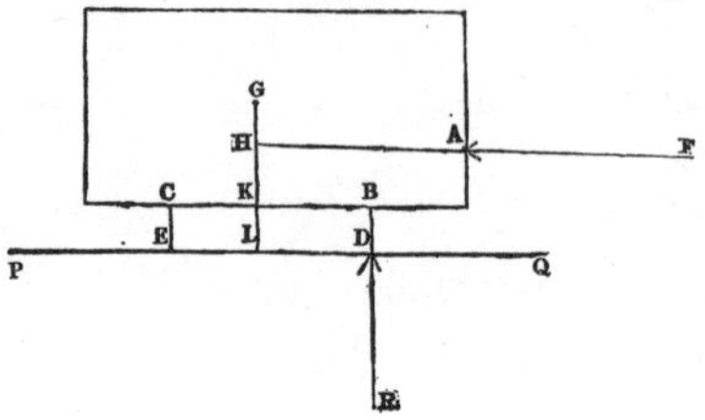

**Let $GH = h$,**

**$DL = a$,**

**$GL = b$,**

**$M$ the mass,**

**$k$ its radius of gyration about $G$,**

**$V$ the horizontal velocity communicated to $D$,**

**$\alpha$ the angular velocity about $D$ resulting from the impact.**

Figure 21 Joseph Power's schematic diagram of a train carriage, with some of his calculations showing the ideal coupling point, from a talk he presented to the Society in November 1841.

Power believed that the connecting chain had been attached too low down on the body of the front engine. He wanted to find the perfect spot—not too high and not too low—at which a chain could be safely attached between two engines. His equation gave a way of working out the most stable way of coupling engines based on their mass, their dimensions, and the distance between them. His calculations are a perfect example of Cambridge mixed mathematics being applied to a real-world problem. Once he had completed his computations, Power asked his friend Robert Willis, Jacksonian Professor of Natural Philosophy, to help him build a small-scale replica of the train so that he could test his theory. Willis, like his predecessor William Farish, was a handy mechanic, easily able to knock up such a model in his workroom. Power demonstrated the reality of his equation with the help of the model, showing it off in the Society's meeting room and letting the fellows see the truth of his theory for themselves. It was an elegant solution to an important problem. But the railway companies tended to deal with problems on a case-by-case basis; they weren't looking for a general theory of coupling and so they paid little attention to Power's academic research, and his suggestions were not implemented.[5]

In the few months between the reading of Power's paper at a meeting and its publication in the Society's *Transactions*, an accident took place on the Great Western Railway. On 24 December 1841, a train consisting of three third-class carriages filled with labourers going home for Christmas and several heavily laden luggage trucks was derailed by a landslide in Sonning Cutting near Reading. Power believed that, had his theory been implemented, the carriages could not have rolled over each other the way they did, crushing the passengers between the heavy luggage trucks and the tender, leaving nine people dead.[6]

Though Power's suggestions were not implemented by the railway companies, they show how the fellows of the Cambridge Philosophical Society were keen to engage with the issues of the day, and how they were willing to address topics that had not traditionally been studied in the university. A few decades earlier, there would have been no forum for a talk like Power's in Cambridge, nor a journal in which to publish it. Power's talk is a nice example of how Cambridge was becoming more outward-looking and more practical as the nineteenth century wore on. It also shows how the trends of Victorian England made themselves felt in this small university town: the mania for railways was reflected in many of the talks given to the Society during the 1840s. Power's talk about the Brighton accident was followed that same evening by another by John Frederick Stanford—a graduate of Christ's College—on a newly invented locomotive engine.[7] It would be 1845 before Cambridge had a railway of its own. At the colleges' insistence, the station was built more than a mile away from Cambridge because of worries that the noise and smoke would disturb the tranquillity of the town—trains were all very well on paper, but the reality was less popular with Cambridge fellows.

As the 1840s progressed and the railway network expanded, the railway companies and the government began to seek out opinions from academics on ways to improve the safety of their trains. In 1847, Robert Willis was appointed to a Royal Commission to investigate the use of iron in structures exposed to 'violent concussions and vibration'. The Commission had been set up after the Dee Bridge disaster, in which a train fell through an iron bridge into the river below, killing five of its passengers. To test how iron bridges responded to the vibrations of the trains that steamed across them, the Jacksonian Professor set up a miniature railway track, just nine feet long, and drove carriages of different weights over it at different velocities. Even with this simplified model, Willis could see how much the iron bridge was stressed by the passing train. In 1849, the same year he became President of the Cambridge Philosophical Society, Willis travelled to Salisbury, where he observed a seventy-foot bridge in action beneath the passing trains. Having made his observations, Willis turned to his friend George Gabriel Stokes for assistance with the mathematical analysis of the problem.[8]

Stokes, an Irishman, had come to Pembroke College, Cambridge, in 1837 and studied mathematics under the famous coach William Hopkins. Hopkins was known as the 'wrangler-maker': a well-deserved epithet, as more than two hundred of the countless students he coached from the 1830s to the 1860s became wranglers, including seventeen senior wranglers. The coaches were a vital component of Cambridge life in this period; they were essentially private tutors to ambitious students, pushing their young men to dizzying mathematical heights. The word 'coach' had arisen from the stage coaches that raced at ever increasing speeds between Cambridge and London at that time: at the turn of the century, the journey between the two towns took about seven and a half hours; by the 1820s, it was five and a half hours; by the 1830s, it was four and a half hours as the coachmen drove their horses faster and faster. The weary students of Cambridge felt equally pushed, and jokingly began to talk of being 'coached' by their tutors.[9] Hopkins was the most famous coach of all. Though easily clever enough to have won a fellowship, Hopkins operated outside the college system because he had married while still an undergraduate. He was a well-known figure around Cambridge and, after his friend Adam Sedgwick introduced him to the field of physical geology, he became a frequent speaker at the Philosophical Society. Hopkins's papers to the Society usually focused on applying mathematical analysis to geology. This was precisely the kind of thing Hopkins taught to his students; thus, men like Stokes were well versed in practical applications of mathematics by the time of their Tripos.[10]

In 1841, Stokes was senior wrangler and first Smith's Prizeman and he was quickly elected to a fellowship at Pembroke. The following year, Stokes presented his first paper to the Philosophical Society; in it he examined the motion of incompressible fluids. He further developed his ideas about fluid motion in a series of papers over the next few years and soon came to be seen as a leading authority on the subject.[11] But Stokes was not a man to limit himself to just one field; he gave multiple papers to the Society on optics, mathematics, and astronomy, and on practical questions relating to the building of railway bridges.[12] Stokes's approach to the problems of bridges, like Power's approach to the problem of coupling engines, was essentially mathematical. Dealing mathematically with the multiple variables encountered when a train crosses a railway bridge was an extremely complex task, as Stokes realized:

> [T]he forces acting on the body and on any element of the bridge depend upon the positions and motions, or rather changes of motion, both of the body itself and of every other element of the bridge, so that the exact solution of the problem, even when the deflection is supposed to be small...appears almost hopeless.[13]

By simplifying various parts of the problem, Stokes (using data supplied by Willis) was able to come up with a differential equation that shed some light on the issue. His results pointed towards an underlying problem in the design and structure of the iron girders. These results broadly agreed with the report of the Commission, which had relied on more empirical investigations.[14] Willis and Stokes gave a series of joint papers to the Cambridge Philosophical Society in 1849 in which Willis explained his experimental results on railway bridges, while Stokes supplied mathematical explanations to sit alongside Willis's work.[15] Their talks transported their listeners out of the sedate elegance of the Society's meeting room and hurtled them into the drama of the industrial world, where science, engineering, and seemingly abstract mathematics could save real lives. By the time the final paper was delivered, Stokes had been appointed Lucasian Professor of Mathematics and had begun to lecture undergraduates in the sorts of mathematical principles seen in his work on railways.[16]

Railways were not the only newsworthy topic being discussed in the Society mid-century: the success of *Vestiges of the natural history of creation* had given rise to several papers about evolutionary theory, as we have seen. Photography was another recent invention that caused excitement amongst the fellows of the Society. The Reverend W. Towler Kingsley gave multiple papers on the new art of photography, speaking, for example, on how photography might be used in conjunction with the microscope to produce new kinds of scientific results.[17] Kingsley's photographs were much praised by his Cambridge colleagues and were also exhibited at the Photographic Society in London, where one critic wrote that they were 'the most remarkable specimens . . . the whole of [Kingsley's] collection from collodion and waxed-paper negatives are exquisitely delineated, and appear to me the most successful, especially of semi-opaque objects, that have yet been taken'.[18]

The Society even addressed one of the great humanitarian disasters of the century—the Irish Famine. Though there were many factors behind the famine—political and social as well as agricultural—the easiest aspect to address scientifically was the blight that had led to the failure of several successive potato harvests. George Kemp, a fellow of Peterhouse, analysed diseased tubers, hoping to find the cause of this blight. Kemp observed potatoes as they progressed through the three stages of the disease: first, dark brown patches would appear beneath the skin; next, dark lines would begin to move towards the centre; finally, the whole tuber would morph into 'a soft, pultaceous, blackish, and offensive mass, in which all traces of organization are lost'. Kemp attacked the problem with chemical analysis. He worked out the percentages of carbon, hydrogen, nitrogen, oxygen, and ash within both healthy and diseased vegetables as a way to understand the fundamental effects of the disease. In papers delivered to the Society in February and March 1846, Kemp gave the result of his analysis and described how the effects he saw may have been caused by premature germination of the plants. Kemp even suggested a practical solution to the blight—planting in autumn to ensure germination at the correct time. Had Kemp's solution worked, it might have saved millions from starvation, but the blight resisted all attempts to eradicate it for many years, and debate about its true cause continued for decades.[19]

These fellows threw themselves into finding solutions for the problems that were plaguing nineteenth-century society. As they grappled with the rising number of railway accidents, Power, Willis, and Stokes reimagined the idealized abstracted knowledge long associated with University learning to suit more practical concerns, while Kemp, trying to figure out the cause of potato blight, found himself at the cutting edge of applied chemistry. But, while some fellows rushed to engage with such real-world issues, others looked beyond our world altogether: at the same time that some fellows were busying themselves with the hurly-burly of the modern world, one fellow of the Society was quietly discovering a new planet.

From the earth, there are five planets easily visible to the naked eye: Mercury, Venus, Mars, Jupiter, and Saturn. In 1781, a sixth planet had been discovered by the astronomer William Herschel (father of John Herschel). He had looked through his telescope and noticed that a particular star appeared to be moving relative to the stars around it. At first he thought it was a comet; but, as he watched it over weeks and months, he came to realize that it might be a true planet—a newly discovered part of our solar system.

This new planet—now called Uranus—takes more than eighty years to complete one orbit of the sun. In 1821, forty years after it was discovered, a French astronomer called Alexis Bouvard published astronomical tables predicting the position of Uranus in the sky as it completed its journey around the sun. By the 1840s, it was apparent that Bouvard's predictions, though mathematically sound, did not match up with the actual observations of Uranus's position. The planet wobbled. There were two possible explanations for this: first, that Newton's law of gravitation was wrong;

second, that there was a hidden body whose gravitational pull was affecting Uranus—another planet. In Cambridge, even though *Principia mathematica* was more than 150 years old by then, it was rare to assume that Newton was wrong; and so one young astronomer began to search for another new planet in order to explain the wobble of Uranus.

This astronomer was John Couch Adams and, unlike William Herschel, he did not seek the new planet through the lens of a telescope. Instead, schooled in the Cambridge mathematical tradition, he used pen and paper. Adams, the son of a poor Cornish farmer, had come to St John's College as a sizar in 1839. This modest and diffident young man had quickly proved himself, excelling in his studies of classics and becoming senior wrangler and first Smith's Prizeman in 1843. When he had time to spare from his own studies, Adams would tutor other students to earn whatever money he could to support his brothers' education, and he even found time to teach his college bedder—the housekeeping staff who made the beds—how to read. Adams was fascinated by the problem of the misbehaving planet and, as soon as the Tripos and Smith's Prize exams were finished, he began his calculations to solve the mystery. Uranus's mass was known, as was its deviation from its expected position: Adams used these data in his attempt to compute the orbit, mass, and position of the body that was disturbing Uranus. By September 1845, he believed he had an accurate solution to the problem and he approached James Challis (Professor of Astronomy, Head of Cambridge Observatory, and President of the Philosophical Society from 1845 to 1847) with his figures. Challis in turn gave Adams a letter of introduction to George Biddell Airy, the former Cambridge Professor of Astronomy who had by now gone on to become the Astronomer Royal at Greenwich. Though Adams shared his findings with both Challis and Airy, he did not do so very forcefully, and there were several miscommunications between the three, with the result that Adams's calculations were largely ignored for several months.[20]

Meanwhile, in Paris, the mathematician Urbain Le Verrier had also realized that there was something wrong with the orbit of Uranus and had begun computations of his own. He reported his calculations to the

Paris *Académie des sciences* in a series of papers. One, given in June 1846, caught Airy's eye: it reminded him of Adams's work lying forgotten beneath a pile of papers on his desk. Anxious that Britain, not France, should be the first to find the new planet, Airy asked Challis to use Cambridge Observatory's Northumberland Telescope to look for it. For Airy, a Cambridge man through and through, it seemed natural to seek assistance from his *alma mater* and from the observatory he himself had once run: he believed that Cambridge not only had the most suitable telescope, but also the best observers. He did not contact any other observatories about Adams's calculations; he would later be criticized for what was termed a 'Cambridge snuggery affair'.[21]

Challis agreed to help and began his search in July. The Northumberland Telescope, a gift from the Duke of Northumberland in 1833, was one of the world's largest refracting telescopes at the time (see Figure 22). Tucked

**Figure 22** Cambridge Observatory's Northumberland Telescope, housed under a wooden dome specially designed for it by George Biddell Airy.

away in a little building of its own in the grounds of Cambridge Observatory, beneath a wooden cupola which rotated on six old cannon balls, the telescope had an extremely accurate clock-driven equatorial mounting which allowed it to follow particular stars with great precision as they travelled across the sky.[22] Challis and his assistants got to work, methodically scanning the heavens, looking for a previously ignored pinprick of light.

In the short summer nights, the astronomers had only a few hours to find what they were looking for before their work was interrupted each morning by the rising sun. As they raced against the clock, Le Verrier in Paris continued his calculations and, in August 1846, he presented his final prediction of the position of the new planet to the *Académie*. In September, Le Verrier wrote to the Berlin astronomer Johann Galle, asking him to look for something unexpected in a certain quadrant of the sky. Within a few hours of receiving Le Verrier's letter, Galle had found it—a speck of light almost exactly where Le Verrier had said it would be.

The Cambridge men were devastated to have lost priority to the French, but the rest of Europe responded jubilantly to the new discovery. Neptune (as the planet came to be known) caused tremendous excitement and was much celebrated in the popular press. It was called 'one of the most wonderful scientific achievements of our time', a 'beautiful discovery', and a 'noble achievement'. The mathematical method of its discovery caused as much interest as the new planet itself, as one journalist enthused:

> [H]e seeks to behold in its obscurity of dimness and distance a heavenly orb on which the eye of man has never yet been fixed: but his eye never seeks the heavens; and night after night the stars come out in the fulness of their glory and the kindliness of their love, yet fail to woo him from the complex records and figured papers over which his pale and aching forehead droops.[23]

Prince Albert (recently elected both Chancellor of the University of Cambridge and Patron of the Cambridge Philosophical Society) even attended a meeting of the British Association at which Adams, Le Verrier, Challis, Airy, and others discussed the new planet.[24]

At the first meeting of the Cambridge Philosophical Society after the discovery had been confirmed, Challis gave an account of his observations.[25] The diarist Joseph Romilly reported:

> Professor Miller came to tea and gave us an account of Challis's talk this evening at the Philosophical about his own observations in search of the newly discovered planet:—it is very hard that Adams and Challis should seem likely to lose the éclat attached to this triumph of science: Adams made calculations of the orbit earlier than Le Verrier, and Challis actually saw the planet before Galle (tho' he did not know it was the planet).[26]

As John Herschel wrote in a private letter, 'Neptune ought to have been born an Englishman, and a Cambridge man every inch of him'[27] (see Figure 23). Challis and Airy admitted that they had been too slow to begin their search, and many thought that Adams had been too shy about publicizing his results. Determined to have Adams's contribution recognized, Herschel and the astronomer Richard Sheepshanks—another Cambridge-trained man living in London—successfully campaigned for Adams to be officially declared as a co-discoverer of Neptune.[28]

Thanks to the efforts of Herschel and Sheepshanks, Adams became a celebrity, and the discovery of Neptune came to be seen (in Britain, at least) as an important triumph for Cambridge science.[29] A poem in the *Cambridge chronicle*, grandly comparing Adams's discovery to the works of Newton, conveyed some of the pride inspired by the discovery:

> Gentle mother, upon thee
> Smiles a new prosperity.
> Science shall no longer brood,
> In neglected solitude.
> Oe'r the hidden worlds which lie
> In the deep infinity
> Of the boundless sky;
> But with speculation true
> Soar beyond our mortal view

**Figure 23** A French cartoon showing the Englishman John Couch Adams losing out to Frenchman Urbain Le Verrier in the race to discover the new planet Neptune (1845).

> Into regions ever new.
> Until the envious world against its will
> Confess that Granta boasts a youthful Newton; still.[30]

Adam Sedgwick also saw a comparison between Isaac Newton and John Couch Adams, writing that

> [i]n the early part of the last century Queen Anne visited this University; and, after dining in the Hall of Trinity College, conferred, at a public academic levee, the honour of Knighthood upon Newton—a man (I need not tell the reader) whose name stamps a glory upon Cambridge, and who, by his labours, changed the whole face and form of natural science. In 1847, our present Sovereign [Queen Victoria]...came hither to grace and honour the Installation—visit of Prince Albert [as Chancellor of the University]; and after a similar festivity, and at a similar academic levee, the honour of

> Knighthood was offered to Mr Adams for the part he had taken in the most striking astronomical discovery that had been made by any Englishman since the days of Newton.[31]

Unfortunately, Adams felt that he had to decline the knighthood, as he was not sufficiently wealthy to support himself without a college position, and the idea of a knight tutoring undergraduates was quite unthinkable at the time. Perhaps as a consolation, Adams's college—St John's—instituted a prize in his honour: the Adams Prize for work in mathematics or the mathematical sciences. In addition to the prize money, the recipient of the Prize won the opportunity to have their winning essay published in the *Transactions of the Cambridge Philosophical Society*.[32]

Adams was a quiet unassuming man who seems never to have resented Challis and Airy's initial lack of interest in his analysis of the orbit of Uranus. But many in Cambridge remembered the incident with regret. More than twenty years later, William Peverill Turnbull, a young Trinity fellow and an assistant tutor, described attending a meeting at the Philosophical Society:

> It was gaslight. The general audience—among these were Professors Cayley and Miller, and a few ladies who would lend countenance to philosophy—were seated on tiers of benches, and below this general audience sat an audience fewer and more select; a more vital part of the Society, I suppose... Behind these worthies there is an array of large diagrams intended by Adams to illustrate his discourse. Challis also has something written out on a blackboard, and both the lecturers are accompanied (silently) by instruments. Distance and the careful closing of doors and windows keep out the noise and air of the world.

Adams, who had been appointed the Lowndean Professor of Astronomy and Geometry in 1859 and Head of Cambridge Observatory in 1861, spoke that evening about the Leonid meteor shower. Every thirty-three years, the Leonids put on a particularly spectacular show, as hundreds of thousands of shooting stars fill the night sky. Adams discussed five possible reasons for the thirty-three-year period of the meteor shower, and outlined why he felt that a link between the Leonids and the comet Tempel was the most compelling explanation. Challis disagreed strongly with Adams's theory. In considering which theory might be correct, Turnbull let history cloud his judgement:

> I prefer Adams's authority on this point because all men judge that Adams's prediction with regard to Neptune was correct; and I remember against Challis that he was not altogether propitious to the efforts of Adams on the former occasion. Had it not been for Challis, I suspect that England, and not France, would have had the glory (slight as it was) of priority in that discovery.[33]

Regardless of some lingering disappointment that the discovery of Neptune had not been solely a Cambridge affair, it was generally agreed that Cambridge science was in a much healthier state than it had been a generation earlier. Through the middle decades of the nineteenth century, the Society maintained a membership of at least 500 fellows; the meetings remained vibrant and the publications were still much sought-after.[34] The University and colleges were paying more attention to the sciences. The Observatory had been a great success, and the University's next major scientific project was the creation of the new University Botanic Garden. There had been a small botanic garden in Cambridge since the 1760s, located on about five acres of land in the area now occupied by the New Museums Site. By the 1820s, the site was described as a 'deserted wilderness', 'completely unsuited to the demands of modern science'.[35] John Stevens Henslow began campaigning for a bigger site for the garden shortly after he became Professor of Botany in 1825. Finally, in 1831, the University bought forty acres just south of the town, near the road to Trumpington, with the idea of creating an expanded garden which would include a significant number of trees as well as systematic beds. There was little progress on the garden during the 1830s, as negotiations with the tenants farming the land inched along. But, by the 1840s, Henslow, Whewell, and others were meeting in the house of the Philosophical

Society to plan the layout and management of the garden.[36] When the garden finally opened in 1846, it was hailed as an important addition to the scientific resources of the University, and it continued to expand and develop through the rest of the century. Several university museums—such as the anatomy museum under the care of William Clark—were also expanding in this period; the Woodwardian Cabinet and the University Library found new homes in the purpose-built Cockerell Building in 1844 and the Fitzwilliam Museum finally opened on its present site in 1848 (more than thirty years after the collection was given to the University)—another sign of the University being willing to engage with new subjects and new audiences[37] (see Figure 24).

Since the founding of the Philosophical Society in 1819, there had been a slow shift towards the sciences in Cambridge. More people were engaged in research, more science was being taught to undergraduates, and facilities

**Figure 24** Image of the Cockerell Building under construction (1840s). The new building would house the expanding University Library, and some University collections such as the Woodwardian Cabinet. It is now the library of Gonville and Caius College.

were being improved. Welcoming these changes, Adam Sedgwick compared the 'unhealthy death-like stagnation' that had gripped Cambridge at the beginning of the century to the current state of affairs in which 'chemistry, mineralogy, geology, botany, anatomy and physiology, and other natural sciences...all flourish well together'.[38] Sedgwick firmly believed that this rising interest in science in Cambridge was a direct consequence of the creation of the Philosophical Society:

> Thirty years are gone since the first formation of the Cambridge Philosophical Society; and in the *Transactions* of that body (which are a good record of the severe labours of Cambridge men) there is hardly a subject which has engaged the attention of the great Mathematicians of Europe that is not discussed in original papers of very great value. All the powers of high analysis have, in our *Transactions*, been brought to bear on the most severe and knotty questions of physics:—such, for example, as the theory of light and electricity —the theory of undulations in connexion with the great practical question of the movement of the tidal waves—the planetary perturbations—the figure and internal structure of the earth—the theory of sound and many other practical applications of profound mechanical philosophy. Nor has analysis, as an implement of pure reason, and without any reference to its immediate use or application, been overlooked by our younger members. It continues to be pushed beyond its former limits, and it carries its rewards in the delights of discovery, and the enjoyment flowing from the pure unmixed love of truth; and sometimes, perhaps, in the hope of its future application to the business of life and the practical good of our fellow-men. We need only mention the names of Cumming, Christie, Clark, Herschel, Babbage, Whewell, Henslow, Jenyns, Willis, Power, Lubbock, Airy, Challis, Miller, Hopkins, De Morgan, Murphy, Earnshaw, Kelland, O'Brien, Ellis, Cayley, Green, Stokes, and many others—all of whom were trained amongst us, have partaken of our honours, and been contributors to our *Transactions*—and we do enough to justify what I have said of the philosophical spirit that has animated our University for the last thirty years.[39]

Of course, science wasn't just growing in popularity in Cambridge. Across Britain, scientific societies were increasing in number and activity; the British Association for the Advancement of Science was still touring the country; practical centres of learning such as mechanics institutes and schools of mines opened in Liverpool, Manchester, London, and many other cities and regions; and new universities were growing up in London and elsewhere. People were becoming more interested in the sciences, and the sciences were becoming more relevant to everyday life. This could be seen in every part of the country; and yet, perhaps Sedgwick was right to say that the Cambridge men were, particularly in the more mathematical parts of science, leading the way.

The Cambridge Mathematical Tripos remained unique in the country: *every* student in the University was obliged to study mathematics before they could graduate. Even after 1851, when two new degrees were created in Cambridge University—one in natural sciences and the other in moral sciences—the Cambridge system remained fundamentally mathematical. There had been much controversy about reforming the University curriculum, and it had taken the intervention of Prince Albert, acting in his role as Chancellor of the University, to bring change to the ancient institution. In 1847, shortly after his election, Albert had approached Henry Philpott, who was Vice-Chancellor of the University, Master of St Catharine's College, and President of the Cambridge Philosophical Society, to discuss the state of the University curriculum. Philpott drew up an elaborate chart of all of the teaching and examinations going on in the colleges and the University. Albert showed the chart to his advisor Robert Peel, the former prime minister, who instantly saw the disjunction between the professors' lectures and the students' examinations.[40]

The college and University teachers worried about displacing mathematics from its central place in the Cambridge curriculum because it was seen as such a pure and truthful subject; it dealt in facts, not speculation, making it extremely useful for training young minds. It was about learning a way of thinking, rather than just learning facts off by heart. Some of the more progressive members of the University agreed that reform might be beneficial, but it must be gradual. For example, William Whewell thought that scientific facts might be included in University examinations when such facts had been accepted for a century or more. Robert Peel was horrified by the idea of such slow change, as he wrote to Albert:

> I think Dr Whewell is quite wrong in his position—that mathematical knowledge is entitled to *paramount* consideration, because it is conversant with indisputable truths—that such departments of science as chemistry are not proper subjects of academical instruction because there is controversy respecting important facts and principles, and constant accession of information from new discoveries—and danger that students may lose their reverence for Professors, when they discover that the Professors cannot maintain doctrines as indisputable as mathematical or arithmetical truths.
>
> The Doctor's assumption, that *a century should pass* before new discoveries in science are admitted into the course of academical instruction, exceeds in absurdity anything which the bitterest enemy of University Education would have imputed to its advocates. Are the students at Cambridge to hear nothing of electricity, or the speculations concerning its mysterious influence, its possible connection with the nervous system and with muscular action, till all doubts on the subject are at an end?[41]

Prince Albert invited Philpott to Windsor to discuss the problem. Philpott agreed that reform was needed, and saw that new examinations were necessary if they wished to entice more students into attending scientific lectures and to better align the research interests of the professors with the course of studies followed by the undergraduates. Together with Robert Phelps (who was Master of Sidney Sussex College and was about to succeed him as Vice-Chancellor), Philpott put together a syndicate in Cambridge's Senate House to discuss reform. The syndicate included several prominent members of the Philosophical Society—Whewell, Professors James Challis and John Haviland, and 'the wrangler-maker' William Hopkins—as well as tutors, the professors of civil law and divinity, and several heads of colleges. The syndicate recommended the creation of the two new Triposes. The Moral Sciences Tripos would include compulsory lectures and examinations in moral philosophy, political economy, modern history, and law, while the Natural Sciences Tripos would include anatomy, physiology, chemistry, botany, and geology. It took several years of arguments, debates, petitions, votes, and eventually a Royal Commission before the new Triposes officially came into being.[42]

To sit either of the new examinations, a student already needed to have passed the Mathematical Tripos; therefore, all students studying natural sciences in Cambridge had a good grounding in mixed mathematics.[43] In the early years of the new degree, a student would sit the Mathematical Tripos after three years of study; he would then have just one additional year to study multiple scientific subjects before sitting the examinations for the Natural Sciences Tripos. Each paper within the Natural Sciences Tripos covered the full spectrum of knowledge in that subject, as one student complained after sitting Sedgwick's first-ever geology examination:

> Geology seems to have been tolerably done by all, brilliantly by none. If the paper had been a quarter of the length, it would have been more satisfactory to all parties. Fuller says Sedgwick boasted of having made it a 'very complete' paper, and got *all* geology into it, to be written out in four hours![44]

Because of the enormous workload coupled with the lack of career prospects in the sciences, few students actually sat the new Tripos in the 1850s, but the reforms showed that the University recognized the increasing interest in science, and proved its willingness to make changes. Slowly, as the Tripos was amended to make it more appealing to students and numbers began to increase, the need to offer science tutorials to students forced the colleges to create new teaching posts. The availability of these additional teaching positions created career paths for young scholars and encouraged more people to pursue the natural sciences after graduating.[45]

One of the students to sit the new Tripos examination in its first year was George Liveing, the son of a Suffolk surgeon. He had sat the Mathematical Tripos in 1850 and been placed eleventh wrangler; in 1851 he came top of the Natural Sciences Tripos. He then set up a little laboratory in a cottage on Corn Exchange Street. Liveing had only limited means and so his laboratory was primitive and under-equipped, but it was still one of the best laboratories in Cambridge at the time, and it was there that many medical students came for private tuition in practical chemistry. After a year of teaching students about reactions and reagents and explaining the basics of titrations, distillations, and electrolysis, Liveing was offered a fellowship at St John's College in 1853. Mirroring the increased University interest in sciences, the colleges too began to take practical steps to support the new Tripos: St John's College made Liveing a college lecturer in chemistry and built him a laboratory in the college, just behind New Court and much grander and better equipped than the one he had made for himself in Corn Exchange Street. Many years later, Liveing reflected on the importance of this step: 'the College built for me the Chemical Laboratory, which was the first seed sown towards the growth of a large Chemical School.'[46] Liveing was extremely active in the Philosophical Society and became its president in 1877. He also went on to become University Professor of Chemistry and campaigned for (and got) better laboratories and lecture rooms.

Cambridge was moving away from the 'death-like stagnation' which had so worried Sedgwick. Attendance at lectures began to rise, more students took the scientific papers, and science became more visible in the University. Following the success of the Cambridge Philosophical Society, many new societies were founded with focus on particular disciplines within the sciences. The Ray Club was founded in 1837 by Charles Cardale Babington in honour of the seventeenth-century naturalist John Ray. Babington had studied at St John's College from 1827 to 1830 and, having shown an interest in botany, he was invited by Henslow to act as his assistant. So began a lifelong interest in the field of plant taxonomy. After graduating, Babington remained in Cambridge and continued to live in college (though he was not awarded a fellowship until 1874). He established the Ray Club 'for the cultivation of Natural Science by means of friendly intercourse and mutual instruction'. Its members (originally limited to just twelve, though the membership later expanded) met each Wednesday evening during term time to present and examine natural history specimens. Babington was the Ray Club's secretary for fifty-five years; he was also Secretary to the Philosophical Society from 1851 to 1870 and its president from 1873 to 1875, by which time he had succeeded

Henslow as Professor of Botany.[47] Some new societies in the mid-nineteenth century were particularly aimed at students; these included the Natural Sciences Society and the Medical Sciences Society. There was also a host of non-scientific societies springing up in Cambridge throughout the century, including the Antiquarian Society, the Camden Society, the Eranus Society, the Grote Society, and the famous Cambridge Apostles (the Antiquarian Society and the Camden Society held their meetings in the house of the Philosophical Society).[48] There was even a new journal devoted to publishing mathematics: *The Cambridge mathematical journal* was founded in 1837 in Trinity College by three recent wranglers and was judged to be 'full of very original communications'.[49]

Though there were many new forums for science in Cambridge, the Philosophical Society remained central. It was not just a place for resident members to present their latest research; there were also many speakers from outside Cambridge. Some of these were old members returning to their *alma mater*. This was the case with George Biddell Airy, the former Plumian Professor of Astronomy and an early director of Cambridge Observatory. In 1835, after being appointed Astronomer Royal, he had left Cambridge for Greenwich. Though London had its own Astronomical Society from 1820, and though that society was highly regarded, Airy chose to present much of his research at Cambridge. He regularly travelled to Cambridge to give papers on optics, solar eclipses, new telescope designs, the workings of the eye, and mathematics.[50] The Philosophical Society very much counted Airy as one of their own, as Sedgwick made clear when he wrote that

> two great discoveries have [in the last thirty years] been made by Cambridge men: and it deserves remark, that these two are the only great discoveries in physical astronomy made by Englishmen since the days of Newton. I here, of course, allude to Professor Airy's discovery of the long period of perturbation in the Earth's orbit by the action of the planet Venus; and to Mr Adams's theoretical discovery of a new planet external to Uranus.[51]

Airy's paper on the effect of Venus on Earth's orbit had actually been presented to the Royal Society in London, but Sedgwick saw it as an essentially 'Cambridge' paper, one that would not have been possible without Airy's early training in the mathematical thinking of Cambridge, and the influence of the Cambridge Philosophical Society.[52]

Airy was not the only Cambridge man to maintain a connection to the Philosophical Society after departing the city. Augustus De Morgan, a student of Trinity College, had been fourth wrangler in 1827 and was much respected by his tutors William Whewell and George Peacock. Because he lacked strong religious beliefs, he was never ordained and did not seek a Cambridge fellowship; instead, he was elected Professor of Mathematics at the new non-denominational London University in 1828. Finding that London lacked a proper place to present his mathematical paper, De Morgan regularly returned to Cambridge over the next decades to present his momentous work on algebra.[53] Likewise, Philip Kelland (the senior wrangler in 1834 who became Professor of Mathematics at Edinburgh), David Ansted (a former student and fellow of Jesus College who later became Professor of Geology at London's new King's College), and Matthew O'Brien (third wrangler in 1838 and later Professor of Natural Philosophy and Astronomy at King's College, London) all returned to Cambridge to speak at the Philosophical Society, which they considered as important as any of the learned societies of London or Edinburgh.[54]

It wasn't just former Cambridge students and fellows who saw the Society as a desirable place to speak; many eminent natural philosophers with no previous link to Cambridge travelled to the house on All Saints' Passage to present their work. Richard Owen, the great comparative anatomist, had been proposed as an honorary member of the Philosophical Society in 1842—the same year he became resident curator of London's Hunterian Museum and coined the word 'dinosaur'. He spoke several times at the Society about the fossils he studied, including a beaked lizard that had been found near Shrewsbury (see Figure 25). The lizard was significant because it gave evidence that four-footed animals, perhaps also

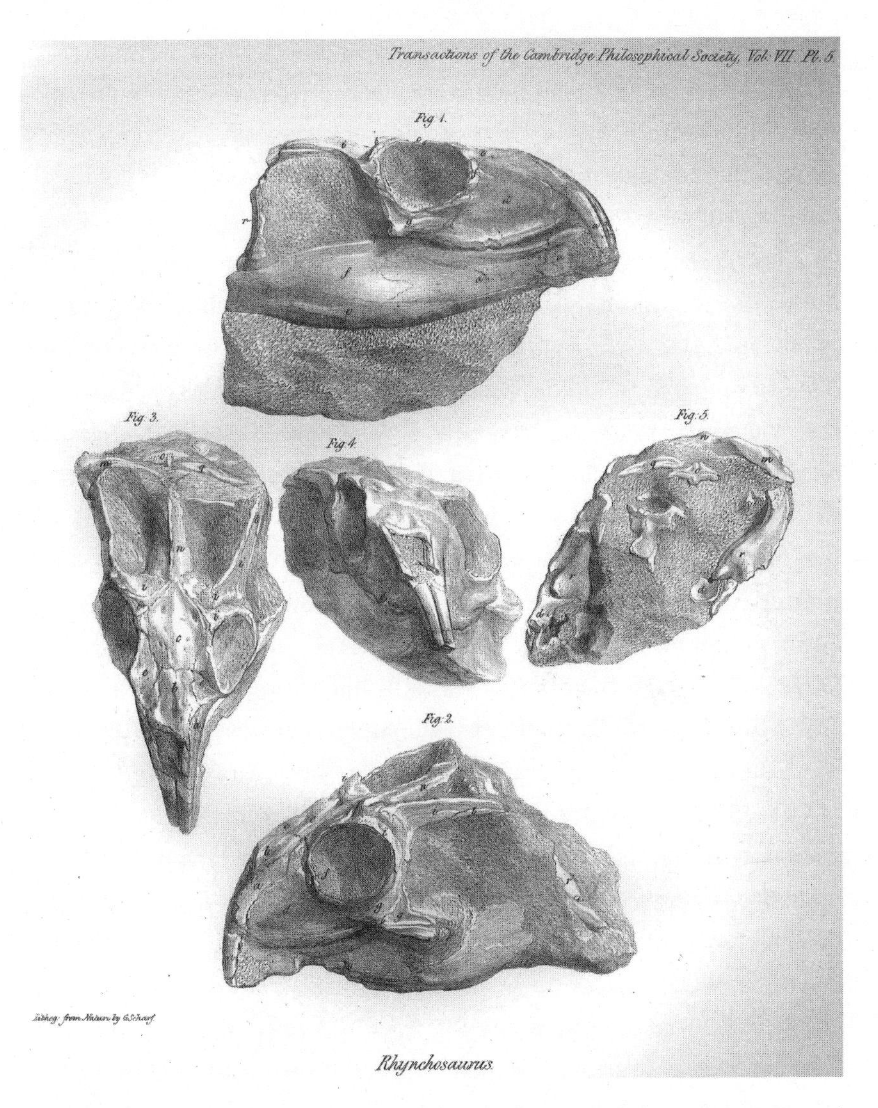


**Figure 25** Richard Owen (the man who coined the word 'dinosaur') presented these images of a prehistoric beaked lizard to the Society in April 1842.

'warm-blooded', existed when the New Red Sandstone had been laid down more than 200 million years earlier.[55]

Through the middle of the nineteenth century, the Society also maintained its programme of publication and periodical exchanges—showcasing its work for the world, and keeping abreast of new developments elsewhere. The *Transactions* continued to peer-review and publish the best papers that had been read before meetings of the Society. By the mid-nineteenth century, the journal was leaning more heavily towards mathematics and mathematical physics. The first volume of *Transactions,* published in 1822, had had a roughly equal spread of articles across mathematics, chemistry, physics, geology, astronomy, medicine, anatomy, meteorology, and engineering. The eighth volume, which appeared in 1849, had 36 papers on mathematics or mathematical physics, with just a handful on geology, biology, medicine, and other topics like clock-making. A few within the Society, notably William Whewell, attempted to retain the philosophical breadth of the Society by writing articles outside the increasingly narrow bounds of 'science'. Whewell's articles for the *Transactions* covered topics including Hegel and German idealism, Aristotle's theory of induction, political economy, the history of science, and Plato's notion of dialectic, as well as occasional papers on pure mathematics.[56] Though the Society was still open to papers on a range of topics, the increasingly mathematical nature of the journal reflected the teaching and research interests of the Society's fellows.

The unique nature of the *Transactions* as the record of a university society was highlighted by John Herschel when he declared that they were 'full of variety and interest, and such as no similar collection, originating as this has done in the bosom, and, in great measure, within the walls of an academical institution, can at all compare with'.[57] The *Transactions* acted not just as a repository for the papers presented to the Society, but as an advertisement for the University as a whole, as Whewell saw it:

> We think we are doing a service to our readers in calling their attention to [the *Transactions of the Cambridge Philosophical Society*], both on account of the valuable and curious memoirs on various subjects which it contains, and also because it may assist them in forming a correct opinion of the condition of one of our English Universities with regard to the cultivation of physical science.... [the English Universities] have of late been incessantly assailed with charges of the neglect of modern knowledge and improvements....

> [H]aving shown what [the *Transactions*] contain, we really cannot but feel tempted to ask—where is the evidence of the sloth and prejudice, the want of habits of inquiry and of sympathy with the advances of modern knowledge, which are so plentifully attributed to the English Universities, whenever they are mentioned by various classes of writers? What branch of knowledge is uncultivated? With what modern discoveries are they unacquainted? To what corresponding class of their countrymen are they inferior? What set of men can be pointed out as doing more with means such as they possess?[58]

Despite the quality of the papers in the *Transactions*, the Society struggled to make them pay. With each new volume, the council of the Society would approach the Syndics of the University Press to ask for help covering the costs of printing the journal. The University Press was unceasingly generous, usually paying half of the printing costs throughout the 1840s, and increasing their funding to pay the full printing costs and providing paper in the 1850s. Attempts to reorganize finances and renegotiate arrangements with booksellers did little to alleviate the poor financial state of the journal.[59]

Though the Society was struggling to keep one journal afloat, its council took the unusual step of launching a second journal. This was called *Proceedings of the Cambridge Philosophical Society*. In summer 1843, it was decided that all papers given at general meetings of the Society should be recorded, even if they were not formally published in the *Transactions*. The *Proceedings*, compiled by the Society's three secretaries, listed most (but not all) meetings that had taken place and summarized the contents of the papers, particularly those that would not be published in full. They also included information about the election of new fellows, the passing of new bye-laws, or additional notes from the speaker. And they began to record tantalizing fragments of the discussions that followed papers, or mentioned some of the questions that had been asked of the speaker: the *Proceedings* show that, by the 1850s, most papers were followed by a lively discussion, though the details are often sketchy.[60]

The Society also maintained and expanded the programme of periodical exchange that had been going on since the publication of the first volume of the *Transactions*. They established new ties with bodies in Munich, Metz, Vienna, Bordeaux, Berne, Cherbourg, Washington, Boston, and Victoria, sending copies of the *Transactions* and receiving journals from around the world in exchange. By the end of the 1860s, the Society was swapping journals with over eighty institutions in Britain and further afield, ensuring its continued expansion.[61]

The Society was reaching new highs in the mid-nineteenth century. Comfortably ensconced in a house of their own, they hosted meetings attended by the great and the good, gave papers on topics of national importance, made world-changing discoveries using the most modern methods, pushed the agenda of science in the University, and reached out to other societies across the globe. Membership of the Society remained above 500 through these decades. It was a given that all of the University's professors in scientific subjects would join and take an active part in the Society's daily life, but there were also hundreds of fellows just beginning their scientific careers, and many who did not seek to be professional 'scientists' but took an interest in natural philosophy as part of a well-rounded life. The Society also attracted prominent natural philosophers as honorary members. These included the expert on heat and energy James Joule, the famed botanist and scientific traveller Joseph Dalton Hooker, the geologist Charles Lyell, the mathematician and logician George Boole, the German chemist Justus von Liebig, the physicist and mountaineer John Tyndall, and the philosopher and political economist John Stuart Mill.[62] The library and reading room were popular as ever, and the museum was still expanding its collection. Science was on the rise in Victorian Britain, and the Cambridge Philosophical Society had been central to that rise. It was hard to believe that the Society was about to come crashing down.

# 5

# THE MISDEEDS OF MR CROUCH

On 24 February 1851, Augustus De Morgan was readying himself to set off for Cambridge. This former wrangler was now Professor of Mathematics at University College London—the radical 'Godless institution of Gower Street'—and, as he made his way north towards his *alma mater*, his head was filled with $\psi$s and $\varphi$s. He was to give a talk that evening to the Philosophical Society on one of his favourite topics: integral calculus.[1] As De Morgan ran through differential equations and their solutions in his mind, and the train from London raced along the track, the Society's council were rushing to prepare the meeting room for their guest. The house on All Saints' Passage—normally so orderly—was in disarray. For, as De Morgan's thoughts scaled the lofty heights of pure mathematics, one man faced a rather more mundane problem. John Crouch, the curator of the Philosophical Society, stood nervously before the County Court of Cambridgeshire. He had been declared an insolvent debtor.

Crouch had been a larger-than-life figure around Cambridge for many years. As a younger man, he had been a keen cricket player, playing for many clubs in the town—the Union, Chesterton, the Hoop, and the Royal George Inn—and was a star player of many a town-versus-gown cricket match, described by his teammates as 'a great wonder for his skill with the willow'. Crouch was also a yeoman bedell of the University (see Figure 26). The bedells performed ceremonial and administrative duties for the University, including assisting at graduation ceremonies, collecting fines from students, enforcing certain University rules in the town's businesses, and acting as criers. As a crier—publicly proclaiming pronouncements

**Figure 26** An esquire bedell (left) and yeoman bedell (right), c. 1815. John Crouch was a yeoman bedell of the University for many decades, performing administrative and ceremonial duties.

from the University—Crouch would have cut a striking figure in a black gown and top hat, carrying a silver mace over his shoulder, his booming voice rising above the din of Cambridge's busy streets. The one surviving image of Crouch, painted in 1847, shows him in middle age posing with the town and gown cricket teams (see Plate 6 and Figure 27). Seated in the foreground, his dark suit in sharp contrast to the young men's whites, Crouch is a solid and imposing figure, seemingly a bastion of respectability.[2]

Crouch had been the Society's curator since 1820. He lived with his family almost rent-free in the Society's house, he was given a generous salary, and he received regular pay rises. His work at the Society was not very taxing, meaning that he had time for his responsibilities as bedell as

**Figure 27** A town-versus-gown cricket match on Parker's Piece, the town's gaol is visible in the background (1840s).

well, for which he was paid a modest wage. But, despite having two reliable sources of income, extremely cheap accommodation, and a reputation as an upstanding Cambridge man, Crouch somehow found himself falling into debt.

His first appearance at the County Court was to examine the extent of his debts and to identify his creditors.[3] On that same day, at a council meeting of the Society, a committee of four fellows was appointed to look into Crouch's insolvency. They found that he had debts of over £700 (approximately ten times his annual income) and assets of only £78. The fellows looked not just into Crouch's finances but into his character, and found him wanting: 'the committee regrets to observe that [Crouch's debts] are of such an amount and of so long standing as to indicate habits of expense incompatible with the curator's station.'[4] One of Crouch's duties in the Society had been to collect library fines, of which he was allowed to keep half for himself; he was immediately suspended from this task, and his salary was reduced accordingly. The committee then considered 'the complaints which from time to time have been made of his inadequate performance of his duties'. A few days later, Crouch was given three months' notice.[5]

Though the fellows no longer saw Crouch as employable, they recognized his long tenure at the Society—he had, after all, served them for more than thirty years—and so voted to award him a pension of £25 per year for the remainder of his life. But Crouch's bankruptcy aroused suspicions amongst some fellows and they began to scrutinize their account books more closely. They started to see discrepancies and soon proved that Crouch had stolen at least £80 from the Society. His pension was stopped immediately.[6] Crouch had had plenty of opportunity to steal from the Society—as well as collecting library fines, he played a role in collecting membership subscriptions, charging visitors an entrance fee to the museum, selling copies of the *Transactions*, and selling on old newspapers from the reading room.[7] Though the Society had a treasurer, he did not oversee each transaction Crouch entered in the ledger, which meant that, for thirty years, Crouch had been able to quietly fiddle the books. No doubt the total missing from the Society's accounts was considerably more than £80. It is very probable that Crouch also cheated the University out of money, since part of his duty as yeoman bedell was to collect fees from market-stall holders. He had also been responsible for collecting subscriptions for the town cricket club, and perhaps this served as another source of personal income. Some of Crouch's debt had accrued when he tried to help out a relation who had become bankrupt, but that accounted for only a small portion of the total—what Crouch spent the rest of the money on is still unknown.[8]

Crouch was ordered to pay £10 per annum to the County Court for five years, but no criminal charges were ever brought. The University continued to employ Crouch as a bedell until his death in 1858, and even increased his salary in 1856.[9] Other than stopping his pension, the Society, perhaps fearing a scandal, took no further action against Crouch, but they were to feel the effects of his actions for many decades to come.

Though the Society gave every appearance of prosperity, with its fine building, and its elegant museum and reading room, it had built its house with borrowed money. Most of the money had been raised by selling bonds worth £50 to fellows of the Society; interest was to be repaid on these bonds at a rate of 4 per cent.[10] Membership subscriptions comfortably covered the repayments on interest, but the Society struggled to save enough to begin making repayments on the capital. Added to this was the problem of paying church rates; after lengthy correspondence and debate, the Society had been compelled to pay rates to their parish church from 1852.[11] Following the discovery of Crouch's misdemeanours, the fellows became more aware of their financial situation and began trying to cut their expenses: they stopped using gaslights in the council room; they cut back on the number of journals in the library and curtailed the buying of new publications because 'the state of the finances would not justify any additional expense'; they even began to think about closing their beloved reading room.[12] At the end of 1854, the fellows resorted to taking out a loan of £200 just to keep on top of daily expenditure.[13]

By 1856, it was apparent that the Society could not continue to function financially in its present state. It had debts of £2050, plus it needed £200 for its upcoming lease renewal. The fellows calculated that they required a surplus of £50 to £60 per year to keep afloat. A subcommittee of fellows was appointed to consider the situation. As well as the problems caused by Crouch, and by the additional expense of £40 per year on church rates, the fellows saw that subscriptions had been gradually dropping off in recent years. As the sciences had grown up in Cambridge, so too had new societies and discussion groups, new laboratories, and new publications. The Philosophical Society was no longer the only important scientific forum in town, and men who would once have joined as a matter of course now had other options available to them. This falling-off of subscription income meant that the Society's expenditure exceeded its income by about £70 per year. The subcommittee's principal recommendation to address the problem was to separate the reading room from the Society.[14]

The beautiful high-ceilinged reading room had been a place of inspiration for many fellows over the years; there they would lounge for hours, riffling through the metropolitan newspapers or poring over the latest journals from Germany and France. It was a meeting place for fellows who were from different colleges and whose paths might not otherwise cross during the working day. It would be a heavy blow to lose it, but perhaps it was a price worth paying if it would save the rest of the house. Dissociating the reading room from the Society would mean eliminating its running costs; furthermore, renting the room out would mean a new income stream. Though the reading room was still popular, it was no longer unique. The colleges had been playing catch-up with the Society, and now many fellows were able to access quality reading material in the combination rooms of their own colleges; newspapers and periodicals were also becoming more affordable, which meant that individuals were more likely to buy them for their own use. There was concern that the subscription fee to become a fellow of the Society, a fee which automatically included use of the reading room, was too high, especially since the reading room was no longer the only such place in Cambridge. Getting rid of the reading room would mean the subscription fee could be lowered, hopefully attracting more fellows and ultimately raising revenues. At a general meeting of the Society in April 1856, the question of what to do with the reading room was put to a vote. The Society's wooden ballot box was placed in the meeting room, and each fellow took one of the little black beans offered (see Figure 28). One by one, they reached their hands into the dark recess of the box and secretly cast their vote. Later, when the box was opened up by the three secretaries—the botanist Charles Cardale Babington, the chemist George Liveing, and the astronomer John Couch Adams—they saw a pile of beans heaped up in the left-hand chamber of the box, while just a few desultory ones rattled about in the right. The fellows had voted to let the reading room go.

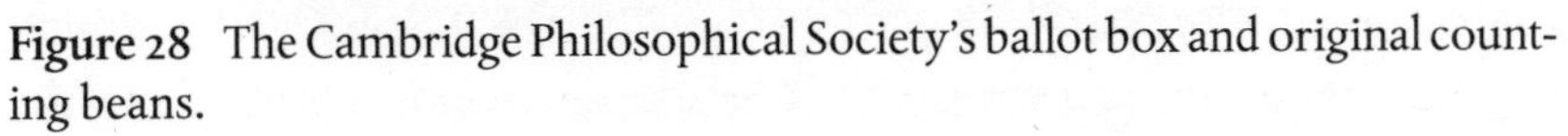
**Figure 28** The Cambridge Philosophical Society's ballot box and original counting beans.

Individual fellows were given the option to take on the running of the reading room themselves, but none did.[15]

And so it was that the old reading room was rented out for a fee of £45 per annum to William Walton, a Trinity alumnus and mathematician. With sadness, the fellows began to dismantle what they had built. James Challis and William Hallowes Miller were put in charge of finding new homes for the instruments and astronomical clock that had adorned the room. The newspapers were sold off, while an auction was arranged for the literary periodicals (the books in the library and the scientific periodicals were not affected). All the furniture, save two bookcases and a glazed cabinet, was valued and sold. These carefully gathered items were scattered to the winds. And the curator, Mr Punchas, who had replaced

John Crouch, was informed that his reduced duties would mean a corresponding reduction in his salary.[16]

Fellows who had paid subscriptions in advance were offered compensation for their loss of access to the reading room, but many waived that right, including Joseph Romilly.[17] This devotion was echoed many times in the coming months, as fellows with long-standing loyalty to the Society tore up their bonds, renouncing all financial claims on the Society. George Peacock, Adam Sedgwick, and Charles Cardale Babington were amongst the fellows who did this. Many also gave donations to the Society, with some of the donations specifically intended for the upkeep of the museum.[18]

These measures afforded the Society some relief from their financial predicament but were not enough to deal with the problem of repaying the loans. The Society still struggled with everyday costs. In 1859, the Secretaries agreed that they would print the next issue of the *Transactions* 'so soon as the funds of the Society should be capable of bearing the expense'.[19] In the event, it was 1864 before the next volume appeared. There had been an eight-year gap between this, Volume X, and the preceding one; earlier in the century, the average gap between volumes of the *Transactions* had been three and a half years. Though Cambridge University Press was still making contributions towards the cost of paper and printing, the Society was unable to cover its share of publication costs.[20] Nor was the Society able to maintain its staff. In 1863, they discontinued the office of curator and let Mr Punchas go. His salary of £65 per year had become too much for the fellows to pay. In his place, they hired a housekeeper with an annual salary of £24.[21] Without a resident curator, the upkeep of the library, the museum, and the house itself became more difficult.

Just as the Society's reading room had suffered falling popularity as the colleges competed with similar offerings of their own, so the Society more generally was a victim of its own success. The Society had led the way on offering facilities for the sciences in Cambridge and had inspired its fellows to push for more resources; and they had been successful in this. Through the 1850s and 1860s, the sciences became much more visible in Cambridge. The new Natural Sciences Tripos was attracting more students each year, college fellowships in the sciences were becoming available, and the University was beginning to think of serious investment in buildings to house scientific laboratories and collections. All of this meant increased competition for the Society as Cambridge men found alternative spaces for science. This made it more difficult for the Society to maintain membership levels and subscriptions.

In the 1850s, the site of the old Botanic Garden (where the Society's earliest meetings had taken place) was proposed by the University as a possible location for new University museums and lecture rooms. The architect Anthony Salvin was asked to draw up plans for these new buildings, creating spaces for students to attend scientific lectures, and designing a home for the University's growing zoology and comparative anatomy collections. Salvin's plans were approved in 1854 but, due to University politics, building work did not begin until 1863. The buildings were ready for use by 1865 (see Figure 29).

Coincidentally, 1865 was also the year that the Society finally conceded that it could no longer afford to keep its much-cherished house. The fellows had tried everything they could to hold on to it but it was not

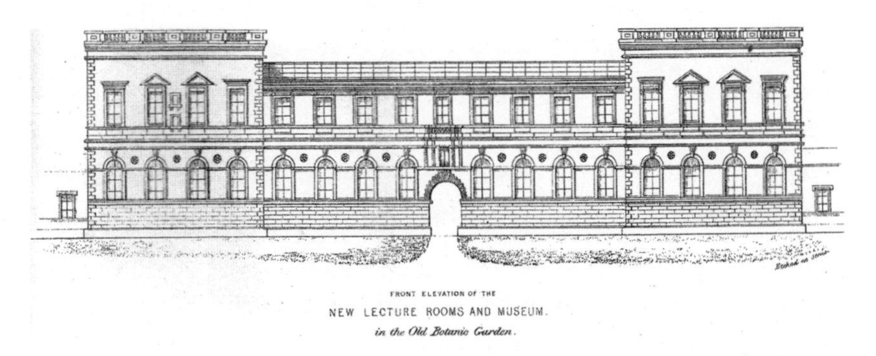


**Figure 29** Anthony Salvin's design for the New Museums, 1865. The Philosophical Society occupied rooms in the central part of the building, near the arch. Most of the building has now been demolished, with only a small portion of the western end remaining.

enough. The house was draining their resources and beginning to threaten their mission. Bravely looking forward, the Society's council began to consider their options. One option in particular jumped out at them. With the completion of the University's New Museums Site, as the set of buildings came to be known, an idea began to form in the minds of several fellows of the Society. In February 1865, it was agreed that the Society's president, William Hepworth Thompson, Regius Professor of Greek, should write to the vice-chancellor of the University to sound him out about the possibility of the Society obtaining some rooms in the New Museums. The vice-chancellor, Henry Wilkinson Cookson—Master of Peterhouse and a long-standing fellow of the Society—was open to the idea, though he could not give formal confirmation until the administrative structure of the new site was finalized.[22]

A few weeks after Thompson's exchange with Cookson, the Society's council began to consider the possibility of offering the Society's museum to the University as a gift. The collection that had begun four decades earlier with a few treasured boxes of insects and shells given by John Stevens Henslow had grown into something extraordinary—a vision of life on earth, held motionless and preserved forever behind the glass of countless vitrines. Through careful curacy, the Society's museum had become the most important scientific collection in Cambridge. It was unique. But, if the Society did not have a house of their own, what would become of the museum?

Henslow had died in 1861, a few years after his dear wife Harriet in 1857. But the other early curator of the museum, Leonard Jenyns, was still alive and living happily in Bath, running a natural history field club he had set up, when the Society approached him and sought his opinion on the fate of the collection. Ever a pragmatist, Jenyns gave his blessing to the idea of gifting the collection to the University, so long as his donations, and particularly a set of Cambridgeshire insects, were kept intact in perpetuity[23] (see Figure 30). The Society also needed permission from another significant donor—Charles Darwin. Darwin had continued to use the Philosophical Society museum as a repository for his *Beagle* collections decades after his return from South America. He would send parcels of

**Figure 30** Some of the specimens of Cambridgeshire insects donated by Leonard Jenyns to the Society's museum in the 1850s, which were later given to the University's new zoology museum.

preserved animals, care of Mr Crouch, to be examined by the Cambridge naturalists, and some of those specimens would end up permanently part of the collection.[24] Darwin too gave his blessing.

Clearly, there were benefits for both parties: if the Society no longer had a house, it would need to make arrangements for its collections to be sold, stored, or displayed elsewhere. From the University point of view, the acquisition of a good-quality natural history collection would be a boon—a proper centre piece for their new museum building. Though the University had a small zoology museum, its collections were more anatomical than natural historical, and very limited in scope. The Society's collection could act as an object of exchange, allowing the fellows to swap their ready-made museum for a new home, and rescue them from their financial crisis.[25] In May 1865, at a general meeting of the Society, the ballot box was produced again and the fellows voted unanimously that their natural history collections be offered to the University and placed in the New Museums (see Figure 31). A formal University Grace was passed a few weeks later, accepting the collection. Another Grace followed shortly afterwards, granting rooms in the New Museums to house the Society and its library.[26] The exchange was complete.

That summer, the precious objects were bundled up in swathes of paper and twine. Spirit jars, exotic birds and animals, and fragile fossils all had to be packed individually—a labour of many hours. They were then loaded up and carted the half mile from All Saints' Passage to the New Museums Site. In their new home, the collections were carefully arranged, regrouped, and reset. Many of them can still be seen in the University's zoology museum today, an echo of an older tradition that lives on in modern Cambridge. The books and periodicals too were packed up, hauled across town, and restacked in the New Museums Site; but these would remain the property of the Society. Now that the contents of the library and museum had been relocated, the fate of the house itself could be decided. There was an attempt

**Figure 31** The University's new Zoology Museum (incorporating many specimens from the Philosophical Society's Museum) in the late nineteenth century.

to lease it out to the conservative student group known as the Pitt Club (which met in another building on All Saints' Passage at that time) but, when that fell through, it was decided to sell the house instead. In November 1865, an auction was arranged; the following February, the fellows accepted George Ralph Carpenter's bid of £1,200 for the house—far less than it had cost them to design and build it, but enough to stave off disaster. And so the Society's association with All Saints' Passage came to an end.[27]

When the academic year restarted in the autumn of 1865, the Society began to hold its meetings in the mathematical lecture room on the New Museums Site. This chamber seemed bare compared to their former meeting room, which had been adorned with so many wonderful specimens from the museum. But, as Alfred Catton began to speak about the synthesis of formic acid, conjuring the pungent aromas of his laboratory bench, and Arthur Cayley unfolded the intricacies of a new theorem for his listeners, the place began to feel like home.[28]

That same night, the fellows elected a new council, to be led by Henry Wilkinson Cookson as President—the same man who had facilitated the Society's move to a University premises. In the following year, William Hepworth Thompson, who had been president of the Society when the move had been proposed, was elected Vice-Chancellor of the University. This cosy relationship, which saw men slide in and out of the Society's presidency, the University's vice-chancellorship, and various senior college positions, highlights the many interrelations between the constituent parts of Cambridge. Though the Society was officially independent of the University, its members needed to hold a Cambridge degree to be elected and its council tended to be made up of senior figures from the colleges and university. Now that the Society had moved into a University building, and the Society's museum had become a university collection, the line of independence became even more blurred.

Though it still had debts to pay off, its new residence in the New Museums Site allowed the Society a period of stability. After the distractions of the financial crisis and the move, it was a relief to return to the normal business of science. And, as the Society settled in, Cambridge science began a renewed ascent. The Salvin Buildings represented a willingness on the part of the University to make large-scale investments in facilities for science. This need for extra facilities was tied in part to the success of the Natural Sciences Tripos. In 1860, the Tripos had been reformed so that students did not have to sit the Mathematical Tripos before studying natural sciences. Prior to that, the Natural Sciences Tripos had attracted a maximum of six candidates each year. Students had only a year to prepare for an intense four-day examination: comparative anatomy and geology on the first day; physiology and botany on the second; chemistry and mineralogy on the third; and general scientific knowledge, including history of science, on the fourth, topped off with a *viva voce*. From 1861 onwards, students could spend three years preparing the same range of subjects.[29] This quickly led to an increase in the number of students sitting the Natural Sciences Tripos; by 1870, almost twenty students were taking the Tripos each year; by 1880, the number had increased to sixty. For the first time, the number of students reading natural sciences equalled the number reading mathematics.[30] There were other reforms too: experimental physics was added to the curriculum of the Mathematical Tripos in 1868 and the colleges began to create more fellowships in the natural sciences at around the same time.[31]

The inclusion of experimental physics in the syllabus was a key moment in the history of Cambridge science. In the late 1860s, the Senate House appointed a syndicate to consider the state of the science in the University. Oxford had just begun building its new Clarendon Laboratory for physics, and Cambridge was concerned about falling behind its old rival. The syndicate, which included George Stokes and George Liveing, recommended building a similar laboratory, hoping to compete with Oxford. But University funds were low and a second syndicate had to be appointed to work out how to raise the estimated £7,000 needed to build, equip, and staff such a laboratory. In 1870, William Cavendish, the seventh Duke of Devonshire—a former second wrangler and Smith's Prizeman, Chancellor of the University after the death of Prince Albert, and Patron of the

Cambridge Philosophical Society—backed the project to the amount of £6,300. This was the same year that Cavendish was appointed chairman of a Royal Commission on Scientific Instruction and the Advancement of Science, which came to be known as the Devonshire Commission. Its task was to assess the state of science teaching at all levels across the nation and, through his work with the Commission, Cavendish began to see how British science was falling behind its European neighbours. The Devonshire Commission would go on to publish eight reports between 1872 and 1875 but, even before the first report was completed, Cavendish had decided to take the practical step of funding a new laboratory for Cambridge himself. His donation would be enough to build the laboratory and buy the necessary apparatus for it, just leaving the University to fund the salaries for a professor, a demonstrator, and a lecture room attendant. A plot on Free School Lane, part of the New Museums Site, was chosen for the Cavendish Laboratory, just yards away from the Society's rooms[32] (see Figure 32).

As with any expensive University project, there had been prolonged wrangling about the funding, design, status, and purpose of the new laboratory. Early plans to obtain some of the necessary funds from the Cambridge colleges had fallen through as the colleges objected to opening their own coffers for this newfangled scheme. There were also more fundamental objections to the creation of the laboratory: at a meeting of the Senate House, several members questioned the necessity of such an institution at all. Several of the University registrars 'urged the superior claims of ecclesiastical history and pastoral theology', while the master of Corpus Christi College 'deprecated exaggerated statements in favour of physical science as a disparagement of classics and mathematics', and Henry Cookson—former President of the Philosophical Society—stressed the potential dangers of allowing moral and spiritual studies to be overborne by the physical sciences, which related only to 'what is material and perishing'.[33] While some feared that the physical sciences were gaining too much ground at the expense of the more established subjects, others thought that even the building of a modern and well-equipped laboratory would

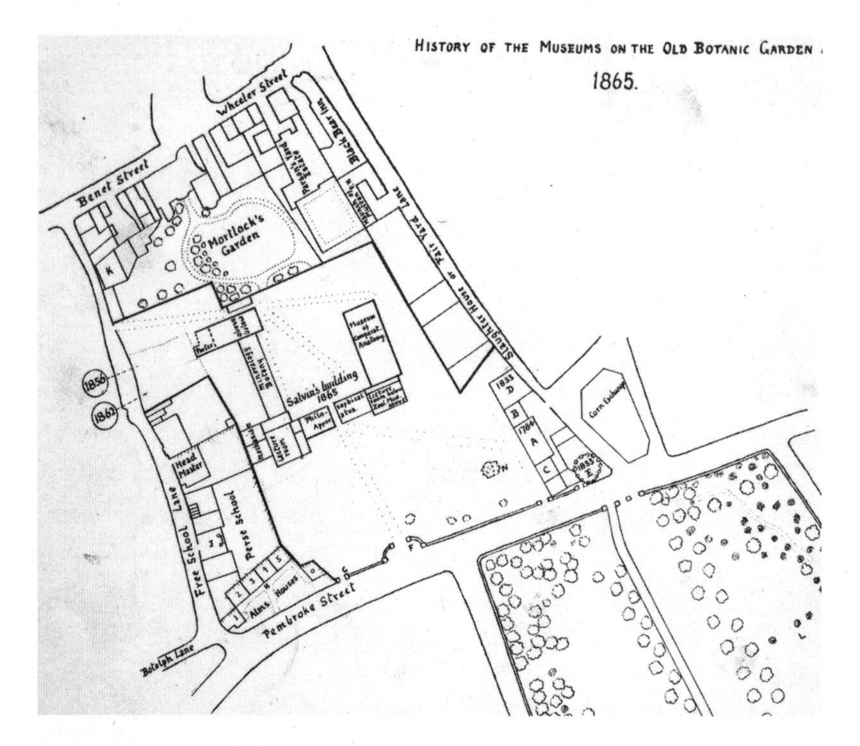


**Figure 32** A map of the New Museums Site in 1865, showing the location of some of the collections and departments.

not be enough to resuscitate Cambridge science. In 1873, just as construction was nearing completion, the young journal *Nature* declared that

> [i]t is known to all the world that science is all but dead in England.... It is also known that science is perhaps deadest of all at our Universities. Let any one compare Cambridge, for instance, with any German university; nay, even with some provincial offshoots of the University in France. In the one case he will find a wealth of things that are not scientific, and not a laboratory to work in; in the other he will find science taking its proper place in the university teaching, and, in three cases out of four, men working in various properly appointed laboratories.[34]

The debate went much deeper than just the state of science in the universities; it went to the heart of what the universities were for. In a follow-up piece in *Nature*, three possible purposes of a university were examined:

> The first regards the University as an ecclesiastical nursery. This was the original view, but now-a-days is passing out of mind, though tenaciously clung to by some resident members at either University. . . .
>
> The second looks upon Oxford and Cambridge as places where the young Tartars of modern English society are covered with a varnish of 'culture', and polished into gentlemen. Dr Lyon Playfair said in the House [of Commons] the other day that the Scotch University taught a man how to make a thousand a year, the English University how to spend it. . . . [The English universities] emphatically discard the idea that it is the duty of the University to equip a man for the struggle for a livelihood, to train him for business, for the arts, for the professions. . . .
>
> The third view, which at present has but few advocates, teaches that the University is a place where anyone and everyone may be trained for any and every respectable path of life, and where at the same time all the interests of higher learning and science are cared for.[35]

*Nature* clearly took the view that the third approach was the best. Founded in 1869, *Nature* was a campaigning instrument as much as a scientific journal; the astronomer Norman Lockyer as editor, and its many contributors, frequently used their leaders to raise awareness of topics including funding for the sciences, reform of science education and careers, and the state of English science compared to her European neighbours. Lockyer deliberately set the letters page up as a site of controversy, and so he was pleased to be able to publish a response from Thomas George Bonney to the two articles about Cambridge.[36] Bonney was a geologist, a fellow of St John's College, and an extremely active member of the Cambridge Philosophical Society. Like many fellows of the Society, he supported the idea of reforming Cambridge and he was aware of the many reforms that had taken place there in recent years: as well as studying classics and mathematics, students could now study natural and moral sciences to a high level; scholarships and fellowships were becoming available to students of the sciences; and the University was already aiming to make itself a place where 'anyone and everyone may be trained for any and every respectable path of life'. In Bonney's mind, the real block to more extensive reforms at Cambridge was not the incumbent fellows or professors, but the Senate House—a body made up of all old members who had been awarded an MA. This meant that thousands of former Cambridge students, awash with nostalgia for their dreamy student days but with little or no interest in academia, held the final vote on how the University was run. This traditionalist group often slowed or halted progress altogether.[37]

But, despite some conservative opposition, the Cavendish Laboratory was completed in 1873 and opened with much excitement in 1874. The laboratory was to be run by a newly appointed Cavendish Professor of Physics. Possible candidates for the role included William Thomson (later known as Lord Kelvin)—a former second wrangler and Smith's Prizeman who had become Professor of Natural Philosophy at Glasgow. Thomson was much celebrated for his role in creating the theory of thermodynamics and for his work on transatlantic telegraph cables. Another suggestion was Hermann von Helmholtz, the German natural philosopher, physiologist, and philosopher, well known for his work on field theory. But both Thomson and Helmholtz were already established in modern laboratories; the promises of the untried Cavendish Laboratory were not enough to lure them to Cambridge.[38]

Attention turned instead to the Scottish physicist James Clerk Maxwell. Maxwell, after a few years' study in Edinburgh, had matriculated at Cambridge and studied under William Hopkins. Even before beginning his studies at Cambridge, Maxwell had been engaged in original research. At Cambridge, Maxwell established himself as an innovative thinker in the fields of physics, mathematics, and philosophy. He spoke frequently at the exclusive Cambridge Apostles club, and gave his first paper to the Cambridge Philosophical Society in March 1854, at the age of just twenty-two.[39] He was second wrangler and a joint Smith's Prizeman in 1854 and briefly became a fellow of Trinity College before being appointed Professor of Natural Philosophy at Marischal College, Aberdeen, in 1856. In 1857, Maxwell was awarded the Adams Prize for his work on the rings of Saturn; he used the theory of dynamics to explore the possibility that the rings were not rigid but fluid. His winning essay was to be published in the Society's *Transactions*, though, perhaps because of its length, it was published as a stand-alone book instead. It was this that cemented his reputation as a leading thinker in mathematical physics.[40] In 1860, he departed Scotland to take up the post of Professor of Natural Philosophy at King's College, London, and, by the end of that decade, having resigned his professorship at King's, Maxwell became an examiner for the Mathematical Tripos in Cambridge. He began to introduce new subjects, asking questions about heat, electricity, and magnetism, and this led to calls for these subjects to be permanently included in the Tripos.

Maxwell was well positioned for consideration as the first Cavendish Professor; in 1871, he put his name forward and was appointed to the chair. He was in post three years before the laboratory opened, which meant that Maxwell could play a central role in its design, and in choosing instruments for it; he even helped to pay for some of these. In his inaugural lecture, which he chose not to present to the great and the good in Senate House but to a small number of students in an unimportant lecture room, Maxwell emphasized the move from the theoretical physics which had dominated in Cambridge for so long to a more experimental physics:

> The familiar apparatus of pen, ink, and paper will no longer be sufficient for us, and we shall require more room than that afforded by a seat at a desk, and a wider area than that of the black board.... We should begin, in the Lecture Room, with a course of lectures on some branch of Physics, aided by experiments of illustration, and conclude, in the Laboratory, with a course of experiments of research.... [B]y opening at once all the gateways of knowledge, we shall ensure the associations of the doctrines of science with those elementary sensations which form the obscure background of all our conscious thoughts, and which lend a vividness and relief to ideas, which, when presented as mere abstract terms, are apt to fade entirely from the memory.[41]

Maxwell's lectures began before the laboratory was open, meaning that his vision of moving freely between the blackboard and the workbench had to be delayed a few years. But his early lectures were still significant; he focused much of his attention on heat, electricity, and magnetism, and these topics were reintroduced to the Tripos in 1872–3. In addition, 1873 was the year that his momentous *Treatise on electricity and magnetism* was published. In this book, Maxwell expressed physical quantities without linking them directly to a mechanical model—something that would go on to be hugely influential in mathematical physics. The *Treatise* increased Maxwell's personal profile, and also that of Cambridge physics.

The Society celebrated its fiftieth anniversary in 1869 with a jubilee dinner. Adam Sedgwick—one of its three founders—was in attendance. Sedgwick had lived to see his goal of a scientific society for Cambridge realized and, though he was to die a few years after that anniversary dinner, he knew that it was now secure enough to live on beyond him. In February 1873, at the first meeting of the Society following his death, the fellows expressed 'deep regret at the great loss' of Sedgwick, then settled down to listen to two papers by Maxwell on equations of motion and the calculus of variations.[42] It was exactly the tribute that Sedgwick would have wanted.

From the beginning, the Cavendish was intended as a place of both teaching and research (see Figure 33). This was common practice in Europe, but a newer idea in England, where the newly coined word 'researcher' was initially intended to be derogatory: there were many who saw the new breed of researcher (and those who campaigned for more funding for research) as being at odds with the true purpose of a university.[43] The curious students who came to the laboratory in its first year found something quite unlike what they had known before. Sunlight streamed into the large open-plan laboratories from enormous windows, illuminating the large sturdy workbenches round which the students gathered. Undergraduates were taught how to use basic equipment, becoming adept at setting up their apparatus, and measuring, calibrating, and adjusting until their simple experiments were running just so. These young students were given a surprisingly free hand in the Cavendish. There was no regimented practical course; rather, the students were allowed to follow their own interests, usually under the benign supervision of the demonstrator, William Garnett. There was, however, a practical examination in which students had to demonstrate that they could perform basic tasks such as finding the focal length of a lens or the electrical resistance of a wire.[44]

**Figure 33** The interior of the Cavendish Laboratory in the early twentieth century.

Students who had already graduated from their first degree came too, undertaking more advanced experiments. Many of these graduates had completed the (purely theoretical) Mathematical Tripos and had never worked in such a laboratory before. Like the wide-eyed undergraduates, they had to be trained in the basic skills of performing experiments. Then the research could begin. Maxwell encouraged a hands-on approach to research, with plenty of room for trial and error. The Cavendish's first graduate student, William Hicks, who had been placed seventh wrangler in the Mathematical Tripos, recalled how Maxwell's research left him 'fired with the desire of measuring experimentally the velocity of propagation of electromagnetic waves'. Hicks designed a piece of apparatus to attempt the measurement but it was practically impossible at the time and with the equipment available to him. 'Of course nothing came of it,' he wrote later, 'but the practice was worth a great deal to me, and it is interesting as showing the kind of atmosphere in which one was working round Maxwell.'[45] Maxwell explained his reasoning to a friend: 'I never try to dissuade a man from trying an experiment; if he does not find out what he is looking for he may find something else.'[46] This spirit of open-minded research was to guide much of the work at the Cavendish even after Maxwell's time there.

As well as encouraging his students, Maxwell was busily involved in original research of his own. He had always been an original thinker: in December 1855 and February 1856 (shortly before departing the city for Aberdeen), Maxwell had read to the Cambridge Philosophical Society a pair of papers which dealt with the problem of understanding electricity. Electricity had caught the imagination of natural philosophers in the eighteenth century, and these elites conducted a huge number of experiments to try to understand its mysterious nature. The public too was fascinated and flocked eagerly to lively demonstrations of the seemingly magical powers of electricity, even queuing up to experience the novelty of an electric shock. Though many advances were made, the essence of electricity was still an enigma when young Maxwell began thinking about it. Maxwell approached the problem using analogies from seemingly disparate areas of physics like studies of the propagation of light, and the concept of action at a distance. Though physically different, there were intriguing mathematical overlaps in the theories behind these phenomena. Maxwell published his early thoughts on the subject in an article in the Society's *Transactions* in 1864. Throughout the late 1850s and early 1860s, the ideas that he had begun to formulate in those first talks persisted in his mind, seeding more ambitious ideas about the interconnections of different branches of physics. These recurring thoughts led to several more publications. A theory began to emerge, and a set of equations.[47]

Maxwell's equations were derived theoretically from work previously done by Michael Faraday of London's Royal Institution on the idea of a magnetic field (or 'lines of force' as Faraday called them). He also drew on experimental work by two German researchers—Wilhelm Eduard Weber and Rudolf Kohlrausch—who had shown that there was a mathematical ratio between electric charge and magnetic force. The genius of Maxwell was to combine elements of theory and experiment to draw a conclusion that would change physics forever: he showed that light is a form of electromagnetic radiation.

The excitement of the moment and the implications of Maxwell's result were summed up many decades later in the words of Albert Einstein:

> Imagine [Maxwell's] feelings when the differential equations he had formulated proved to him that electromagnetic fields spread in the form of polarised waves, and at the speed of light! To few men in the world has such an experience been vouchsafed... it took physicists some decades to grasp the full significance of Maxwell's discovery, so bold was the leap that his genius forced upon the conceptions of his fellow-workers.[48]

Part of the reason that it took some years for the full import of Maxwell's equations to permeate the scientific community was the way in which he had expressed them. The equations were all contained in his papers of the early 1860s, but not in the form they are commonly written today. It was not until 1884 that another physicist, Oliver Heaviside, reformulated them using vector calculus notation, thus making the equations more useable. As the decades advanced, new mathematical formulations of Maxwell's equations were developed and new uses found for them.

On his return to Cambridge in the 1870s, Maxwell also made a return to the Cambridge Philosophical Society. He had never let his fellowship lapse and immediately he threw himself into their meetings and events. Maxwell's new laboratory was located just next to the premises of the Society on the New Museums Site so he could easily visit the Philosophical Library during the working day, leafing through the latest journals from around the world. He was also drawn to the Society's meetings, for there he found a set of kindred spirits—a group whose interests ranged over the whole landscape of the modern sciences. Over the next few years, in numerous papers and in the discussion following other fellows' papers, he would touch upon an enormous range of topics: he would speak about electricity, of course; he would discuss Darwin's theory of pangenesis, making several objections to it from a molecular point of view; he would talk about motion and dynamics; he would analyse papers from the Mathematical Tripos; he would propose improvements in the design of the *camera lucida*; he would describe a machine he had seen for measuring the tides; and he would speak about geometrical optics, the motion of the eye, techniques for sounding lakes, and the importance of diagrams.[49] Though Maxwell was a professor of physics, he was a natural philosopher at heart, allowing his interests to range widely and fitting perfectly into the Philosophical Society with its eclectic mix of fellows and huge variety of papers. In 1872, Maxwell was awarded the Society's Hopkins Prize (an award instituted in 1861 for the 'best original work in mathematico-physical or mathematico-experimental science' and named in honour of the famed coach William Hopkins).[50] In 1875, Maxwell was elected the Society's president. In many ways, Maxwell was the embodiment of the ideal relationship between the Society and the University: he found both great inspiration in the Society's meetings, and an outlet for his own ideas. Even those ideas that did not pertain directly to his role as a professor of experimental physics could be brought to the Society, examined, debated, and elaborated upon—making his years in Cambridge full of intellectual stimulation.

But, sadly, Maxwell's time in Cambridge was brief. He died in 1879 at the age of 48. This presented a problem, as the original terms of the professorship stated that the post would cease when the first incumbent vacated the role. The University may have been willing to back the building of a modern physics laboratory, but they weren't committed to funding a physics professor indefinitely. Still, Maxwell had done enough in his eight years as professor to convince the Senate House of the value of the chair, and a Grace was passed shortly after his death, allowing a new professor to be elected. After William Thomson again declined the chair, John William Strutt, third Baron Rayleigh, emerged as the next most popular choice. Rayleigh had studied under George Stokes and George Liveing; he had been senior wrangler in 1865, and first Smith's Prizeman. He was briefly a fellow of Trinity College, but resigned his fellowship to marry Evelyn Balfour. After leaving Cambridge, Rayleigh set up a laboratory in his family home at Terling Place in Essex and spent many years conducting independent research there.[51]

On his return to Cambridge in 1879, Rayleigh began to lecture immediately, expanding the programme that Maxwell had begun. Under Maxwell, there had been limited undergraduate teaching, but Rayleigh lectured extensively on subjects including colour vision, scattering, sound, electricity, magnetization, and the density of gases. He formalized a system of practical work for undergraduates. More advanced research also continued in the laboratory. Rayleigh's pet project during his Cavendish years was the redetermination of the absolute units of the ohm, the ampere, and the volt. Maxwell had worked on such measurements too, but Rayleigh pioneered new, more precise apparatus and worked painstakingly for many years to complete the work. Knowing these absolute units was critical to the success of the most exciting technological innovation of the time: submarine cable telegraphy. Countries around the world were being connected for the first time by copper wires laid across the seabed; where once messages had travelled by ship and could take weeks or months to reach their destination, now communications could be sent at dizzying speeds, racing from country to country in just hours or even minutes. But, in order for the cables to work efficiently, the way electricity travelled through the metal wires had to be truly understood. The powers behind the British Empire saw the importance of this work, knowing that their ability to be informed of what was happening in their overseas territories, and to control them, was dependent on 'the two or three slender wires that connected the scattered parts of [Queen Victoria's] realm'.[52] For the researchers in the Cavendish, there were more fundamental reasons for wanting to determine the absolute units. Maxwell had once explained it to his students like this: 'Those aspirations after accuracy in measurement . . . are ours because they are the essential constituents of the image of Him who in the beginning not only made the heaven and the earth but the materials of which heaven and earth consist.'[53]

Many of the graduates who worked alongside Rayleigh in the laboratory investigated aspects of these problems; these included Rayleigh's demonstrators, Richard Glazebrook and William Napier Shaw. Rayleigh's closest collaborator, however, was not a graduate of the Tripos. Rayleigh's sister-in-law Eleanor Sidgwick (née Balfour, and known as Nora) was not, as a woman, permitted a Cambridge degree, but she worked many hours in the Cavendish and in Rayleigh's private laboratory and co-published several papers with Rayleigh. Sidgwick was a brilliant experimenter who worked painstakingly with Rayleigh and the others to set up the great spinning coils of wire, the scales and magnetometers, the Argand lamps, the looking glasses, and the telescopic eyepieces needed to record their measurements. The researchers often worked overnight, toiling away when the laboratory was silent and still, exhausting themselves in pursuit of the elusive numbers.

Rayleigh had been elected a fellow of the Cambridge Philosophical Society in February 1880 and gave his first paper there in April of the same year.[54] Over the next few years, he would speak before the Society many times on a huge array of topics: sensitive flames, electromagnetism, optics, the use of telescopes, and improvements in battery design.[55] Like Maxwell before him, his work was extremely wide-ranging. But Rayleigh's tenure as Cavendish Professor was short, as he wished to pursue independent research; he left Cambridge in 1884 and returned to his private laboratory. There, amongst other work, he continued research begun in Cambridge on the density of gases, work which ultimately led to the discovery of a new element—argon—and to the Nobel Prize for Physics.

When Rayleigh left the Cavendish, it was taken for granted that he would be replaced. This time, there was competition for the chair. Five candidates ran, and a young Cambridge graduate named Joseph John Thomson (commonly known as J.J.) emerged victorious. Thomson had studied at Owens College in Manchester as a teenager before coming up to Cambridge. He was second wrangler in the Mathematical Tripos in 1880 and became a fellow of Trinity College in 1881 and an assistant lecturer in mathematics in the college in 1882, the same year that he won the

Adams Prize. He also worked in the Cavendish after graduating, contributing to Rayleigh's project of redetermining electrical units. Though Thomson was enjoying an impressive career trajectory, he was, in one way, an unusual choice for Professor of Experimental Physics: he was known as a clumsy experimenter with poor intuition for the workings of apparatus. But Thomson was a brilliant theoretician and a popular fellow known for his collegiate values and, even though the Cavendish had been operational for a decade at the time of Thomson's appointment, experimental physics still had a way to go in Cambridge; Thomson's lack of practical nous was not seen as an impediment to him becoming a professor.[56]

Despite his shortcomings, Thomson turned out to be an inspired choice to run the laboratory. He left much of the undergraduate teaching to Glazebrook and Napier Shaw and, as before, researchers were given freedom to choose their own projects. Thomson himself, in stark contrast to Rayleigh's work on precision measurement, turned to experiments on the discharge of electrical currents through gases. Patiently, he would watch and wonder as the electricity sprang through the glass vessels and made the gases glow eerily. One of the key attractions of these experiments was that their results were more qualitative than quantitative. Thomson's lack of experimental skill, though it may have seemed like a disadvantage, was key to the laboratory's success in this period. It allowed for more latitude than Rayleigh's approach had; researchers were not particularly encouraged to join Thomson in his research programme, but could pursue whatever avenue of research appealed to them.

This study of electric current through gas was not especially fashionable in Cambridge at the time, but it was perfect for Thomson, as it linked his various theoretical interests, allowing an investigator to consider the interplay of gas molecules with an electric field. The basic apparatus consisted of a thin-walled glass tube with a negative electrode at one end, and a positive one at the other. Inside was a gas at very low pressure. When a voltage was passed between the electrodes, the negative electrode (called the 'cathode') would emit rays that caused the tube to glow and become warm. Thomson's first paper on discharge tubes after his appointment to the chair was delivered to the Cambridge Philosophical Society in May 1886, and later published in the *Proceedings*. Using two specially designed electrodes in a vacuum tube, Thomson created an electrical discharge and watched it change shape as he slowly dropped the pressure in the tube. At higher pressures, the discharge took the shape of 'an Indian club' but, as the pressure was lowered, the 'neck' of the club lengthened; as the pressure fell further, the discharge changed form again and a bright disc appeared near the negative electrode (see Figure 34). The colour of the light emitted in the experiment was rather pleasingly described by Thomson as 'a pale Cambridge blue'.[57] But what caused this puzzling phenomenon? Thomson considered various possible explanations, but found none convincing. Many in the 1880s, including Thomson, believed that the discharge might be explained as the disruption of vortex molecules due to an electric field. This was tied to the concept of the ether (a medium considered to fill all space), and to the popular theory that atoms were vortices in the ether.[58]

It would be some years before Thomson and others could satisfactorily explain what they were seeing in the discharge tubes. In the meantime, Thomson continued to experiment widely. His work at the Cavendish was not limited to these electrical experiments, as the many papers he presented at the Philosophical Society show. He spoke there frequently on topics including conductors, magnetization, surface tension, the polarization of light, the theory of telephones, and absorption of energy.[59] Like his predecessor Maxwell, Thomson was elected President of the Philosophical Society very soon after his appointment as professor. This

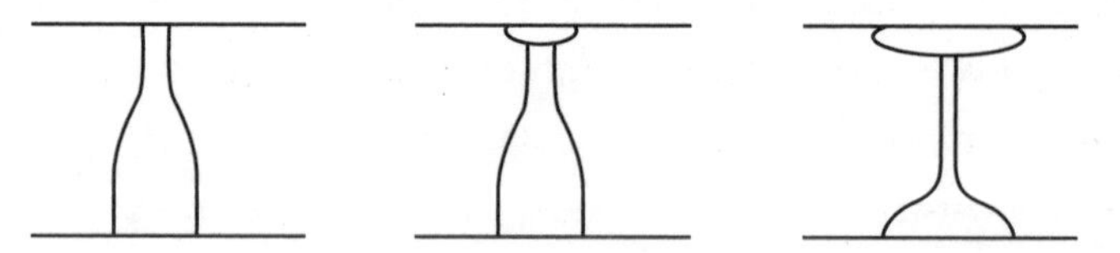

**Figure 34** A diagram from one of J.J. Thomson's early experiments with cathode ray tubes, showing the shape of the discharge, which glowed "a pale Cambridge blue". From a talk presented to the Society in May 1886.

cemented the already strong relationship between the two bodies and it became increasingly common for researchers at the Cavendish to present their findings at meetings of the Society or to publish them in *Transactions* or *Proceedings*. The link to the Cavendish strengthened the reputation of the Society, particularly in the eyes of readers who were outside Cambridge but were avidly following developments in the new laboratory. Probably Thomson's experiences at the Philosophical Society also prompted the creation of a new scientific society in Cambridge: the Cavendish Physical Society. This group was instituted in 1893 and met fortnightly to discuss the current work of staff and students in the physics laboratory. Though much narrower in scope than the Philosophical Society, and intended as a venue for presenting work in progress rather than polished papers, it provided another outlet for scientific research in Cambridge.[60]

It was coincidence that the Cambridge Philosophical Society relocated to the New Museums Site just as the Cavendish Laboratory was being created, but these two simultaneous events were mutually beneficial for both bodies. The Society could ally itself with the most modern and innovative laboratories in the University, while the researchers were given a natural place to present and publish their findings. From 1887, the Society began to hold its meetings in the Cavendish whenever there were physical papers to be presented, and apparatus belonging to the laboratory was frequently employed at Society meetings, whirring into life before audiences who might not otherwise get to see inside this now world-famous laboratory.[61] The Society's two journals were amongst the most popular places for an aspiring young physicist to publish his papers; only the Royal Society's *Philosophical transactions* and the independent *Philosophical magazine* held the same appeal. Even papers that were not destined for the Society's journals were often read at a Society meeting before publication elsewhere.

The rise of the Cavendish Laboratory was felt very keenly within the Society, but it also had an unprecedented impact across the University as a whole. The laboratory repositioned Cambridge in the scientific landscape of late Victorian England. Across the country, there had been rising interest in the sciences in the middle decades of the nineteenth century. Science education in schools was improving, more publications were becoming available to the general public, and scientific institutions were multiplying. There were campaigns for the reform of science teaching and improved career structures at both the national and the university level. On the national level, the reforming spirit of the time was most famously embodied by the X Club. This was a group of nine prominent men of science—including Thomas Henry Huxley, John Tyndall, and Herbert Spencer—who met regularly in London from 1864 to discuss each other's work and the state of science in Britain. One member of the Club wrote that 'the bond that united us was devotion to science, pure and free'; in particular, they shared an interest in the theory of evolution by natural selection, and in reforming the major scientific societies. Significantly, not one of the nine held a degree from Cambridge or Oxford; furthermore, several had studied abroad and they maintained connections with European colleagues.[62] They were acutely aware of how British science compared to its continental counterpart: in the 1870s, there were six times as many papers being published by German chemists as there were by their British peers; and half of all papers in British chemistry journals were by German researchers.[63] Like a leader-writer in the journal *Nature* (to which several members of the X Club had strong links), they worried that 'England, so far as the advancement of knowledge goes, is but a third-rate or fourth-rate power'.[64]

Over the next few decades, members of the X Club held most of the key posts in the Royal Society and the British Association for the Advancement of Science, as well as taking part in various government commissions. They used these positions to further their agenda on the professionalization of science. In order to professionalize science, science education and training for science teachers needed to be rethought. Huxley was particularly involved with this: he lectured to aspiring men of science and science teachers at the Museum of Practical Geology in London's Jermyn Street, at the government-backed Science and Art Department in South

**Plate 1** A selection of objects from the Woodwardian Cabinet.

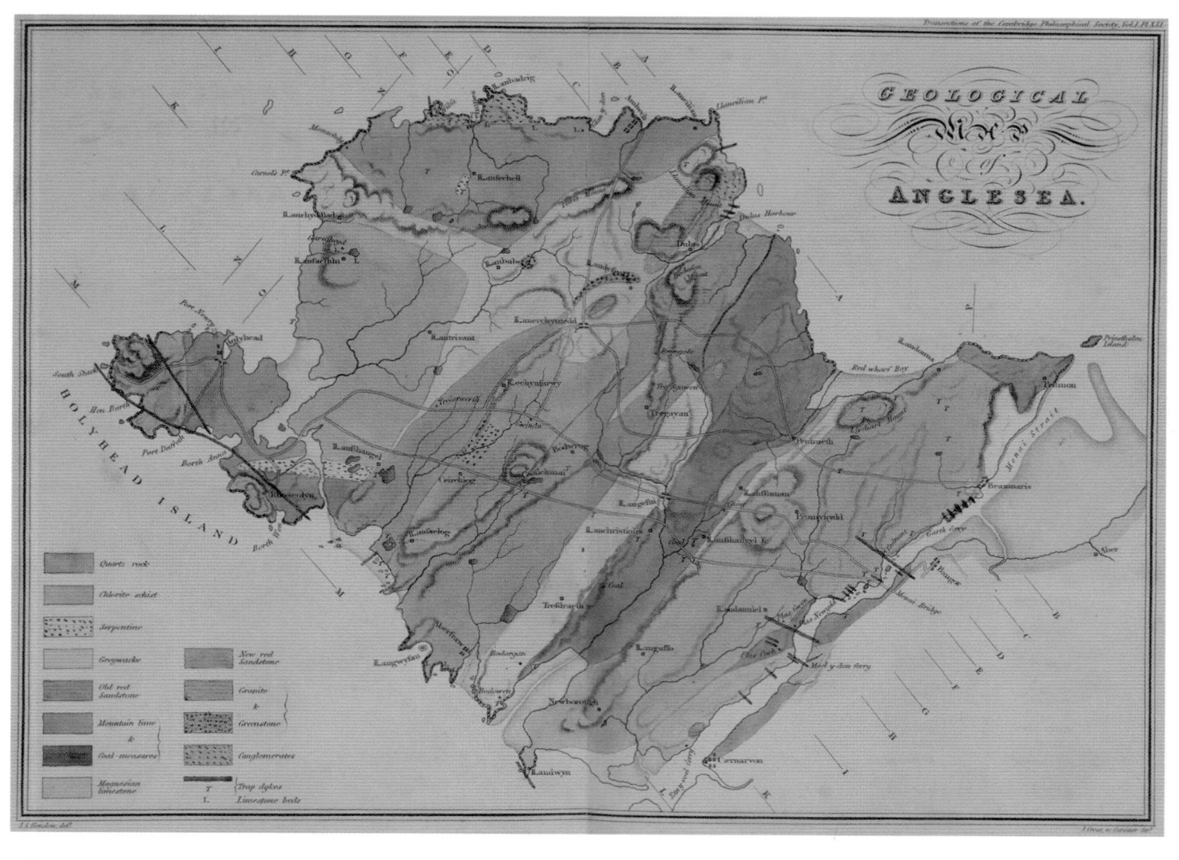


**Plate 2** John Stevens Henslow's geological map of Anglesea (Anglesey) in North Wales, from his first scientific paper, which was presented to the Society in November 1821 and published in 1822.

**Plate 3** Cambridge observatory, showing the director's apartments in the east wing (c. 1829).

**Plate 4** *Disa cornuta,* an African orchid, from *Flora Herscheliana,* Margaret and John Herschel's joint project to record the plant life of South Africa.

**Plate 5** An image of the New England Cutting, near Brighton, showing crowds gathering to celebrate the opening of the line in the early 1840s.

**Plate 6** John Crouch (seated, far left) with the town and gown cricket clubs (1847).

Kensington, and at the London Institution.[65] The proliferation of such establishments in the Victorian period reflected not just the growing appetite for the sciences, but also their increasing viability as a career option: there were now new opportunities for schoolteachers, lecturers, and researchers.

Not everyone was pleased about such developments. In Oxford, the creation of the Clarendon Laboratory led to protests about the 'Germanizing' of universities as they introduced more science teaching and created new professorships.[66] In some quarters, there was a general concern that teaching research skills would mean that 'the prospects of [students] would be so injured that it would be difficult for them after to find congenial employment'.[67] Furthermore, debates about evolution had caused some critics to become more vocal about the negative impacts that science might have upon religion. In an effort to quell such worries, there was an attempt to have members at the 1864 meeting of the British Association for the Advancement of Science sign a declaration stating that science and scripture were not in conflict.[68] But even this powerful lobby could not quiet those who believed that science was overstepping its boundaries. Yet, despite these dissenting voices, science in England generally was more visible, more accessible, and better funded than it ever had been: additional government funding became available in the second half of the century, and the number of members of metropolitan scientific societies doubled between the years 1850 and 1870.[69]

Meanwhile, Cambridge was undergoing reforms of its own. Many of these reforms were, like those proposed by the members of the X Club, tied to the professionalization of science. The colleges were beginning to recognize that their system of life fellowships did not lead to the highest standards of teaching and research. This ancient system awarded a job for life to young men based on their results in the Mathematical Tripos. Life fellows did not have any particular responsibilities. Many of them did some tutoring in mathematics, but the best mathematics tutors were generally agreed to be the private coaches who operated independently of the colleges. Life fellows were not required to produce original research, nor did they have to justify how they spent their time. Many fellows also had church livings and were allowed to split their days between their parish and their college. The only catch was that the fellowship would have to be given up in the event of marriage. This meant that some men left the fellowship in their twenties or thirties, but many took the idea of *life* fellowship at its word and remained bachelors in perpetuity.

From the 1870s, the colleges began taking steps to reduce the number of life fellowships or to demand that life fellows take a greater part in academic life. It was also in the early 1870s that the 'religious tests' were finally fully abolished. For centuries, the ancient English universities had only admitted students who were members of the Church of England. These rules about students were relaxed slightly in the 1850s, but all members of the Senate House (effectively anyone who wanted a degree, as well as all college and University staff) still had to declare themselves members of the established church. This placed a severe limitation on who could study or teach at Cambridge. The famous case of Numa Edward Hartog, a Jewish student who was placed senior wrangler in 1869 but could not take up a college fellowship on account of his religion, highlighted the absurdity of the situation. There were fierce debates throughout the 1860s and early 1870s between those who wished to open up college and University positions to non-Anglicans, and those who saw this as an unspeakable threat to the very purpose of the University. Eventually, even Parliament became involved in the debates. Testimony from Hartog and others before a committee of the House of Lords swayed opinion away from the tests and the then prime minister, William Gladstone, proposed an act to abolish them in 1871.[70]

Another major reform of this period saw colleges begin to allow their fellows to marry. A small number of colleges began this practice from the 1860s and, by the 1880s, it had become the norm. This meant that promising young scholars did not have to commit to a life of celibacy in order to ensure a fellowship. Together, the abolition of religious tests and the introduction of married fellows completely changed the landscape of

Cambridge; positions now became available to new sections of society, and became compatible with normal family life. Careers in Cambridge began to resemble those in the outside world, and the idea of 'professionalizing' Cambridge science became more viable. Funding improved in the 1880s when the colleges were obliged to make financial contributions towards the University—this further reform allowed the University to become more ambitious about funding science teaching and research.[71] For the first time, the majority of young college fellows planned careers for themselves in research, rather than the church.[72]

Quite quickly, the Cavendish came to be seen as a symbol of the new English attitude towards science, and of the new mood in Cambridge. This new breed of 'research school' set the pattern for many others that were to follow. The Cambridge Philosophical Society allied itself to this modern institution and enjoyed a renewed intellectual prosperity in the late nineteenth century, thanks partly to this link, despite its financial crisis and the loss of its house and museum. Many philosophical societies in the late Victorian period suffered financial troubles, as well as loss of status, as the modern idea of 'science' began to overtake the older, more gentlemanly ideal of 'natural philosophy' in the public imagination. In Yorkshire, the new Bradford Scientific Association flourished while the older Bradford Philosophical Society collapsed in 1905; likewise, many other philosophical societies folded or curtailed their activities while more modern-sounding 'mechanics institutes', 'field clubs', and 'microscopical societies' grew up in the same towns. The philosophical societies had often fostered educational organizations and supported them in their early years, but now these became independent and cut their ties to the older societies: the Yorkshire College of Science in Leeds, and Firth College in Sheffield, had been supported by their local philosophical societies but now they became autonomous and began to assume some of the functions once performed by their parent organizations (these institutions were later assimilated into Leeds and Sheffield Universities, respectively). Societies that did not have their own premises were hardest hit.[73] The Cambridge Philosophical Society suffered from some of these problems:

certainly, it had much competition from more modern-sounding clubs and societies in Cambridge, and its membership dipped in the late nineteenth century. But the loss of its house—which could have been disastrous—had allowed the Society to reposition itself in the cutting-edge New Museums Site; and its journals and meetings were reinvigorated by the influx of Cavendish researchers and other young fellows who were taking advantage of the reforms sweeping through their University. Many regional philosophical societies were beginning to create links to educational institutions, but the Cambridge Philosophical Society had always been uniquely positioned within a university, and it used this position very much to its advantage, while carefully maintaining independence.

Despite this positive association between the Society and the New Museums Site, the Society had still not completely recovered from its mid-century financial troubles. The ramifications of John Crouch's actions were still being felt late into the century. Though the Society had given up its reading room and museum in the 1850s and 1860s, they had retained their library and moved it with them to the New Museums Site. The library continued to grow through purchases, donations, and, of course, the extensive programme of periodical exchanges. The growing library was becoming increasingly cramped in its new home, and the costs of maintaining it—including staffing, cataloguing, binding, and arranging—were becoming ever greater.[74] The Society's fellows knew that their library required more space and more funding. At around the same time, the University began to realize the advantages that could be gained from having an official scientific library on the New Museums Site. In 1880 Alfred Newton, Cambridge's first Chair of Zoology and Comparative Anatomy, and President of the Philosophical Society, wrote to the University vice-chancellor to point out that 'the want of a central scientific library in the New Museums for the use of the professors, lecturers and students has long been felt'. Newton had a solution: 'if such a library were founded and placed in a suitable room, the Council of the Philosophical Society would be prepared to recommend to the Society that their library should be deposited in it... It would then form a nucleus for such a collection of books as is required.'[75]

This mutually beneficial solution was welcomed by the University and, following negotiations, the Philosophical Library was made available to members of the University from 1881. On the University side, the negotiations were handled by the mathematician and astronomer George Darwin, the zoologist and embryologist Francis Balfour (brother of Nora Sidgwick), and the physicist William Garnett—all of whom were fellows of the Society. On the Society side, the negotiations were conducted by Alfred Newton, John B. Pearson, John Willis Clark, and Coutts Trotter—Newton, Clark, and Trotter all held University posts.[76] These seven men arranged that the University should provide a larger room for the library as well as a full-time librarian, while the Society was required to continue its system of periodical exchanges so that the library would be stocked with the most up-to-date journals from around the world. All University members would be able to access the books, though only fellows of the Society would be able to borrow them. The library was managed by a committee of six—half from the Society, and half from the University.[77] A Grace confirming these terms was passed in June 1881 and, with it, the relationship between the Society and the University shifted once more. Unlike the museum, which had been fully handed over to the University (though it was superintended by John Willis Clark who was Secretary and later President of the Philosophical Society), the Society maintained some official control over the library and continued to stock it with periodicals and to contribute to the costs of running it.[78] This joint asset symbolized the complex entanglement of Society and University: each needed something the other had, and each benefitted from their close relationship. Though the unscrupulous Mr Crouch may have cost the Society its house and museum, his actions ultimately led to a closer bond between the Society and the University—something that saved the Society from the obscurity that overtook so many other philosophical societies in this brave new world of science.

# 6

# A WORKBENCH OF ONE'S OWN

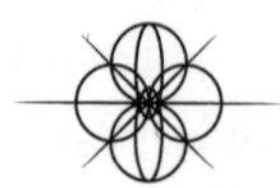

The question of whether birds were descended from dinosaurs had been debated since the 1860s. Could the innocent feathered creatures fluttering benignly through English towns and countryside really be related to the strange monsters that were being quarried out of ancient rocks? The discovery of *Archaeopteryx*—a fossilized reptile with feathers—in Germany in 1861 excited much curiosity amongst both geologists and the public. In 1862, a specimen arrived in the British Museum, and the debate in England truly took off. The comparative anatomist Richard Owen cautiously described it as a bird in a talk before London's Royal Society, but could not fully explain some anomalous features. The much brasher Thomas Henry Huxley pounced on the specimen as proof that birds were 'merely an extremely modified and aberrant reptilian type' and went on to propose that dinosaurs probably had had hot blood and bird-like hearts and lungs. In the sixth edition of *On the origin of species*, published in 1872, Darwin discussed *Archaeopteryx* and other 'linking fossils' which might be used to understand the relationship between ancient and modern species.[1]

By the 1870s, many (even amongst those who were sceptical about Darwin's theory) believed that there was some kind of connection between dinosaurs and birds, but the details were still to be worked out. And so it was that Alice Johnson stood up before a meeting of the Cambridge Philosophical Society one February evening in 1883 and delivered a paper likening bone development in chicks to that in dinosaurs. Johnson compared the pelvis of dinosaurs with that of embryo birds, adult birds, reptiles, and mammals. Her results showed the greatest similarities between dinosaurs and embryo birds, and she concluded that there was a close relationship between these two classes of animal. Johnson's paper was noteworthy not just for its subject matter—the fashionable field of dinosaur studies—nor for its bold conclusion about the connection between such different creatures, but also because it was the first paper ever presented to the Society by a woman.[2]

Johnson spoke as a guest. She was not a fellow of the Society; she was not eligible for membership because only graduates of the University of Cambridge could become fellows and women would not be awarded Cambridge degrees until 1948. The idea of allowing women some kind of membership of the Society had been debated earlier in the century when the celebrated mathematician and science writer Mary Somerville had paid a visit to Cambridge. Somerville, who had taught herself mathematics as girl despite being forbidden to read mathematical texts by her father, had made a name for herself in 1831 with the publication of *The mechanism of the heavens*. This book was initially envisaged as a condensed translation of Pierre Simon Laplace's five-volume *Mécanique céleste*. In that book, the French *savant* Laplace had explained his nebular hypothesis of the solar system using mathematical analysis and it quickly became a key work for astronomers and cosmologists. Somerville soon realized that, in order to make her version of this book useable for her English-speaking audience, she would have to include a huge amount of original material explicating Laplace's ideas. What began as a simple translation project turned into a three-year effort involving much original thinking and consultation with other experts, including the Cambridge men John Herschel, Charles Babbage, and Augustus De Morgan. When the book was finally ready, Somerville sent copies to William Whewell, Trinity College Library, and the Cambridge Philosophical Society—three recipients that would welcome a book on continental analysis. Whewell was impressed; he called the book 'one of the most remarkable of our age'. He was particularly grateful that she had sent a copy to the college library, writing: 'I am glad that our young mathematicians in Trinity will have easy access to the book, which will be very good for them.' But perhaps the extent of Whewell's esteem for Somerville and her book was best expressed in the closing lines of the sonnet he wrote for her:

> [D]ark to you seems bright, perplexed seems plain,
> Seen in the depths of a pellucid mind,
> Full of clear thought, pure from the ill and vain
> That cloud the inward light[.] An honoured name
> Be yours; and peace of heart grow with your growing fame.[3]

George Peacock also expressed his admiration for Somerville, writing that *The mechanism of the heavens* was 'a work of the greatest value and importance' and that he was introducing it to his students at Cambridge—'I have little doubt that it will immediately become an essential work to those of our students who aspire to the highest places in our examinations'.[4] Thanks to the support of Whewell, Peacock, and others, most of the print run of *Mechanism* was sold in Cambridge. Adam Sedgwick, who was President of the Philosophical Society when the book was published, was also struck by the genius of Mary Somerville and declared that 'it is most decidedly the most remarkable work published by any woman since the revival of learning'. Indeed, Sedgwick was so impressed that he proposed that the Cambridge Philosophical Society elect her as a member. But, as Sedgwick confided to his friend Charles Lyell, 'some objected to the manner of doing it'. The details of the proposal and its defeat are not known, but there must have been strong opposition to counter the will of Adam Sedgwick—founder and President of the Society.[5] The category of 'honorary fellow' had existed since the inception of the Society, and honorary fellows did not require a Cambridge degree; indeed, honorary fellowships could only be awarded to those who had *not* studied at Cambridge. No mention of a requirement for honorary fellows to be male was ever mentioned in the Society's minutes or regulations, so it is not clear why Somerville could not have been made one.

There was also some consternation about what to do with Somerville at the 1832 Oxford meeting of the British Association for the Advancement of Science. William Buckland, who was presiding that year, felt that the presence of women would add an unwelcome note of frivolity to proceedings, writing to a confidant that 'ladies ought not to attend the reading of the papers—especially in a place like Oxford—as it would overturn the thing into a sort of Albemarle dilettanti meeting instead of a serious philosophical union of working men'.[6] In the end, Somerville made it easy for the Association by choosing not to attend, which Buckland took as a sign that she agreed that such meetings were not proper places for women. The following year, at the Cambridge British Association meeting, Henslow was so impressed by a seventeen-year-old called Paulina Jermyn, who diligently wrote reports of the evening lectures, that he began to campaign for the admission of women to the Association's meetings, but it would be another twenty years before he saw that hope realized.[7] Other societies were more forthcoming with their commendations, and Somerville received honorary fellowships from the Royal Astronomical Society (where she became the first female member alongside Caroline Herschel in 1835), the Royal Irish Academy, and the Bristol Philosophical and Literary Society.[8] Though she was never a fellow, Somerville published some of her experimental results (on the magnetizing power of sunlight) in the Royal Society's *Philosophical transactions* and she was the first and only woman to have a marble bust in the hall of the Royal Society.[9] Somerville herself, always demure and perfectly mannered, never made any comment about her exclusion from certain societies, and graciously accepted an invitation to Cambridge from her admirers there in 1832. It was Sedgwick who took the lead on issuing this invitation and making the arrangements for Somerville and her husband's visit. No preparation was too elaborate for Sedgwick's honoured guest; he even arranged for a four-poster bed—'a thing utterly out of our regular monastic system'—to be brought into Trinity College for the duration of her stay. Somerville's diary was quickly filled up by the Cambridge men: 'On Tuesday you will, I hope, dine with Peacock; on Wednesday with Whewell: on Thursday at the Observatory [with the Airys]. For Friday, Dr Clarke [sic], our Professor of Anatomy, puts in a claim.' Sedgwick could barely contain his excitement at the prospect of such a guest: 'We have no cannons at Trinity College, otherwise we would fire a salute on your entry.'[10] The visit was a great success, marred only slightly by Sedgwick's inability to persuade the Society to have Somerville as a fellow.

Many years later, in 1872 (the year of Somerville's death), the Cambridge Philosophical Society proposed a new membership category that might also have accommodated women: the associate member. This was aimed at those residents of Cambridge who were interested in the sciences but did not have a Cambridge degree. Associates would be able to attend meetings and use the library, but would not have a vote on issues affecting the running of the Society. Following a debate at a general meeting, the associate membership of the Cambridge Philosophical Society came into being in 1873 and the first associates were elected that February. As with honorary fellows, there was no stipulation about gender in the regulations governing associates, yet no women were elected in the nineteenth century.[11]

This is particularly notable because, by 1873, Cambridge was home to two women's colleges—Girton and Newnham—and both had many staff and students who were actively engaged in scientific studies and research (see Figure 35). These residential colleges were run along similar lines to the men's colleges: their students attended University lectures, and high academic standards were the norm; yet the University did not officially recognize either college, and their students were not regularly allowed to sit Tripos examinations until 1881. Even after women were permitted to sit the Tripos, they were not awarded degrees. In 1890, there was much press coverage of the Newnham student Philippa Fawcett, who was awarded the highest mark in the Mathematical Tripos but did not receive the title of senior wrangler, or a degree. Yet, despite much sympathy for Fawcett and other women in similar positions, ballots in the Senate House in the late nineteenth and early twentieth centuries consistently denied degrees to female students.[12]

Though women had no official status in the University, or in the many clubs and societies around it, they did play an active role in the

**Figure 35** The teaching staff of Newnham College, 1896. Upper image: Nora Sidgwick is seated in the centre of the front row; the unofficial senior wrangler Philippa Fawcett is seated on the far right; Marion Greenwood, a demonstrator and later Director of the Balfour Laboratory, is fourth from the left in the front row, seated on the ground; and the physicist Helen Klaassen is second from the left in the back row. Lower image: From left to right, Marion Greenwood is second, Philippa Fawcett is fourth, Nora Sidgwick is seventh, and Helen Klaassen is eighth.

intellectual life of the town. Women had been attending meetings of the Cambridge Philosophical Society since its foundation and, following the appearance of Alice Johnson in 1883, research conducted by other women began to be presented there. Like Johnson, many of these women were involved in research in the life sciences, and particularly in the fields of comparative anatomy, embryology, and physiology.

There were several reasons for this strong female representation in the life sciences. First was the general rise in the life sciences in Cambridge from the 1870s. Though there had been professors of botany, physic, and anatomy in the University for centuries, the teaching of these subjects had been patchy until the nineteenth century. The creation of the Natural Sciences Tripos and of the New Museums Site had given impetus to improve the teaching of biological subjects, and allowed more room for specialization and for laboratory-based studies of living things. This meant that more students overall began studying experimental sciences in preference to preparing for the Mathematical Tripos. Second, biological subjects like botany had long been considered 'appropriate' for women in a way that mathematical subjects were not. The Botanical Society of London had been the first nationally important scientific society to admit women as members (in 1836), and scientifically minded women were often pushed towards botanical pursuits. Third, more specifically, the abolition of religious tests had allowed Trinity College to employ Michael Foster as their praelector in 1870, and he became a fellow there in 1871.

Foster, a religious nonconformist, had studied in the medical school of University College London and had excelled in anatomy, physiology, and chemistry. After graduating, he completed his clinical training in Paris, and then worked as a naval surgeon and private physician for some years. In 1867, he was offered a position as an instructor in physiology and histology at University College London and quickly rose to become a lecturer and later assistant professor. He was also appointed Professor of Physiology at London's Royal Institution and worked as a demonstrator for Thomas Henry Huxley in his South Kensington laboratory.[13]

Trinity College recognized the need for more scientific fellows amongst their number; they also recognized the potential advantages of electing men on their track record of original research rather than Cambridge Tripos results and so they invited Foster to join the college on the recommendation of Huxley. When Foster was elected to the fellowship of Trinity in October 1871, he became the first non-Anglican fellow in Cambridge or Oxford; when he married in 1872, he became the first married man in Trinity. He represented a real departure for the college, both socially and academically. When Foster arrived in Cambridge, there was little organized physiology teaching at the University; England was still considered a backwater in the science, especially in comparison to France and Germany where it was a sophisticated and professionalized pursuit, or even Scotland, which had a long tradition of physiology research. England had no physiology journal; the subject wasn't represented at meetings of the British Association for the Advancement of Science; and, apart from the tiny laboratory at University College London, there were no physiology facilities in the country.[14]

Foster's position was a college one, but he determined from the beginning to teach students from across Cambridge, including the women's colleges. The University agreed to provide some space on the New Museums Site (in a room which had been intended for displaying philosophical instruments), and Trinity College provided equipment.[15] Upon arrival, Foster went about setting up a laboratory-based biology class for students—the first in an English university. Previously, the subjects of botany and zoology had been treated separately in Cambridge, but Foster believed that all branches of biology were united by fundamental principles, and that the Darwinian theory of evolution could be applied across the plant and animal kingdoms to explain the workings of life. Foster was heavily influenced by his old mentor Huxley in such beliefs, and he modelled his teaching on Huxley's practical classes in South Kensington. Foster's new syllabus saw first-year students studying elementary biology while second years studied practical physiology and third years studied advanced physiology. Initially, Foster taught from just half a partitioned room but, despite the cramped conditions, he insisted on welcoming undergraduates from across the University. Though this might have been seen as controversial, the University Professors of Botany and Zoology—Charles Cardale Babington and Alfred Newton, respectively—were supportive of Foster, recognizing the importance of modernizing the life sciences in Cambridge. Soon after beginning his practical biology classes, Foster was attracting forty to fifty students each term.[16]

Demand for physiology teaching was far outstripping supply. Initially, Foster was assigned to share a partitioned room with the Plumian Professor of Astronomy, James Challis. The chamber was cramped and Foster had to fight for additional space in the New Museums Site to accommodate his growing classes; one of his students described how Foster went about this:

> The accommodation was all too small, but Foster did not complain: with a wise humour he suggested to one of his pupils a research into some of the rarer of the dissociation products of proteid metabolism; if I remember aright, the substance was uroerythrin. It was a research which, in order to obtain an adequate amount of material, demanded the boiling down over several days and weeks of many gallons of excrementitious fluid. The Plumian Professor did not merely vacate the premises, he spread about so dire a report of the disadvantages of close propinquity to the physiological laboratory that by a natural process Foster's laboratory accommodation was doubled.[17]

By 1873, Foster had two rooms on the New Museums Site, but it was still not enough to cope with his growing number of students. As one German visitor to the city remarked around this time: 'there are plenty of riches here, why are not more laboratories built for your University?'.[18] Throughout the 1870s, there were campaigns to increase the amount of space available for physiology. Foster himself, who was not a graduate of the University of Cambridge, did not have a vote in Senate House and had little clout in debates on the subject; but there was evidently much support for physiology, as a new building was agreed upon and completed in 1879. Foster's previous set of rooms became the new home to the expanding Philosophical Library, while Foster and his students moved into their spacious new premises.[19]

By 1880, Foster had several keen graduates working in his new laboratory, and chief amongst them was Francis Maitland Balfour (known as Frank). Balfour was the brother of Nora Sidgwick, who had collaborated with Rayleigh in the Cavendish (another sister, Alice, was a respected naturalist, and their brother Arthur would become prime minister in 1902). He had entered Trinity College in 1870 and studied under Foster, conducting original research on chick embryos and publishing several papers while still an undergraduate. The story was later told of how an uncertain young Balfour, 'sitting in the little room of the philosophical library at Cambridge . . . asked Foster to advise him as to his future career'.[20] Foster suggested pursuing embryology and so Balfour took a placement at a zoological research laboratory in Naples which ultimately led him back to a fellowship at Trinity. He rose to become a college lecturer, published several successful books on embryology, was elected a fellow of the Royal Society, aged just twenty-seven, became President of the Cambridge Philosophical Society a few years later, and was appointed to run Cambridge's new morphological laboratory. Worried that another university might poach their young star, Cambridge created a chair in animal morphology for Balfour in 1882. In the following year, Foster was made Professor of Physiology—a university position which finally gave him some official status in the New Museums Site—and, in 1884, he was elected President of the Cambridge Philosophical Society.

Foster and Balfour were notable not just for their groundbreaking research, but also for welcoming women into their laboratories. Foster, who had come from outside Cambridge, was very open to the idea of sharing laboratory space with women. His *alma mater*, University College London, had been set up as a progressive institution and it began awarding degrees to women in the 1870s. Balfour too, whose sister had collaborated closely with the Cavendish Professor and who would go on to become Principal of Newnham College, was fervently in favour of higher education for women. While some Cambridge professors excluded women from their lectures and practical sessions, Foster and Balfour allowed students from Newnham and Girton attend theirs. As one student later recalled:

> Michael Foster allowed us women to sit up in a gallery overlooking his big lecture room, full of men . . . May I add a word about Mr. Frank Balfour? I attended his lectures on embryology one May term, in a tiny room, where men and women were squeezed together . . . I remember specially his long, delicate hands and beautiful manipulation of sections, but also his marvellous and stimulating teaching.[21]

The question of whether or not to permit women to attend lectures was left up to individual lecturers. Adam Sedgwick had allowed women into his geology lectures from the beginning, though he clearly thought they were a little distracting, as he wrote to a female friend:

> Do you know that the Cambridge daughters of Eve are like their mother, and love to pluck fruit from the tree of knowledge? . . . And, do you know, it is now no easy matter to find room for ladies, so monstrously do they puff themselves, out of all nature, in the mounting of their lower garments, so that they put the lecture-room quite in a *bustle*. Lest they should dazzle my young men, I placed them with their backs to the light, on one side of my room. And what do you think was the consequence? All my regular academic class learnt to squint, long before my course was over.[22]

By the 1870s, twenty-two out of thirty-four Cambridge professors allowed women to attend their lectures, but there were marked differences between different disciplines. Physics was seen by many as an especially male pursuit. James Clerk Maxwell had effectively banned women from his lectures and from workspaces in the Cavendish Laboratory throughout his tenure there. The idea that excessive study (particularly the study of physics) and examinations could actually damage women was quite common in this period.[23] Debates about the differing natures of males and females were often informed by the most-up-to-the-minute scientific research: Charles Darwin's 1871 book *Descent of man* was frequently cited by those wishing to demonstrate the natural superiority of the male brain and physique. Indeed, it came to be believed that social structures in which women assumed inferior positions simply reflected evolutionary truths.[24] In 1863, Emily Davies, a campaigner for women's education and one of the co-founders of Girton College, had summarized the views of those who opposed degrees for women thus:

> [They believe] that women ought not to pursue the same studies as men; and that they would become extremely unwomanly if they did. A woman so educated would, we are assured, make a very poor wife or mother. Much learning would make her mad, and would wholly unfit her for those quiet domestic offices for which Providence intended her. She would lose the gentleness, the grace, and the sweet vivacity, which are now her chief adornment, and would become cold, calculating, masculine, fast, strong-minded, and in a word, generally unpleasing.[25]

Twenty years later, in 1884, the satirical magazine *Punch* was still expressing much the same sentiment in verse:

> The Woman of the Future! She'll be deeply read, that's certain,
> With all the education gained at Newnham or at Girton;
> She'll puzzle men in Algebra with horrible quadratics,
> Dynamics, and the mysteries of higher mathematics;
> Or, if she turns to classic tomes a literary roamer,
> She'll give you bits of Horace or sonorous lines from Homer.
>
> . . . . . . . . . .
>
> O pedants of these later days, who go on undiscerning
> To overload a woman's brain and cram our girls with learning,
> You'll make a woman half a man, the souls of parents vexing,
> To find that all the gentle sex this process is unsexing.
> Leave one or two nice girls before the sex your system smothers,
> Or what on earth will poor men do for sweethearts, wives, and mothers?[26]

Earlier in the century, an occasional woman had been permitted to enter scientific circles, and her femininity had even been seen as an advantage. As William Whewell had once written in a review of Mary Somerville's book *On the connexion of the physical sciences*: 'one of the characteristics of the female intellect is a clearness of perception . . . where women are philosophers, they are likely to be lucid ones.'[27] But now that hundreds of women were agitating for increased access to higher education, it became more common to believe that 'we should have half the young women in the country in brain fever or a lunatic asylum, if they were to make up their minds to try for [a degree]'.[28] Mary Somerville had been seen as unproblematic partly because she was a rarity, and partly because her femininity was not in question: she was married, she was a mother, and she had never done anything as presumptuous as study in a university. But in the later nineteenth century, the women of Cambridge faced enormous prejudice and had to deal with commentators questioning whether they truly *were* women at all. Some, like the physicist Helen Klaassen, deliberately chose masculine-looking clothing so that their appearance could not be considered a distraction in the laboratory, while others, like the unofficial senior wrangler Philippa Fawcett, dressed in as feminine a manner as possible as a way to downplay potential controversy and avoid accusations that 'the women's colleges are peopled by a sort of impossible race of eccentrics'[29] (see Figure 36).

Little by little, these pioneering women began to gain access to more parts of the University. Though Maxwell would not allow women into the Cavendish, in the summer vacation of 1879 the demonstrator William Garnett began teaching physics to women in the laboratory while Maxwell was holidaying in Scotland. In just a few weeks, these women undertook a full course on electrical measurement, a course which would normally take male students several months to complete[30] (see Figure 37). When Rayleigh took over as Cavendish Professor, he was far more relaxed about allowing women into the laboratory and, in 1882, women began attending classes there on the same terms as men. But Rayleigh's reign was brief. When J.J. Thomson was appointed to the chair, he did allow women to continue in the laboratory, as Rayleigh had, and he admitted that they often did quite well in the elementary part of the course, but he expressed doubts about the wisdom of allowing women to study advanced physics. He once wrote to a friend: '[women] always do very well in the first [part] the Tripos, but make

**Figure 36** Terrible result of the higher education of women! 'Miss Hypatia Jones, Spinster of Arts (on her way to refreshment), informs Professor Parallax, F.R.S., that "Young men do very well to look at, or to dance with, or even to marry, and all that kind of thing!" but that "as to enjoying any rational conversation with any man under fifty, that is completely out of the question!"' From Punch 1874.

**Figure** 37 Women in the Cavendish Laboratory, c. 1900.

an awful hash of the second, in fact I think in nineteen cases out of twenty they had better not attempt it.'[31] And, in another letter, he wrote:

> I think you would be amused if you were here now to see my lectures—in my elementary one I have got a front row entirely consisting of young women . . . and they take notes in the most painstaking and praiseworthy fashion, but the most extraordinary thing is that I have got one at my advanced lecture. I am afraid she does not understand a word and my theory is that she is attending my lectures on the supposition that they are Divinity and she has not yet found out her mistake.[32]

This attitude is surprising since Thomson's own wife, Rose Paget, a former student of Newnham College, had trained as a physicist and researched at the Cavendish. But her research came to an abrupt halt on her engagement to Thomson; instead, she became a sort of 'hostess' in the laboratory, welcoming new staff and students, and presiding over social occasions.[33]

Meanwhile, over in the physiology laboratories, Foster and Balfour were encouraging their female students to undertake independent research. But, as physiology grew in popularity through the 1870s and into the 1880s, and as more women began to take the Tripos from 1881, Foster struggled to find more laboratory space. The new building that had been completed in 1879 quickly filled up, and an extra storey had to be added in 1882.[34] Because this was a University building, it was generally agreed that space in it should be reserved for University students. Neither of the women's colleges was officially affiliated to the University and so their students could not be prioritized. Concerned that overcrowding might lead to their students being excluded from laboratories, the women's colleges began considering how best to ensure that their members could continue their studies and research unhindered. Each of the women's colleges had its own on-site laboratory, but these were small,

**Figure 38** Women in Girton College's laboratory, c. 1900.

were not particularly well equipped, and had, in any case, been designed primarily as chemical rather than biological laboratories (see Figure 38). Some teachers got around the problem of restricted laboratory access for women by teaching their students out of hours, like the chemist Philip Main who taught an early morning women-only practical class in the laboratory of St John's College.[35] But more was needed. The two colleges took very different approaches: Girton wanted the University to provide a separate laboratory for women, part-funded by the women's colleges; Newnham, knowing how slowly the University made decisions, decided to set up a biology laboratory themselves. The students of Newnham instantly began fundraising, and the college put together a committee including Nora Sidgwick (by now Newnham's vice-principal (see Figure 39)), Frank Balfour, and several others to begin planning the venture.[36]

**Figure 39** Nora Sidgwick.

The Newnham committee found a home for the new laboratory in an abandoned chapel on Downing Place, a few hundred metres from the New Museums Site. They invited Girton to become involved, but Girton preferred to continue its policy of pushing for a University laboratory for women. Though the two colleges had different visions, the fellows of Newnham assured their counterparts at Girton that the laboratory would be 'as fully available for the use of Girton students as if a more complete combination had been effected' and Girton offered some funds towards equipment for the laboratory.[37] The chapel was purchased thanks largely to donations from Sidgwick herself and her sister, Alice Blanche Balfour. During 1883, the chapel was renovated and fitted for purpose, ready to be opened in 1884.

It had been decided to call the new laboratory the Balfour Laboratory for Women at Cambridge. This reflected the many contributions of the Balfour family, but especially honoured the laboratory's great champion, Frank Balfour, who had been tragically killed in an alpine climbing accident while the laboratory was still being planned. Frank died only two months after being appointed Professor of Animal Morphology, and while he was still President of the Cambridge Philosophical Society. His loss was deeply felt in the scientific circles of Cambridge. His female students and colleagues were especially aware of how much Frank Balfour had supported their work. It had been Balfour who had suggested the idea of researching the connection between dinosaurs and birds to Alice Johnson—research that had led to that first talk by a woman at the Philosophical Society.[38]

Johnson had studied natural sciences at Newnham and had been awarded a first in the Tripos in 1881. She then began researching in the zoological laboratories of the New Museums Site, thanks to a Bathurst scholarship. This scholarship had been established in 1879 by a former Newnham student to enable young women who had 'passed the Natural Sciences Tripos with credit, and who wish to carry their studies further, independently, but under the advice of the Cambridge teachers'.[39] Johnson had studied under Balfour as an undergraduate and continued to work with him as well as with Adam Sedgwick Jr as a Bathurst scholar. Sedgwick Jr—the grandnephew of the geologist Adam Sedgwick—had studied under Foster and sat the Natural Sciences Tripos in 1877. He then became a demonstrator in Balfour's laboratory and co-published a number of papers with him, as well as publishing independently. Following Balfour's death, Sedgwick Jr had been appointed to Balfour's college lectureship, and became a University lecturer in animal morphology.[40] Though Johnson had space in Sedgwick's laboratory, she transferred her research to the Balfour Laboratory for Women as soon as it opened, conducting much of her research at Frank Balfour's old workbench, which had been donated to the laboratory after his death.[41] She became the laboratory's first director as well as the demonstrator in animal morphology. A former Girton student, Marion Greenwood, became the first demonstrator in physiology and botany.

Though it was not in a modern building like those on the New Museums Site, the Balfour Laboratory had been carefully set up according to the most modern ideas. The Newnham College magazine proudly described how

> [t]he gallery round three sides of the laboratory is now chiefly used for microscopical work, the windows having been enlarged to increase the amount of light. The furniture consists of tables fixed against the wall running nearly all round the gallery, with all the necessary fittings of gas, water, shelves, &c., a row of tables standing further back, and a line of cupboards fixed to the edge of the gallery all round. A demonstrator's room is partitioned off at one corner. The gallery is used for demonstrations in Comparative Anatomy and Physiology, to which the liberal allowance of light and space renders it admirably adapted. It is capable of accommodating in a luxurious manner eighteen students at once, and double that number could be taken in fairly comfortably.[42] (see Figure 40)

The laboratory filled up quickly with eager students and, though it was intended as a biological laboratory, there was also some elementary teaching on physics and chemistry. Crucially, all three subjects were taught *practically*. There were lectures, but the laboratory sessions were considered more essential, and students who were short on time

**Figure 40** The Balfour Laboratory for Women at Cambridge which operated from 1884 to 1914.

were told to 'come only to the demonstrations, not only to the lectures'. This practical approach to the teaching of biology came directly from Michael Foster, who had taught many of the early demonstrators at the Balfour. It meant that the Balfour Laboratory was in line with the University laboratories, and also had similar values to the London institutions where Foster had once trained under Huxley. Students were encouraged to see for themselves and to think for themselves: 'the hand and the eye are educated as well as the brain', as one student wrote. The students worked hard, but they also made time for fun. As Edith Rebecca Saunders, one of the later directors, fondly recalled in an obituary for her colleague Marion Greenwood (who herself served as the director of the laboratory from 1890 to 1899):

> after lunch the short interval before an afternoon class was spent in battledore and shuttle-cock [an early form of badminton], a game with queer hazards played around a stove and among tables and chairs. Another recreation occasionally indulged in was whist—bridge was still in the future.[43]

In the 1880s and 1890s, the Balfour Laboratory attracted about forty students each year. Of those, an average of fourteen intended to sit the Natural Sciences Tripos (some sat other Triposes, and some chose not to sit examinations as they would not be awarded a degree). For more advanced work, permission was sometimes obtained for students to use the University laboratories, which were better stocked with specialized equipment. In 1897, numbers in the Balfour soared when Adam Sedgwick Jr banished women from his classes. Sedgwick had been faced with massive overcrowding in his laboratory. This was a real problem, and others such as Francis Darwin (botanist, son of Charles Darwin, and President of the Philosophical Society), who were supportive of higher education for women, were occasionally driven to exclude female students from their practical classes.[44] But Sedgwick fundamentally opposed the idea of awarding degrees to women, and he worried that, should the Senate House change its policy on female graduates, 'the glorious career of this

University as a producer of great men will receive a serious check'. By this time, Sedgwick was the head of the School of Morphology, so his decision affected a huge number of students. The Balfour, with limited resources, built itself a new lecture room, appointed a new lecturer in morphology, and managed to provide all instruction in biology and morphology to the female students of Cambridge—a considerable feat.[45]

Johnson's talk to the Philosophical Society in 1883 had been an important indicator of the growing visibility of women in Cambridge. Following Johnson, more women spoke at the Society and began to publish in the *Proceedings*; much of this research was associated with the Balfour Laboratory. Florence Eves described her experiments on the liver ferment; Anna Bateson and Dorothea Pertz each presented multiple times on plant physiology; Rachel Alcock spoke about the digestive systems of lamprey larvae; and Elizabeth Dale discussed her botanical work.[46] There were also papers on the physical and chemical sciences, some by women working at the Cavendish (which had not banished women as the morphology laboratories had) and other University laboratories; and some by women within the Balfour, which had its own physics demonstrator from 1891. Helen Klaassen spoke on the conductivity of sulphuric acid, while Florence Martin (who worked alongside J.J. Thomson) talked about her work on discharge tubes.[47] Some of these women read their papers themselves, but those who worked with male collaborators usually sat to one side while their male partner read the paper aloud: Anna Bateson's papers were often read by her brother William Bateson; Dorothea Pertz worked frequently with Francis Darwin, who tended to read on her behalf; and papers by the female Cavendish researchers were often read by J.J. Thomson.[48]

The predominance of women giving papers on physiology is not only indicative of the importance of the Balfour Laboratory but also shows how physiology had grown in popularity across Cambridge in the late nineteenth century. Just as the creation of the Cavendish Laboratory led to a rise in the fortunes of experimental physics, Michael Foster's arrival in Cambridge presaged a new dawn for the life sciences.[49] Since the 1870s, each meeting of the Philosophical Society had been devoted to a particular branch of science, the meetings rotated between mathematics, physics, and chemistry; biology and geology; and literature, history, law, and the moral sciences. While the last of these three categories was often undersubscribed and eventually petered out, the physical and biological strands were extremely popular.[50] The meetings began to take place in specialized locations too. Now that the Society no longer had a meeting room of their own, they were free to roam between venues. Talks often took place in the mathematical lecture room of the New Museums Site but, when there was a need to include a scientific demonstration, the meetings would move to the Cavendish or to one of the biological laboratories or museums.[51]

Throughout the later decades of the nineteenth century, not just the meetings but also the publications of the Philosophical Society came to be more dominated by physiology and physics. Mathematics, traditionally the mainstay of the scientific elite of Cambridge, was still an important topic for papers at the Society but, as the Mathematical Tripos began to be overtaken by the Natural Sciences Tripos in popularity, mathematics lost some of its former supremacy. The *Transactions* still carried a large number of papers on mathematics and mathematical physics but the *Proceedings*, which was evolving from a sort of minute book into a more standard journal, began to give much more space to the life sciences. Volumes of *Proceedings* in the mid-1870s typically contained a handful of papers on biological subjects; for example, Volume III, which covered the years 1876 to 1880, contained a total of eighty-four papers, of which nine were on zoology, physiology, medicine, or anatomy. It also contained forty-one papers on mathematics or physics, fourteen on geology, and a handful each on astronomy, chemistry, and scientific instruments, plus a scattering of papers on other topics such as archaeology. In contrast, by the time Volume IV appeared in 1883, its twenty-seven papers on mathematics and physics were matched by twenty-two on the life sciences—approximately a quarter of the total papers. In every volume of the *Proceedings* published in the remainder of the century, about a third of its papers were on the life sciences, with a heavy emphasis on physiology.[52]

In this, the Society reflected a wider trend in English science. When Foster had arrived in Cambridge in 1870, England had no physiology journal. In 1873, Foster had established the first: *Studies from the physiological laboratory in the University of Cambridge*. This short-lived journal was superseded by the *Journal of physiology* (also founded and edited by Foster) in 1878. As physiology laboratories multiplied in Britain, new places to publish physiological research were needed; by being flexible, established journals such as the *Proceedings* could play an important role, alongside the new specialist journals.[53] There was new funding for physiology too as scholarships were set up, such as the George Henry Lewes Studentship, which was inaugurated in 1879 by the novelist George Eliot (Mary Anne Evans) in memory of her former partner; it was notable for being open to women as well as men.[54]

Specialization was the hallmark of late Victorian science. Research laboratories like the Cavendish, Foster's physiology laboratory, or Frank Balfour's animal morphology laboratory drove narrowly focused research and trained up students in specialist techniques. Along with research laboratories devoted to particular fields came dedicated journals, societies, and meetings. In Cambridge, Thomson set up the Cavendish Physical Society, while Foster set up the Physiology Society; nationally, the British Association created new divisions to deal with the disciplines and sub-disciplines that were springing up; and philosophical societies, which had catered for audiences with eclectic tastes, were giving way to newer scientific societies focusing on single subjects.

Cambridge undergraduates were drawn into this trend for specialization with the splitting of the Natural Sciences Tripos into two parts in 1880: part I was an elementary course covering a range of topics, while part II was an advanced course requiring students to develop deeper knowledge and particular research skills.[55] The effect of the new part II courses was significant not just for the students sitting the papers but for the University. In the field of geology, for example, the creation of part II led to the creation of a host of new University posts: in addition to the Woodwardian Professor, there was appointed a lecturer (1883), a demonstrator (1883), a demonstrator in palaeobotany (1892), a demonstrator in palaeozoology (1892), an assistant to the professor (1892), a lecturer in petrology (1904), and a demonstrator in petrology (1910).[56] Within two decades, the solitary University geologist had been joined by a host of other professionals, all working together in the same building, to achieve the same aims: this was the birth of the University department. Across the disciplines (not just the scientific ones), this was being repeated. As each Tripos split into part I and part II, more specialized teaching and research grew up in dedicated centres. By 1910, almost all Triposes had a part I and a part II, which not only allowed the traditional studies of mathematics and classics to evolve but also facilitated the growth of departments of history, law, theology, modern languages, economics, and mechanical sciences.[57] The new Triposes encouraged original thinking amongst students. In the 1870s (following Adam Sedgwick's death in 1873), students of geology had begun to find the syllabus dull, as it focused on simple descriptions and lists of geological features; but, from 1880 onwards, the new part II syllabus encouraged students to link geology to other sciences like physics and chemistry, connect field work to laboratory work, and engage with new sub-disciplines like petrology.[58] This reinvigorated not just the students but also their teachers, especially the young researchers who were graduating and beginning academic careers.

While to many in the colleges and the University, reorganizing the curriculum and employing more specialist teachers seemed like a positive change, not everyone approved of the new departments and faculties. As the classicist and fellow of St John's College, Terrot Glover, put it:

> For centuries the centre of academic life had been the College (and a very good centre, too, with its diversities of types); now it was to be the 'faculty!' (a group of people of one interest). We owed this to the scientific departments; they were each of them centred in huge buildings, more like government offices or factories than the old-time Colleges; their staffs were from every College; buildings were kept up and staffs paid largely by the

> taxation of College revenues which had been given for no such purposes. The 'lab' was really more to the new type of man than the College. . . . A College came to be a place where science men from the labs had free dinners, men very often who had been brought in from outside, and could not be expected to understand College feeling.[59]

Glover also complained of the 'mathematical and natural science men who have been nowhere and seen nothing and read nothing', singling out the physicist Paul Dirac for criticism because he had needed to be told 'who Erasmus was and when'.[60] It was true, of course, that, as curricula had become more specialized, students' education had narrowed; where once all students would have had a smattering of classics, theology, languages, literature, mathematics, and natural philosophy, now many were restricted to a single technical subject.

In addition to undergraduates studying narrow fields, a new kind of student arrived in Cambridge: the research, or graduate, student. They came to the new research schools: young men, and occasionally women, who had studied at other universities or institutions and who were attracted to Cambridge because of its modern laboratories and tight research communities. Previously, not holding a Cambridge degree would have barred these scholars from any significant college or University role, but the expanding laboratories needed external researchers—they benefited enormously from an influx of outside ideas. Such students had been arriving in Cambridge from the time of the completion of the Cavendish Laboratory in the 1870s but, in the 1890s, it was agreed that these students needed some kind of formal recognition, and a way into the structured roles of Cambridge. In 1895, a Grace was passed allowing research students to be awarded BA degrees based on their work. Later, the granting of MA degrees based on research was agreed; and, later still, in the 1920s, Cambridge began to award doctoral degrees to students.[61]

The Cambridge Philosophical Society had accommodated the new research students with the creation of associate members in 1873—just as the Cavendish was coming into being. The Society quickly realized the value of these scholars to its mission of promoting research and disseminating scientific results. And the research scholars recognized the potential of the Society as a place to present and publish their work. Associate membership quickly became popular, with many members of the Cavendish, of Foster's physiology laboratory, and of other laboratories applying for election. The new membership category allowed researchers like Ernest Rutherford, Niels Bohr, and J. Robert Oppenheimer to join the Society—men who would otherwise have been excluded for not having a Cambridge undergraduate degree.[62]

At a time when the membership of the Society was falling slightly (due to the many other scientific societies now active in Cambridge, and to the Society no longer being in the unique position of hosting a museum, reading room, and library), the president, John Willis Clark, noted that 'our annual recruits make up in quality for defects in quantity'. He hoped that these new recruits, many of them research students, would help to maintain 'the ancient popularity of our Society' and that they would contribute to its 'vigorous and healthy life'. He stressed that the Society's meetings were still one of the best places in Cambridge to meet persons of similar tastes and interests and to further scientific pursuits. He also pointed out the meetings' many appeals, including an endearingly English dependence on tea 'as a charm to bring our members together, and promote the ends we have in view'.[63] The meetings were still, at the turn of the century, the single largest forum for scientific papers in Cambridge. And, while specialist societies were growing up within the research schools, the Society maintained its diversity, covering a broad array of topics in its meetings. Likewise, the *Proceedings* was seen as a desirable place for young researchers from a variety of backgrounds to publish.

But, though the Society was trying to promote itself to Cambridge's new residents, and though it wished to attract the new research students, the Society failed to welcome women as warmly as it might have. Women could speak as guests, but they were frustrated at not being able to become fellows or associates. A few societies, like the Sedgwick Club—a geological club founded in 1880 in memory of Adam Sedgwick—allowed women to join and even to take part in field trips, but many other societies—such as the student-run Natural Sciences Club—excluded women. And so women began to create scientific societies of their own. Newnham established its Natural Sciences Club in 1883, and Girton established its own in 1884. These clubs were primarily run by students, but they also included papers by advanced students and demonstrators from the laboratories. In later years, the Newnham and Girton scientific societies took to holding joint meetings.[64] The women's societies were essential in providing a forum for female students; this is particularly notable when compared to groups like the Cavendish Physical Society. At the Cavendish, women could attend meetings, but records show that their main role was to make tea; there is no evidence that women ever gave presentations there.[65]

As the century drew to a close, things were slowly improving for the women of Cambridge, and particularly for those who wished to pursue the life sciences. Allied to the rise of biology and physiology, and to the growing acceptance of Darwin's work, came a new interest in the study of heredity. Many women were drawn to this field thanks to the influence of another supportive mentor—William Bateson. Bateson had taken a first in the Natural Sciences Tripos in the early 1880s. He studied under Michael Foster and Frank Balfour and began presenting original research to the Cambridge Philosophical Society in 1884, just a year after graduating. His first paper there was on the development of the *Balanoglossus aurantiacus*—a marine worm that some believed acted as a kind of 'link' between vertebrates and invertebrates. As Bateson became increasingly interested in the origins of vertebrates, in variation, and in heredity, he gradually moved away from the comparative anatomy and embryology that were so popular in Cambridge. He spent some time in North America developing his ideas on heredity before returning to Cambridge to take up a fellowship at St John's College in 1885. In Cambridge, he was prolific, and the *Proceedings of the Cambridge Philosophical Society* throughout the 1880s and 1890s are scattered liberally with his papers.[66]

Bateson did not work alone. He began, immediately upon his return to Cambridge, to gather students and collaborators around him. Bateson's first informal 'research group' had thirteen members, of whom seven were women: Anna Bateson (William's sister), Dorothea Pertz, Edith Rebecca Saunders, Maria Dawson, Elizabeth Dale, Jane Gowan, and Ethel Sargant.[67] Anna Bateson was one of William's most notable colleagues. Two years younger than William, she had studied at Newnham and sat both parts of the Natural Sciences Tripos in the mid-1880s. Their mother, also called Anna, had served on Newnham's governing Council in the 1880s; their sister Mary later became a history fellow at the college. As well as being part of the progressive Newnham College, the elder and younger Anna Batesons were involved in setting up the Cambridge Women's Suffrage Association in 1884.[68] Growing up in such a family, William could hardly have failed to be supportive of a woman's right to higher education. Happily, he and Anna also shared many intellectual interests. After completing her studies, Anna Bateson was awarded a Bathurst scholarship, which allowed her to undertake research in the Balfour Laboratory, where she later became the botanical demonstrator. She worked as a research assistant to Francis Darwin on his botanical work, and co-published several papers with him on plant physiology.[69] She also worked alongside her brother to study botanical variation—they presented their results at a meeting of the Cambridge Philosophical Society in November 1890.[70] Anna, always a practical person, and known as an eccentric figure who enjoyed smoking pipes and drinking beer, left academia in the 1890s to pursue a career in horticulture, leaving her brother to find new collaborators.[71]

He turned first to Anna's friend Dorothea Pertz. Pertz, like Anna Bateson, had studied at Newnham and had achieved the second-highest result in the Natural Sciences Tripos of 1885.[72] She then worked with Francis Darwin, co-publishing important papers on geotropism and heliotropism in plants, on hybridism, and on the physiology of aquatic plants.[73] After Anna Bateson left Cambridge, Pertz began to work with William Bateson on questions of variation and heredity. Building on the Batesons' earlier work, they published several papers that detailed experiments on cross breeding in plants, showing how abnormal variations were passed on through successive generations.[74]

Another important woman who worked with William Bateson to study heredity was Edith Rebecca Saunders. Saunders had studied at Newnham College, was awarded a first in the Natural Sciences Tripos of 1888, become a Bathurst scholar, and acted as botanical demonstrator in the Balfour Laboratory from 1889 to 1899 before being appointed the director of the laboratory. She began working with William Bateson in 1895. Unlike Anna Bateson and Pertz, Saunders had several years of independent research under her belt when she started to collaborate with William Bateson; she was not a junior colleague, but very much his equal. Their experiments focused on plant hybridism, and allowed them to investigate theories of heredity. Saunders used seeds that Bateson had brought back from a trip to Italy to raise specimens of *Biscutella laevigata*, the buckler-mustard, in Cambridge's Botanic Garden. She crossed two forms of the plant: one with hairy leaves and one with smooth leaves. Saunders found that the plant almost always bred true—offspring would have either hairy leaves or smooth leaves, though a small number of specimens did display an intermediate mixture of smooth and hairy leaves. Saunders wanted to test whether repeated cross breeding of the intermediate form would ultimately lead to a 'blending' of the characteristics, or whether 'pure' smooth or hairy variations would occasionally show themselves in future generations. Saunders and Bateson hoped to develop a statistical model that would predict the likelihood of certain characteristics breeding true. They were interested in variation not just for its own sake, but in the hope that it would provide clues to the mechanism of inheritance, and reveal something about the origins of species. They wrote that

> [t]he essential problem of evolution is how any one given step in evolution was accomplished. How did the one form separate from the other? By crossing the two forms together and studying the phenomena of inheritance, as manifested by the cross-bred offspring, we may hope to obtain an important light on the origin of the distinctness of the parents, and the causes which operate to maintain that distinctness.[75]

Their work led them to move away from theories of 'blending' and towards 'discontinuous' inheritance, which would explain why sometimes features seemed to disappear for several generations before reappearing in a 'true' form. Discontinuous inheritance would also allow for large evolutionary jumps rather than gradual accumulation of changes.[76] To gather evidence for this, Saunders and Bateson continued their cross-breeding experiments, adding new species to their trials, and collating large amounts of statistical data. By 1900, they had gathered enough evidence to show that inheritance was discontinuous, but they struggled to find a general pattern that could explain or predict what characteristics would be visible in any given specimen.[77]

In 1900, the work of Gregor Mendel on the inheritance of characteristics was rediscovered. In May 1900, Bateson and Saunders learned of Mendel's results on cross-bred pea plants, results which showed that 'dominant' characteristics outnumbered 'recessive' characteristics in second-generation plants in a ratio of 3:1. By that autumn, Bateson and Saunders had incorporated Mendel's theory into their analysis and redesigned their experiments to test and expand on Mendel's work. Their work brought Mendelian theory to Cambridge and had a heavy influence on the direction of research for decades to come. Newnham College in particular, where so many of Bateson's collaborators had studied, came to be known as a centre of Mendelism.[78] These Newnham women dominated the Balfour Laboratory, and it too came to be linked to the newest scientific discipline: genetics—a word coined by Bateson in 1905.

By the turn of the century, women's contributions to Cambridge science were being more widely recognized. In 1900, the Cambridge Philosophical Society presented a full set of its *Transactions* and *Proceedings* to each of the women's colleges—a tentative step towards acknowledging their contribution to the intellectual life of Cambridge.[79] The possibility of women becoming honorary members of the Society was proposed by Courtney Kenny, a fellow of Downing College and a law professor, in February 1914. Kenny had pointed out that honorary fellows did not have voting rights, and had only limited rights to use the library or attend meetings. That being so, wrote Kenny, 'I see nothing in the policy and spirit of the Charter and Bye-Laws—and certainly nothing in their actual provision—to invalidate the election of a lady as such a member'. The fellows agreed with Kenny and, the following May, Marie Curie—winner of two Nobel Prizes for her work on radiation and the element radium—was included on the list of honorary fellows, alongside other luminaries such as Max Planck.[80]

In addition, 1914 was the year that saw the closure of the Balfour Laboratory. The end of the laboratory's brief era came about because the University finally allowed women back into its laboratories. Adam Sedgwick Jr, having banned women from his laboratory in 1897, readmitted them in 1906. When Sedgwick left Cambridge in 1909, his successor, John Stanley Gardiner, made it clear that he welcomed women into both his lectures and his practical classes, and even appointed a female assistant demonstrator called Kathleen Haddon to work in the University morphological laboratory. By 1910, the University biological laboratories had been enlarged again, and they were more easily able to accommodate female students alongside the men. The Balfour Committee, which ran the laboratory, knowing that the University laboratories were better funded and equipped than theirs could ever be, jumped at the chance for their students to rejoin the larger laboratories. In 1914, the botany and physiology laboratories followed Gardiner's lead and gave full access to women. The Balfour Laboratory had achieved its aim of providing a quality scientific education to women. It had allowed its students and staff to demonstrate that they could perform at the same level as their male counterparts, it had applied a constant but subtle pressure on the University, and eventually it convinced many that women had an important role to play in Cambridge scientific life.[81]

In this story about how women gradually worked their way into the institutions of Cambridge, the Cambridge Philosophical Society acts as a sort of microcosm. While some in the Society welcomed the contribution of women to their meetings, there was enough opposition amongst the fellows to deny even honorary or associate membership to women for decades. Much like the case in Cambridge more widely, there were some professors and fellows supportive of women's rights, while the Senate House repeatedly voted against awarding degrees to women. In the end, the Society beat the University to equality: it allowed women to become full fellows in 1929, two decades before the University gave equal status to women (though women were allowed to hold some University teaching posts from 1926). It was in 1922 that some fellows of the Philosophical Society began to agitate for admitting women as fellows on the same terms as men. The debate was begun by Arthur Berry, who was a mathematician and astronomer from King's College and had been senior wrangler in 1885. Berry first put his proposal to the Society in May 1922, but discussion of it was postponed for many months, and it was November before the fellows actually got around to considering it. Then, Berry's proposal was brusquely dismissed, with Hugh Newall (Professor of Astrophysics) and Arthur Seward (Professor of Botany)—both former presidents of the Philosophical Society—arguing that there was little point discussing the issue until the University decided to award degrees to women.[82] There the matter rested for some years until November 1928, when Joseph Needham, a biochemist who later became distinguished as a historian and sinologist, proposed amending the Society's bye-laws to allow women to become fellows. His proposals were brought before a special general meeting of the Society in January 1929. The president, statistician George Udny Yule, read a statement in favour of the proposal, arguing that 'if [women] were to be admitted as fellows . . . they should be treated as other fellows, and if of the same standing should be eligible for the council'. What had changed in the six years since the matter had last been debated at a Society meeting is unclear but, when the ballot box was produced and the little black beans secretly cast, the fellows, persuaded by Needham and Yule, voted unanimously to support the election of women to the Society.[83]

When the Society had first formed in 1819, a common phrase used to describe practitioners of the sciences was 'gentlemen of science'—a phrase that reflected both the gender and the class of the 'ideal' student of nature. The word 'scientist' had not yet been coined and, even when it was, it was considered ugly and was little used in Britain until the end of the nineteenth century. By 1900, the scientific community had shifted: it now included a significant number of women; it included people from a greater variety of social backgrounds; and it also contained many specialists who focused on narrow scientific disciplines and were happy to embrace the word 'scientist' with its more modern and inclusive connotations. Though the Cambridge Philosophical Society retained its original name, the idea of natural philosophy was being eroded in Cambridge and elsewhere; the fellows who peopled the Society no longer saw themselves as philosophers, or gentlemen of science, but as scientists.

# 7

# THE LABORATORY IN THE LIBRARY

In the Philosophical Library, John Rashleigh sat down and surveyed the wooden and brass instruments laid out before him. Rashleigh, who was studying to become a medical doctor, had come to the library today not to consult its books, but to be the subject of a scientific experiment. The Philosophical Society's librarian, Mr White, was to perform the experiment, methodically undertaking a series of observations and measurements of Rashleigh's body. Rashleigh's skin was described as ruddy, his hair as brown and wavy, his cheekbones as inconspicuous, and his ears as flat, with lobes present. Then came the measurements: Rashleigh's head was recorded as being 190 millimetres long, 152 wide, and 132 high; his nose was 51 millimetres long and 29 across and was described as 'sinuous'. Likewise, the length and breadth of his face were measured, as well as the distance between his eyes; his height, weight, and arm span; his breathing power; the strength of his arms while pulling and squeezing; and his eyesight and colour sense. Rashleigh paid threepence for the privilege of taking part in this experiment and walked away with a small card carrying all his measurements—a duplicate of one that Mr White kept and carefully filed away in an index case[1] (see Figure 41).

Rashleigh's card is one of thousands that survive in the archives of the Cambridge Philosophical Society. There are cards that record the length

**Figure 41** An anthropometric card from the archives of the Cambridge Philosophical Society. There are many thousands of such cards in the archives. This student, John Cosmo Stuart Rashleigh III, went on to become a medical doctor and High Sheriff of Cornwall.

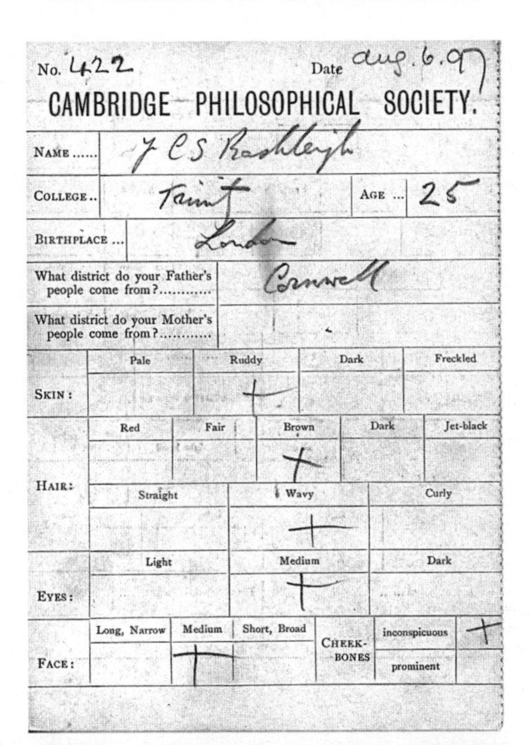

No. 422 Date aug. 6. 97

CAMBRIDGE PHILOSOPHICAL SOCIETY.

| | | | |
|---|---|---|---|
| NAME ...... | J C S Rashleigh | | |
| COLLEGE .. | Trinity | AGE ... | 25 |
| BIRTHPLACE ... | London | | |
| What district do your Father's people come from? ........... | Cornwall | | |
| What district do your Mother's people come from? ........... | | | |

| | Pale | Ruddy | Dark | Freckled |
|---|---|---|---|---|
| SKIN: | | + | | |

| HAIR: | Red | Fair | Brown | Dark | Jet-black |
|---|---|---|---|---|---|
| | | | + | | |

| HAIR: | Straight | Wavy | Curly |
|---|---|---|---|
| | | + | |

| | Light | Medium | Dark |
|---|---|---|---|
| EYES: | | + | |

| | Long, Narrow | Medium | Short, Broad | CHEEK-BONES | |
|---|---|---|---|---|---|
| FACE: | | + | | inconspicuous | + |
| | | | | prominent | |

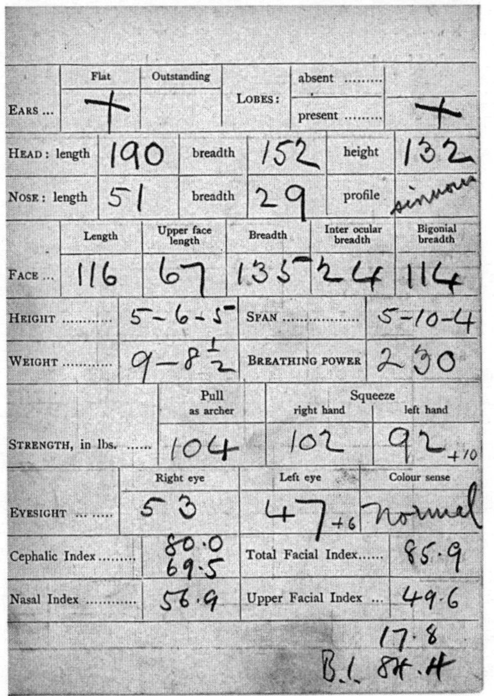

| | Flat | Outstanding | LOBES: | | |
|---|---|---|---|---|---|
| EARS ... | + | | | absent ......... | |
| | | | | present ......... | + |

| | | | | | |
|---|---|---|---|---|---|
| HEAD: length | 190 | breadth | 152 | height | 132 |
| NOSE: length | 51 | breadth | 29 | profile | sinuous |

| | Length | Upper face length | Breadth | Inter ocular breadth | Bigonial breadth |
|---|---|---|---|---|---|
| FACE ... | 116 | 67 | 135 | 24 | 114 |

| | | | |
|---|---|---|---|
| HEIGHT ........... | 5-6-5 | SPAN ................. | 5-10-4 |
| WEIGHT ........... | 9-8½ | BREATHING POWER | 230 |

| | Pull as archer | Squeeze right hand | Squeeze left hand |
|---|---|---|---|
| STRENGTH, in lbs. ...... | 104 | 102 | 92 +10 |

| | Right eye | Left eye | Colour sense |
|---|---|---|---|
| EYESIGHT ... ..... | 53 | 47 +6 | normal |

| | | | |
|---|---|---|---|
| Cephalic Index ......... | 80.0 69.5 | Total Facial Index...... | 85.9 |
| Nasal Index ............ | 56.9 | Upper Facial Index ... | 49.6 |

17.8

B.I. 84.4

of John Maynard Keynes's nose, the distance between Ernest Rutherford's eyes, and the arm span and pulling strength of astronomer George Darwin. They are relics of the work of the Anthropometric Committee, which operated in Cambridge, within the Philosophical Society, from the 1880s to the early 1900s. Anthropometrics—the mathematical study of the dimensions and capabilities of the human body—was a popular science in the late nineteenth century. Its most famous proponent was Francis Galton, a cousin of Charles Darwin's who had studied mathematics at Cambridge before travelling extensively and becoming interested in anthropology. Galton's attraction to mathematics and anthropology, along with an interest in evolution after the publication of Darwin's *On the origin of species*, led him to think about how a mathematical study of a population could feed into studies of heredity.

Today, Galton's legacy is argued about: he inspired much of the research behind fingerprinting and IQ testing but he is also linked to racial and class profiling, and to the eugenics movement of the early twentieth century. In the 1880s, however, Galton was seen as a fairly uncontroversial scientific figure who was attempting to apply mathematical techniques to understanding the human body and brain. While some of his research was unorthodox—such as training himself to do arithmetic by smell, or investigating paranoia by telling himself that horses were watching him—he was viewed by many as a pioneer pushing forward the boundaries of science.[2] In 1884, he was invited to give the annual Rede Lecture at Cambridge. Galton chose as his topic for this public event 'the measurement of human faculty'. With this 'very graceful address', Galton awakened an interest in anthropometry amongst the fellows of Cambridge.[3] In addition, 1884 was the year in which Galton had set up a public anthropometric laboratory at the International Health Exhibition in South Kensington (see Figure 42). Such exhibitions were tremendously fashionable in late Victorian England, and the public flocked to learn about the latest developments in medicine and the health sciences. The most popular exhibits were those, like Galton's, that invited audience participation. In Galton's exhibition laboratory, members of the audience

**Figure 42** Francis Galton's International Health Exhibition laboratory in South Kensington, 1884.

could pay a threepenny bit to have seventeen of their physical and mental characteristics measured, including things like height, weight, pulling strength, and keenness of eyesight. Almost 10,000 people passed through the laboratory in the few months that it operated. The results not only allowed the individual subjects to feel like they were learning about their bodies and contributing to scientific research; it was also the first major data set collected by Galton, and he hoped that it could be used to draw conclusions about the British population at large.[4]

The exhibition laboratory had publicized Galton's ideas and kindled a popular interest in the science of anthropometry. The success of the laboratory, coupled with the warm reception of Galton's Rede lecture in Cambridge, led to the suggestion of a similar laboratory being set up in Cambridge. In January 1886, Oscar Browning (a fellow at King's College, a historian, and a correspondent of Francis Galton's) wrote to the Cambridge Philosophical Society's secretary, Richard Glazebrook, at the suggestion of Galton himself. Browning proposed that the Society should take possession of Galton's anthropometric instruments and set up a laboratory in which to use them. As there was not, at that time, a formal department of anthropology or anthropometrics in Cambridge, the Society seemed like a natural home for this kind of scientific work. The Society's council voted to accept the stewardship of Galton's apparatus, and to set up a committee to superintend it. The committee would consist of the Society's president (in 1886, the office was held by Coutts Trotter—a physicist and the vice-master of Trinity College), Professor of Anatomy Alexander Macalister, Horace Darwin (son of Charles Darwin and co-founder of the Cambridge Scientific Instrument Company), and one Mr Lea, as well as Glazebrook, a physicist by training.[5]

Following a very brief trial in the Cambridge Union Society, it was agreed that the laboratory should be set up within the Philosophical Society's library on the New Museums Site. The library had space for such an endeavour and it had a full-time attendant who could be trained to carry out the measurements. Since 1881, it had been open to all members of the University and thus it was hoped that a large number of people would pass through the laboratory each term. Galton gave £10 towards setting up the equipment, and the ongoing costs of staffing and maintaining the laboratory were to be covered by charging each subject threepence. By February 1886, the laboratory was ready for use. A sign was fixed to the outside of the building, advertising the modern apparatus within and, inside of two months, almost 200 individuals had been anthropometrically measured in the Philosophical Library.[6]

The fellows of the Philosophical Society were eager to hear about the new research going on in their premises, and so Horace Darwin and Richard Threlfall (a demonstrator in the Cavendish who worked alongside J.J. Thomson) gave a presentation on their work to a meeting of the Society in February 1886. It was too early to have produced any notable results, so Darwin and Threlfall, both of whom had a practical interest in

**Figure 43** Typical anthropometric apparatus from the late nineteenth century. The location of the apparatus used in the Philosophical Society's laboratory is unknown.

the workings of scientific instruments, 'exhibited and described some of the apparatus used by Mr Galton for his anthropometric measurements and then gave an account of improvements and modifications which they had suggested in some of the instruments'[7] (see Figure 43). This was one of the principal purposes of the laboratory, and one of the reasons for locating it in Cambridge: as well as it being a place to gather data, Galton saw it as a chance to refine his apparatus in association with Darwin's Cambridge Scientific Instrument Company. In 1887, after the laboratory had been operational for a year, the Cambridge Scientific Instrument Company published a catalogue of anthropometric apparatus designed in collaboration with Galton. Many of these instruments had been adjusted since the time of the South Kensington laboratory in response to feedback from the staff and subjects of the Cambridge laboratory.[8]

The catalogue not only described the instruments for measuring all aspects of the human body—height, arm span, breathing capacity, eye tint, keenness of hearing, reaction times, and so on—but also explained why these measurements might be useful. First, there was the question of personal use for oneself and one's family: the instruments could be used to test for normal development in children. Reminiscent of today's parents being given weight charts and head circumference charts for newborns, the author wrote that

> [the measurements] draw attention to faults in rearing, to be diligently sought for and remedied lest the future efficiency of the child when it grows to manhood or womanhood be compromised. There are hundreds of thousands of cases in which eye-sight has been heedlessly injured beyond repair by pure neglect; of lopsided growth, and of stunted chest capacity, which measurement would have manifested in their earlier stages, and which could have been checked if attended to in time.[9]

Second, there were issues of national importance at stake:

> Anthropometric records, when treated statistically, show the efficiency of the nation as a whole and in its several parts, and the direction in which it is changing, whether for better or worse. They enable us to compare the influences upon bodily development of different occupations, residences, schools, races, &c.[10]

A laboratory like that in South Kensington or Cambridge could fulfil both of these ambitions by giving individual subjects copies of their measurements and encouraging repeat visits and by retaining copies of all results for later statistical analysis.

Of course, there were also many other purposes to these experiments. Galton frequently stressed the medical usefulness of large-scale studies. He cited as an example a study which had been conducted by the War Office during the American Civil War and which showed that light-haired men were far more susceptible to diseases than their dark-haired counterparts were. If something as trivial as hair colour could affect susceptibility to disease, what might more detailed measurements predict?[11] From a mathematician's point of view, the experiment could provide huge amounts of data that would allow them to develop and refine statistical techniques using real-world examples. And then there were other motivations behind the experiments, some of which were not articulated so publicly. Some wished to use the results to 'rank' humans: individual against individual, nation against nation, and race against race. Some blithely argued that these rankings might, for example, encourage sporting competitiveness amongst schoolchildren, or be useful in ordering candidates for civil service positions, but there were also more sinister motivations for this kind of ranking.[12]

In Cambridge, the results produced in the Philosophical Library were primarily used to investigate the question of whether the size of one's head was proportional to one's intellect. There was a general folk belief that a large head was associated with greater brain power, as Galton put it: 'it is well known that the size of the caps worn by university students much exceed that of the uneducated population.'[13] But was this true? And, if so, could it be used to make predictions about the intellectual potential of young people? The first problem was accurately measuring the length, breadth, and height of a living person's head so that the capacity of their skull, and therefore the likely volume of their brain, could be estimated. Together, Galton and Horace Darwin designed a 'horizontal head spanner' in which

> two pieces of wood are arranged to slide one against the other in a longitudinal direction, and a pair of slender steel rods are fixed to the ends of each piece of wood, projecting at right angles to them. Each pair of rods is in a plane at right angles to the direction of motion of the slide. In using the spanner, the slide is moved so that each pair of rods just touches the opposite surfaces of the head and the spanner is moved about until the maximum dimension in the required direction is obtained.[14]

This fairly simple piece of apparatus could be bought from the Cambridge Scientific Instrument Company for £3. Far more complicated to design and use reliably was the 'vertical head spanner'. The vertical height of the skull had to be taken at right angles to a well-defined plane of reference. Galton and Darwin used the 'plane passing through the holes in the ears and through the lowest part of the orbital cavity of the eyes', but the instrument was not perfect and they hoped to improve upon it after further experiments.[15]

The first results from the Cambridge laboratory were presented in April 1888 by John Venn—a fellow of Gonville and Caius College, with interests in logic, philosophy, and mathematics. Venn was already well known in Cambridge for the papers which he had presented to the Philosophical Society in 1880, explaining his diagrammatic method of representing logical propositions—the images that would later become known as 'Venn diagrams'[16] (see Figure 44). Galton and Venn had been aware of each other's work for many years, and they shared a particular interest in studies of probability and chance.[17] Venn's paper on the Cambridge results collated measurements made of about 1,100 students, with the measurements having been taken between 1886 and 1888.

Though the number of subjects measured in the Cambridge laboratory was very much smaller than that measured during the South Kensington exhibition, there were many advantages to the Cambridge sample group. The Cambridge subjects were remarkably homogeneous. Most Cambridge students at this time still came from a small range of socio-economic backgrounds. They belonged, as Venn put it, to 'the upper professional and gentle classes...they had always been fed well and clothed, and had started well by being the progeny of parents who had mostly for some generations enjoyed the same advantages'. Most had similar

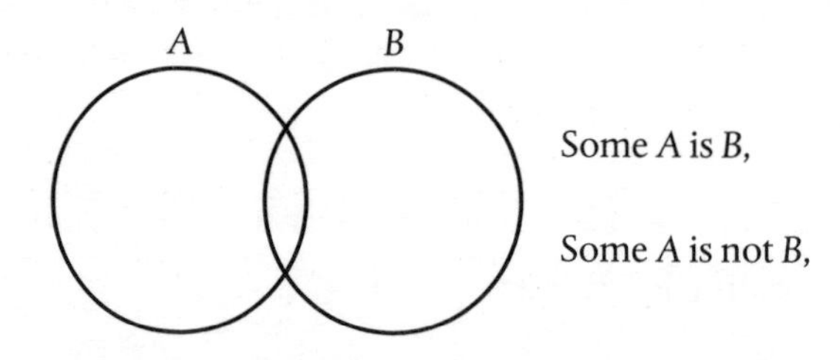


**Figure 44** One of the earliest Venn diagrams. From a paper John Venn presented to the Cambridge Philosophical Society in 1880—his first talk on the subject.

educational backgrounds, most were in their late teens or early twenties, most were active and physically healthy, and most were male (females were omitted from the analysis regarding head size and intellect anyway, as there were not enough female subjects to draw robust conclusions). During the South Kensington exhibition, Galton's samples had been drawn from across a huge spectrum of ages, classes, and educational backgrounds—and so it was harder to make generalizations.[18] And the South Kensington laboratory had had problems with taking women's head measurements, due to the impropriety of asking a woman to remove her bonnet.[19]

Venn presented his first paper on the Cambridge results at London's Anthropological Institute, a body over which Galton presided. Venn did not publish this version of the paper, but Galton later summarized it in *Nature*. For the purpose of this analysis, Venn divided the subjects into three groups: Class A contained students who were predicted by their tutors to get the highest marks in their Tripos examinations; Class B contained those who were predicted to be awarded middling marks; and Class C contained those who were expected to be the 'poll men', who didn't fare well academically and were awarded pass degrees. Tutors from all the colleges assisted Venn in assessing the potential of their students. Next, Venn divided the students into age brackets. He then calculated the 'head product' or relative brain volume for each student by multiplying the length, breadth, and height of the students' heads. Of course, such a calculation would equate to a rectangular box, but it was used as an approximation and Venn believed it would be proportionate to actual skull volume. Venn's analysis gave a clear result: 'men who obtain high honours have... considerably larger brains than others'[20] (see Figure 45).

Not everyone agreed with Venn's analysis. Questions were raised about whether the calculated head products were really proportional to cranial capacity, about the errors introduced during calculations, and about possible errors brought about by the instruments and by the person operating the instruments. One member of Trinity College cited the examples of three friends of his who had each been measured multiple times in the

*Head Products.*

| Ages. | Class A. "High onour" men. | Number of measures. | Class B. The remaining "honour" men. | Number of measure. | Class C. "Poll" men. | Number of measures. |
|---|---|---|---|---|---|---|
| 19 | 241.9 | 17 | 237.1 | 70 | 229.1 | 52 |
| 20 | 244.2 | 54 | 237.9 | 149 | 235.1 | 102 |
| 21 | 241.0 | 52 | 236.4 | 117 | 235.1 | 79 |
| 22 | 248.1 | 50 | 241.7 | 73 | 240.2 | 66 |
| 23 | 244.6 | 27 | 239.0 | 33 | 235.0 | 23 |
| 24 | 245.8 | 25 | 251.2 | 14 | 244.4 | 13 |
| 25 and upwards | 248.9 | 33 | 239.1 | 20 | 243.5 | 26 |
| | | 258 | | 476 | | 361 |

RELATIVE BRAIN CAPACITY OF CAMBRIDGE UNIVERSITY MEN ACCORDING TO THEIR PROFICIENCY AND AGE (FROM DR. VENN'S TABLES).

*The Numerals along the top of the diagrames sinify the product of the three Head measures, viz. :-*
*Length × Breadth × Height (in inches)*

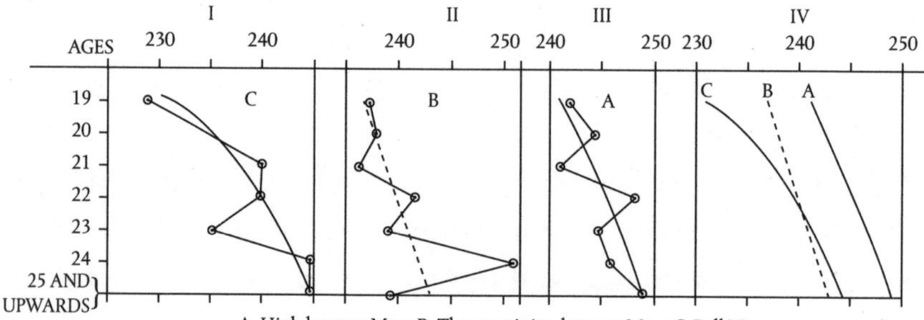


**Figure 45** Galton and Venn's graph showing how larger heads relate to better grades. Group A represents students scoring the highest grades; group B represents students scoring middling grades; group C represents men scoring the lowest grades. Galton and Venn's work shows how a larger 'head product' (the product of head length × breadth × height, measured in inches) is noticeably larger in Group A than in the other two groups.

Philosophical Library: the three men found that the measurements were different on each occasion. 'As anyone would expect who had seen the instrument used,' wrote the anonymous Trinity man, 'the height measurement is most unsatisfactory.'[21] Galton responded immediately, addressing some of these concerns and agreeing that the analysis should be repeated with a larger sample set in order to reduce errors and that the vertical head spanner needed to be improved.[22]

In a much-expanded paper published in 1889, Venn discussed the Cambridge results at greater length. For him, the most interesting part of the experiment wasn't necessarily the question of whether a larger head meant better grades, but rather how large amounts of raw data could be analysed to come up with meaningful conclusions. Venn wanted to grapple with a statistical concept that had been invented by Galton: percentiles. Venn explained to his reader that 'on this scheme the total numbers are supposed to be reduced to the scale of 100, and these to be then divided into ten equally numerous classes'. This would allow an individual to instantly see where he lay in comparison to all other subjects. It could also be plotted quite easily on a graph[23] (see Figure 46).

| Class. | Height in inches. | Strength of pull (lb.). | Squeeze (strongest hand). | Breathing capacity (cubic inches). |
|---|---|---|---|---|
| 1st ten .. | 72 to 77·8 | 102 to 155 | 103 to 125 | 305 to 400 |
| 2nd ten .. | 71 — 72 | 95 — 102 | 98 — 103 | 284 — 305 |
| 3rd ten .. | 70·2 — 71 | 90 — 95 | 94 — 98 | 274 — 284 |
| 4th ten .. | 69·5 — 70·2 | 86 — 90 | 90 — 94 | 265 — 274 |
| 5th ten .. | 68·9 — 69·5 | 83 — 86 | 87·5 — 90 | 254 — 265 |
| 6th ten .. | 68·3 — 68·9 | 80 — 83 | 86 — 87·5 | 248 — 254 |
| 7th ten .. | 67·6 — 68·3 | 76 — 80 | 82·5 — 86 | 235 — 248 |
| 8th ten .. | 66·8 — 67·6 | 72 — 76 | 78·5 — 82·5 | 223 — 235 |
| 9th ten .. | 65·6 — 66·8 | 68 — 72 | 73 — 78·5 | 209 — 223 |
| Lowest .. | 60·1 — 65·6 | 35 — 68 | 45 — 73 | 90 — 209 |

**Figure 46** A percentile chart, from a paper by John Venn. The concept of percentiles was developed to deal with anthropometric data. Here, Venn applies percentiles to anthropometric data about the undergraduates of Cambridge, gathered from the laboratory in the Philosophical Library, to illustrate the spread of heights, strength (pulling and squeezing), and breathing capacity.

The large data set also allowed Venn to explain the theory of probability to his readers. He followed his conclusion with an exhortation to think of the wider implications of this kind of analysis: 'in respect of head-measurement there is a decided superiority of A's over B's and B's over C's. But something more than this must be attempted...appeal must be made to the theory of Probability.'[24] Venn was especially interested in using the data to examine how 'probable errors' could be calculated and understood. This was an interesting exercise in its own right, but it also allowed Venn to show that the variations between the groups were real and not just caused by probable error. In 1890, Venn published another paper on the head sizes and grades of Cambridge undergraduates. He had data from thousands of extra individuals to consider, and the larger sample set confirmed his earlier findings.[25]

The matter rested there for more than a decade until Karl Pearson, a statistician with an interest in heredity, began to re-examine it. Pearson, a former Cambridge wrangler, had studied physics, languages, and law in Germany before settling in London. It was there that he met Galton and began to develop an interest in statistics. Pearson was interested in the question of 'which, if any, physical characteristics are sensibly correlated with intellectual ability' and so he asked the Cambridge Philosophical Society if he could use the data they had been gathering from students in their library to try to answer this question.[26] Working with several students from Girton College (who did much of the tedious data processing), the Registry of the University of Cambridge, and several researchers from University College London, including Alice Lee, who was an expert on the subject, Pearson ascertained the *actual* grades obtained by the students measured and set about reducing the data.

The data gathered in the Philosophical Library was especially useful to Pearson because the sample was so remarkably homogeneous. There had been studies that, for example, showed that 'the head of the Cambridge undergraduate is larger than the head of the criminal population' but these sorts of studies rarely controlled for the social or economic backgrounds of the subjects.[27] Using measurements of the length, breadth, and height of thousands of Cambridge students' heads and correlating these with the students' final academic grades allowed Pearson to conclude that 'there is no marked correlation between ability and the shape or size of the head'. Pearson's results differ from Venn's partly because he treated the three head dimensions separately rather than combining them into a 'head product' and partly because the sample size had increased in the two decades since Venn's analysis, allowing for more accurate calculation of probable error. When Alice Lee extended the study in the following year (looking at characteristics such as height, strength, and fitness as well as head size), she confirmed Pearson's findings, writing that

> [i]n no single case . . . is the correlation between intelligence and the physical characters sufficiently large to enable us to group the honours men as a differentiated physical class, or to predict with even a moderate degree of probability intellectual capacity from the physical characters of the individual.[28]

But, despite the lack of clear-cut answers from the field of anthropometrics, the students of Cambridge continued to flock to the laboratory in the Philosophical Library into the twentieth century. Some were so fascinated by the science of anthropometry that they returned again and again, paying their threepence each time: Pearson commented on 'one senior wrangler who was measured no less than five times!' while Venn remarked that 'one man, who seems to have had a morbid love of this physical inspection, has actually had his various dimensions and capacities tested no less than eighteen times during the course of some three years'.[29] These sorts of laboratories (and Cambridge did not have the only one; others later followed at Dublin, Eton, and Oxford) raised public awareness not only of anthropometrics but also of science generally. They encouraged the idea that ordinary people could participate somehow in large-scale experiments. The results were also intelligible to non-specialists. People could immediately grasp how mathematics was being applied to a study of the human body; they could get their heads around the statistics by seeing concrete examples.

The leaders of this new science often cited its relevance to studies of heredity and to Darwinian inheritance in particular: they were interested in seeing how particular traits were replicated in large populations, and what the social consequences of those traits were. By the end of the nineteenth century, the British Association had set up a subgroup to deal with anthropometrics, and Galton's apparatus was being displayed and used at their meetings. London's Royal Society had also set up a committee which featured several Cambridge men, including Rayleigh, G.G. Stokes, and Michael Foster.[30] Funding became available, both from organizations like the British Association and from the British government.[31]

Beyond the blue-skies research, many in the public realm were interested in the immediately practical uses of the science: could it be used to predict criminal behaviour, rank competitors for jobs, or to finally determine whether men were truly superior to women? The popular press jumped upon anthropometrics as a way to talk about some of the social and political issues of the day. When Galton published some results comparing male and female subjects and showing that the strongest individual female had greater pulling strength (86 lbs) than the average man, the satirical magazine *Punch* immortalized her in verse as 'The squeeze of 86':

> Maiden of the mighty muscles,
> There recorded, you would be
> Famous in all manly tussles,
> And it's very clear to me,
> That if in the dim hereafter
> Any husband should play tricks,
> You would, with derisive laughter,
> Give a 'Squeeze of 86'.[32]

But, though Galton and others hoped for definitive answers to questions about things like male versus female abilities, their conclusions sometimes veered from the path of scientific objectivity, as when Galton wrote:

> I found, as a rule, that men have more delicate powers of discrimination than women, and the business experience of life seems to confirm this view. The tuners of pianofortes are men, and so, I understand, are the tasters of tea and wine, the sorters of wool and the like.... If the sensitivity

> of women were superior to that of men, the self-interest of merchants would lead to their always being employed.[33]

A generation later, Pearson dismissed this reasoning, pointing out 'how many callings at that date were closed to women, without their really being unfitted for them'.[34] Likewise, when anthropometrics was applied to questions concerning race or class, the conclusions drawn were often skewed by prejudices.

In the archives of the Philosophical Society, kept apart from the main run of cards is a little bundle marked 'Indian students'. The cards themselves do not have a place to record race, though they do have one for birthplace. The options for skin tone are limited to 'pale', 'ruddy', 'dark', or 'freckled'. Though there were increasing numbers of Indian (and other overseas) students coming to Britain in the later decades of the nineteenth century, the Anthropometric Committee obviously did not consider there was enough racial diversity to merit a question about it on the laboratory's cards. From the final quarter of the century, and particularly after the creation of the associate membership, many Indian names began to appear in the records of the Philosophical Society, the most famous of which was that of the mathematician Srinivasa Ramanujan, who came to Cambridge in 1914. Whether the pile of Indian cards was the subject of any particular analysis is unknown.[35]

Though Galton and many other practitioners may have had all manner of preconceived notions about gender, race, and class, they had taken some positive steps towards introducing mathematical methods into the human sciences. Pearson, though sceptical of many of Galton's results, wrote admiringly about how

> Galton's new calculus...enabled us to reach real knowledge...in many branches of enquiry where opinion only had hitherto held sway. It relieved us from the old superstition that where causal relationships could not be traced, there exact or mathematical inquiry was impossible. We saw the field of scientific, of quantitative, study carried into organic phenomena and embracing all the things of the mind.[36]

Similarly, the physician Charles Myers, who was sceptical of many of the claims that had come from anthropometrics about race, admitted that Galton had done some important work in pushing the field towards studies of heredity and Mendel's theories. He also emphasized the possibilities of the new science of biometry—the use of statistical analysis to understand biological data—which was growing out of anthropometrical research.[37]

The laboratory in the Philosophical Library closed by 1910, and the cards on which the details of the experiment had been recorded were sent to Karl Pearson in London for further analysis. Throughout the early decades of the twentieth century, such data sets were considered of vital importance to science and to government. There was a new interest in public health, in the fitness of the population, and in the nascent discipline of eugenics. Eugenics, in its simplest form, refers to the idea of improving populations through consideration of theories of heredity. Its proponents argued that the power of science could be harnessed to improve the lot of future individuals, and of nations: the idea was immensely popular, and eugenics societies began to appear around the country, and the world. But the harsh realities of eugenics soon began to emerge. *The Times*, while fundraising in support of a new eugenics laboratory following the death of Galton in 1911, cited the example of recent industrial action as proof of the necessity of exerting greater control over the working classes: 'The state of morals and of intelligence disclosed by the recent strikes . . . [show] the need for the study and the application of Eugenics.'[38] Meanwhile, the geneticist John Burdon Sanderson (J.B.S.) Haldane prophesied a world one hundred and fifty years in the future in which the eugenicists 'succeeded in producing the most violent opposition and hatred amongst the classes whom they somewhat gratuitously regarded as undesirable parents'.[39] Indeed, what had begun with seemingly benign ideas of encouraging healthy people to marry young and have large families quickly spiralled into a much harsher reality as practices of forced sterilization, racial segregation, and even the murder of those deemed unworthy of life received official sanction in many countries. Eugenics, it quickly became apparent, was not simply a way of promoting the good health of future generations, or reducing the effects of genetic diseases; it had become a form of social control to be used by states against some of their most vulnerable citizens.

Anthropometrics fell out of fashion as the twentieth century advanced—it had become too closely associated with the horrors of eugenics that had dominated the mid-century. After Pearson's retirement in the 1930s, the cards were forgotten in a basement at University College London for many decades. In the 1960s, they were rediscovered there, and a statistician kindly wrote to the Cambridge Philosophical Society asking whether the Society would like them back, 'or do you think they may be scrapped?'[40] Not sure of what to do with this strange archive, the Society stored the cards for decades in a tea chest next to the 'big skulls' in the basement of the University's zoology museum.[41] But now, after many years, the cards are being studied again. They provide a rare snapshot of the physicality of a large group of individuals in a single place at a particular period of history: the kind of data set that can tell us something about people's lives, but that also tells us much about the scientists conducting these experiments and the nuances of this lost science—what their concerns were, how they designed and carried out these experiments, what they were hoping to prove, and why they were trying to prove it.[42]

The Cambridge Philosophical Society has supported much research over the two centuries of its existence, but the anthropometric laboratory was an experiment on an unprecedented scale. From the 1880s until the early years of the twentieth century, the library staff made tens of thousands of precise measurements of undergraduates' bodily dimensions and capabilities, seeking statistical correlations between different attributes. The most lasting legacy of this enormous project was not so much the anthropometric data as the methods of analysis that grew from it—the large-scale application of mathematics and statistics to the study of living things.

By 1923, the Philosophical Society's anthropometric laboratory had been inactive for more than a decade and anthropometrics had lost some of its earlier popularity; but the essential method of the science—applying statistics to biological data—had surfaced in a new form. In that year, J.B.S. Haldane arrived in Cambridge. He had come through Eton and Oxford, where he studied mathematics, classics, and philosophy, and he had served in France and Mesopotamia during World War I. He then became a fellow of Oxford's New College, where he began to study physiology and genetics, in 1919. Within a few years, he had established himself in the life sciences and was offered a position in Cambridge. He joined the Philosophical Society in March 1923, and was elected to its council the following year.[43] The first major paper of his Cambridge years was published in the Society's *Transactions*; it was titled 'A mathematical theory of natural and artificial selection'.[44]

This title is striking: Haldane was seeking to mathematize evolution. While the physical, chemical, and earth sciences had become increasingly mathematical in the preceding century, the life sciences were still seen by many as more qualitative than quantitative. But anthropometrics, and its sister science, biometrics, played an important role in changing this perception. Biometrics focused on the large-scale accumulation of measurable data about living things; but, since it did not necessarily deal with humans, it avoided many of the moral and philosophical problems associated with anthropometry. The key figure in the creation of biometrics was Karl Pearson. He founded a school of biometrics at University College London and set up the journal *Biometrika* to present the results of this new discipline. The biometric school in London was the only place in Britain at that time to provide advanced training in statistical theory and it drew many mathematicians into the study of living things.

There were several potential uses for biometrics, but chief amongst them was its use as a tool for the study of Darwinian selection. Though the basic fact of evolution had largely been accepted by the beginning of the twentieth century, Darwin's mechanism of slow continuous change due to natural selection was not so widely believed. Instead, many scientists believed that evolution of species occurred in jumps—saltations—in which offspring which were radically different from their parents might appear suddenly due to mutations in the hereditary material. But the biometricians preferred Darwin's idea of slow continuous change, and the evidence seemed to back them up. Take, for example, the question of height: if the heights of hundreds of men or women are measured and plotted on a graph, a smooth curve will be produced. Most people will cluster around the average height, while smaller numbers will be exceptionally tall or short. The more individuals measured, the smoother the graph will be. This kind of graph supported the biometricians' view of continuity of measurable features in living organisms. Biometrics was seen by some as more 'scientific' than other approaches to biology, relying as it did on quantifiable factors. It was contrasted with older methods of observation and with the sort of work done by animal morphologists who tried to create evolutionary trees by grouping animals based on appearance.

In opposition to the biometricians stood the saltationists. These scientists believed in evolution, but not quite as Darwin had described it. Instead, they looked to the work of Gregor Mendel on inheritance to argue that evolution occurred due to the selection of dramatically different new mutations. The saltationists allied themselves to the emerging science of 'genetics'. In fact, when William Bateson had coined the word in 1905, he had intended it as a way to differentiate between his ideas and those of the biometricians: 'genetics' was created to counter the pure Darwinism of the biometricians. And so a stand-off emerged between, on the one side, those known as Mendelians, geneticists, or saltationists and, on the other, the biometricians, or Darwinians.

The new science of genetics was controversial—so controversial that, not long after the discipline had been established, the journal *Nature* decided that it would no longer publish on the subject. After receiving some articles from Cambridge, its editor explained that he was 'not prepared to continue the discussion on Mendel's Principles and therefore returns herewith the papers recently sent him by Mr. Bateson'.[45] *Biometrika*, admittedly a journal specifically set up to further the work of the biometricians, also began to refuse to publish any articles on Mendelism or genetics. But there were some who were willing to support the radical work being promoted by Bateson, as his collaborator Reginald Punnett later recalled:

> It was a difficult time for struggling geneticists when the leading journals refused to publish their contributions to knowledge, and we had to get along as best we could with the more friendly aid of the Cambridge Philosophical Society and the Reports to the Evolution Committee of the Royal Society.[46]

But, despite some setbacks, the Mendelians were determined to investigate the theoretical model they had developed. Gregor Mendel's famous nineteenth-century experiments on peas and other plants showed that certain characteristics were passed down from parent to offspring without being 'blended'. This meant that a tall parent and a short parent would not produce offspring of intermediate height, but would create either tall or short offspring. The ratio of tall to short plants in the younger generation was generally found to be fixed, which led to the idea that some characteristics were dominant while others were recessive. The Mendelians wanted to test and expand on this theoretical framework so that they could model the process of heredity and understand how parents passed on genetic material (or 'Mendelian factors') to their children. The very idea that there was a genetic material was still contested. Even Bateson, who pioneered much of the early work in the field, did not believe that 'genes' (as Mendelian factors would later be called) were real entities; instead, he saw them as a sort of 'harmonic resonance'.[47]

In contrast, the biometricians took a more empirical approach. They collected large amounts of data and focused on quantifying physical characteristics; then they looked for statistical correlations. The two approaches, which could have been treated as complementary, were seen by their warring proponents as utterly incommensurable. The Mendelians wanted to understand the 'genotype'—the genetic make-up of an organism—while the biometricians wanted to describe the 'phenotype'—its physical characteristics. Discoveries that the Mendelians hailed as breakthroughs were seen as irrelevant by the biometricians, and vice versa. The biometricians' figurehead, Pearson, saw Mendelism as hopelessly narrow, while the foremost Mendelian, Bateson, saw biometrics as unskilled and not really very biological at all. Pearson summed up the Mendelians as being confined by 'confused and undefined notions' while he and his colleagues were guided by 'clear and quantitatively definite ideas'.[48]

But there was one man who saw through the differences of the Mendelians and the biometricians and sought to bring them together. This was George Udny Yule, the future president of the Cambridge Philosophical Society who would argue that women should be allowed to become full fellows. Yule had trained as an engineer at University College London and had met Karl Pearson there. Yule's interest in applied mathematics and statistics led him to a position as Pearson's demonstrator in 1893. He was later appointed as an assistant professor at University College London before moving to Cambridge to become the first lecturer in statistics in 1912.[49] In 1902, Yule published an article which sought to reconcile the views of the biometricians and the Mendelians, writing that he wished to 'institute comparisons between the results obtained by the two schools, and to discuss the bearing of the two classes of observations on each other'. He attributed the gulf between the two schools to a problem of language:

> There has always been a good deal of misunderstanding between biologists in general and those who have done pioneer work in the use of statistical methods, due in great part, I believe, to the fact that the two do not use such terms as *heredity, variation, variable, variability,* in precisely the same signification.[50]

Yule took elements from papers written by Bateson, by Pearson, and by Galton and tried to synthesize them into a coherent whole, showing that the biometricians' approach was compatible with Mendel's laws and not, as Bateson had contended, that the two were 'absolutely inconsistent'.[51] But this paper of Yule's received little attention and the controversy between biometricians and Mendelians raged on.[52]

The possibility of uniting biometrics and Mendelism was largely ignored until 1918, when a Cambridge mathematics graduate named Ronald Aylmer Fisher waded into the debate. It was Leonard Darwin, another son of Charles Darwin, who had encouraged Fisher to try to link biometrics with Mendelism. The two men had met through their shared interest in eugenics: Fisher had co-founded the Cambridge University Eugenics Society in 1911 and Darwin, as President of London's Eugenics Education Society, had been invited to address the Cambridge group in 1912. By 1916, Fisher had worked up a statistical model for linking the two competing sciences, but he could not find a place to publish it. London's Royal Society rejected the paper, as one of their reviewers found it too genetical, and the other too mathematical. But, thanks to Darwin, Fisher managed to have the paper published in the *Transactions of the Royal Society of Edinburgh* and it was instantly acclaimed as groundbreaking.[53] Fisher showed that the inheritance of continuous traits (e.g. as seen in that smooth bell curve of heights) could be explained in Mendelian terms, writing that 'in general, the hypothesis of cumulative Mendelian factors seems to fit the facts very accurately'.[54] In doing so, he successfully linked the slowly changing continuous evolution of Darwin to the saltations of the geneticists.

Fisher's paper opened to door to a new way of thinking about evolution. But much detail still needed to be worked out. Haldane became part of this movement, beginning his mathematical studies of natural selection. Soon after arriving in Cambridge, Haldane started to work out the mathematical implications of natural selection. In a series of papers published by the Cambridge Philosophical Society between 1924 and 1934, Haldane presented his novel ideas to the world. He outlined his project thus:

> A satisfactory theory of natural selection must be quantitative. In order to establish the view that natural selection is capable of accounting for the known facts of evolution we must show not only that it can cause a species to change, but that it can cause it to change at a rate which will account for present and past transmutations.[55]

Haldane set up a simple model using specimens like insects, certain fish, and annual plants—organisms that did not breed across generations. And he began by limiting his studies to single, completely dominant Mendelian factors. As the series of papers progressed, Haldane introduced more complex cases with more variables, until the models became 'as formidable as any in mathematical physics'.[56] Over the course of those ten years, Haldane neatly set out his ideas on the speed with which selection occurred, how dominant and recessive qualities were expressed, how environmental pressures shaped natural selection, how mutations worked, the role of geographical isolation, the impact of mortality rates, and many other aspects of natural selection. Haldane derived equations to explore all of the different facets of evolution, and his models became increasingly complex as the series progressed.[57]

Mathematics was a useful tool for Haldane, but it was not an end in itself. Near the end of the series, Haldane described a set of simultaneous non-linear finite-difference equations. 'Their complete solution,' he wrote, 'is desirable for a discussion of problems raised by eugenics and artificial selection.'[58] Like many other biologists, biometricians, and geneticists of this age, Haldane was fascinated by the possibilities of eugenics—what he called 'the application of biology to politics'.[59] For Haldane, his interest in eugenics was closely linked to his interest in socialism: he saw each as an avenue to the betterment of mankind.[60]

Perhaps it was because of his interest in politics and his belief in the power of science to change society that Haldane was keen to explain his work to as wide an audience as possible. He was one of the earliest writers to explain the complex new mathematical ideas about natural selection to the general reader. In 1932, he transformed the nine intensely technical papers he had published with the Cambridge Philosophical Society into the easily readable *The causes of evolution.* This book focused not on mathematics (most of the statistical work was confined to an appendix) but on the ideas behind natural selection. And, though Haldane was keen to point out that the entire modern science of evolution was predicated on advanced statistical knowledge, he recognized that the most interesting questions about evolution could be asked (and perhaps answered) by anyone. When it came to questions like, 'Is [evolution] good or bad, beautiful or ugly, directed or undirected?' Haldane was no better equipped to answer them than his readers. He wrote:

> I can write of natural selection with authority because I am one of three people who know most about its mathematical theory. But many of my readers know enough about evolution to justify them in passing value judgments upon it which may be different from, and even wholly opposed to, my own.[61]

Thus, the public were invited to take part in the cutting-edge debates of the day.

*The causes of evolution* was a key text in what became known as the 'modern evolutionary synthesis'. The synthesis was, in essence, the outcome of the resolution of the debate between Mendelians and biometricians. The biologist Julian Huxley described how the synthesis relied on 'facts and methods from every branch of [biology]—ecology, genetics, palaeontology, geographical distribution, embryology, systematics, comparative anatomy—not to mention reinforcements from other disciplines such as geology, geography, and mathematics'.[62] It united Darwin's idea of natural selection with a rigorous mathematical and experimental methodology.

When Haldane declared himself to be one of the three people who knew most about the mathematics behind this synthesis, he was counting himself with his peers Ronald Fisher, whose 1918 paper had ignited interest in the field, and the American geneticist Sewell Wright, who pioneered many techniques in the field of population genetics. Working in parallel, these three men not only created a new theory, but popularized a new way of doing science. As well as proving a Darwinian theory using Mendelian results, they showed how biology, long viewed as a primarily qualitative science, could be thoroughly quantitative. Haldane concluded *The causes of evolution* with this thought: 'the permeation of biology by mathematics is only beginning...it will continue, and the investigations here summarised represent the beginnings of a new branch of applied mathematics.'[63] Evolution had been mathematized.

For a long time, biological studies in Cambridge, as elsewhere, had been largely focused on qualitative studies. A typical example of the kind of work that had dominated for so long can be found in a paper presented to the Cambridge Philosophical Society by Charles Marshall, an assistant to the Downing Professor of Medicine. In his paper, Marshall told how, just after lunch one Wednesday in February 1896, he had undertaken two experiments simultaneously. The first was a fairly straightforward chemical distillation, performed at the request of the professor. The second was rather less orthodox: he ingested a not insignificant quantity of cannabis resin. Marshall, who had obtained the drug from some chemical fellows he knew at Clare and Caius Colleges, was interested in the pharmacological effects of the compound. Of course, the drug had been in use for millennia, but it was not particularly well known in Britain at this time, and few scientific studies of its effects had been undertaken. Within forty-five minutes of taking the drug, Marshall began to feel something:

> I suddenly felt a peculiar dryness in the mouth... This was quickly followed by paræsthesia [pins and needles] and weakness in the legs. Gradually my mental power diminished—I was no longer able to control the steps of the operation [i.e. the distillation] and commenced to wander aimlessly about the room. I had the most irresistible tendency to laugh; everything seemed ridiculously funny.

Marshall had little choice but to let the chemical effects play out, but he continued to record those effects in the name of science:

> I was now in a condition of acute intoxication: my speech was slurring, my gait ataxic. I was free from all sense of care and worry and subsequently felt extremely happy. Fits of laughter occurred, especially at first, and sometimes the muscles of my face were drawn to an almost painful degree. The most peculiar effect was a complete loss of time relation: time seemed to have no existence.[64]

Finally, after about three and a half hours of giggling in his laboratory, Marshall drank two cups of coffee and, 'feeling somewhat better', walked home.

Marshall duly related this experiment to a meeting of the Society and wrote it up for their *Proceedings*. Though his topic was unusual and must have raised a few eyebrows at the Society, the paper itself was a classic example of the kinds of papers read to the Society in the late nineteenth century in the fields of pharmacology, physiology, and biology. It dealt with the close observation of a phenomenon within a very small sample set; in Marshall's case, the sample size was one. Obviously, the intention was that other researchers would add their own similar experiments so that a larger body of data would be built up. But, in some areas of the life sciences, that model was slowly becoming less popular, and two new methods became increasingly used as the new century began. The first was the application of mathematics and statistics, which had characterized anthropometrics and biometrics and which ultimately came to be seen as a pillar of the modern evolutionary synthesis. The second was the adoption of techniques from the physical sciences to study biological phenomena.

The discovery of x-ray diffraction by Lawrence Bragg, and its announcement at a meeting of the Cambridge Philosophical Society (which we will come to in the next chapter), had an enormous effect not just on physics and chemistry, but also on biology. Bragg's technique allowed scientists to see inside molecules, including biomolecules. It allowed matter to be understood in a way never previously imagined. The study of life, which had been gradually changing in scale from the study of whole organisms, to the study of their organs, to the study of their cells, now became the study of their constituent chemicals. Older vitalist theories of life were being abandoned as living systems were reduced to chemicals and numbers. But, though the new biology may not have had the holistic integrity of the older system, it had enormous explanatory power, and slowly the secrets of life began to be unlocked.

The Society gave space to all kinds of biological investigation, from the mathematical to the physical or chemical, while still welcoming more traditional observational studies. The seriousness with which the Society took the expanding biological disciplines is evinced in a new journal it began publishing in the 1920s: *Biological reviews*. It was agreed in 1922, after much discussion, that the Society should separate out the disciplines in its various publications, reflecting the reality of twentieth-century sciences. The *Proceedings* became a journal for mathematics and non-biological sciences, while papers about the life sciences were gathered together in *Biological reviews*.[65] The latter began accepting papers in 1923, and the first issue was completed in 1925.[66] But, though the sciences were fragmenting, becoming ever more discrete disciplines, a new interdisciplinarity was growing up, allowing the methods of chemistry or physics to appear regularly in biological journals, while biological data sets popped up in statistical and mathematical papers. The mathematization that the Society's earliest fellows had hoped to bring to the natural world was finally coming to fruition more than a century after they had first declared it a central aim of the Society. They had once written that 'Botany, Zoology and other branches of science... have already been partially illustrated by the application of Mathematical principles, and may perhaps be destined to acquire a still greater portion of the precision and certainty which attend the conclusions of demonstrative science'.[67] Haldane, Fisher, and the many others who contributed to the modern synthesis went beyond even this ambitious vision—they attempted to make biology a branch of mathematics.

# 8

# 'MAY IT NEVER BE OF ANY USE TO ANYBODY'

In the Cavendish Laboratory, late in the year 1896, J.J. Thomson was carrying out his experiments on discharge tubes with a new vigour. Since the 1880s, he had been observing the beautiful glows that lit up the tubes as mysterious rays emanated from the cathode and traversed the vessels. In 1895, Thomson's work had taken on a new significance when the German scientist Wilhelm Röntgen discovered what he called 'x-rays' (and which many others called 'Röntgen rays'). These x-rays were emitted when cathode rays were accelerated through a discharge tube by the application of high voltages; as the cathode rays struck a metal surface, x-rays would be released. Though some earlier investigators had noticed the effects of these rays, Röntgen was the first to pay them any serious attention. Röntgen's 1895 announcement of the discovery caused a sensation and spawned hundreds of research papers as physicists across the world scrambled to explain the true nature of these rays.

In Cambridge, Thomson was uniquely placed to lead the charge. He had spent more than a decade perfecting the apparatus and honing the research skills needed to understand the enigmatic x-rays. Thanks to the University's new policy of granting entry to graduate students who had not previously studied at Cambridge, Thomson also had several new researchers in his laboratory—chief amongst these was Ernest Rutherford, who arrived in 1895 from New Zealand, where he had already completed his first degree (see Figure 47). He became an associate of the Philosophical Society in 1896.[1]

**Figure 47** A group of Cavendish researchers, 1897. J.J. Thomson is fourth from left in the front row, Ernest Rutherford is far-right in the front row, Charles Wilson is second from left in the back row.

X-rays had been discovered just two months after Rutherford's arrival in the Cavendish, and very soon afterwards Thomson invited the young researcher to join him in his discharge tube investigations. Together, they produced one of the most significant papers to emerge from the Cavendish—'On the passage of electricity through gases exposed to Röntgen rays'—which they read to a meeting of the British Association for the Advancement of Science in September 1896.[2] The paper described a number of experiments in which Thomson and Rutherford examined the ability of x-rays to change a gas from an insulator to a conductor and then studied how electricity passed through this charged gas. Thomson used their experimental results to formulate a set of equations and elaborate his theory of ionization.

Though their experimental work yielded some answers about the properties of x-rays, Thomson was still perplexed by more fundamental questions about the nature of the cathode rays that gave rise to x-rays (see Figure 48). He resolved to investigate further and, in February 1897, he presented some of his preliminary findings to a meeting of the Cambridge Philosophical Society. The meeting, presided over by Francis Darwin, heard Thomson describe the experimental set-up that allowed him to show conclusively that there was 'a flow of negative electricity along the cathode rays'. Moreover, Thomson was able to use magnetic fields to deflect the beam of cathode rays, meaning that the beam itself must be carrying a charge.[3] Though he was not yet ready to draw bigger conclusions before this meeting of the Philosophical Society, these properties of cathode rays led Thomson to further speculations about their true nature. By April 1897, following further experiments and calculations, Thomson felt confident enough to draw some major conclusions about cathode rays—that they were made up of tiny particles, that these particles were negatively charged, and that these particles were a constituent of atoms.[4]

**Figure 48** J.J. Thomson during a lecture-demonstration in 1909. There is a glass discharge tube visible on the right-hand side of the image.

This was an extraordinary conclusion. Atoms, by their very definition, were indivisible particles. If Thomson's theory was correct, it would shake modern science to its very core. Thomson persevered with his theory, supporting it by calculating the charge-to-mass ratio of the particles and showing that the negatively charged 'corpuscles'—as he called them—were tiny compared to the mass of even the smallest atom, hydrogen. But though he could prove that the particles were real and could ascertain properties about them, like their charge and their mass, there was much resistance to the idea of a divisible atom. There were also few practical applications for Thomson's theory at the time, and so many of his contemporaries largely ignored his findings. Following a British Association meeting at which Thomson had spoken of his work on the corpuscles, the influential journal *The electrician* reported that 'no scientific discovery of prime importance has been announced during the recent meeting'.[5] It would be almost a decade before Thomson's idea that tiny, negatively charged corpuscles existed within atoms received widespread recognition.

In 1906, Thomson was awarded the Nobel Prize for his work on discharge tubes. The particles in cathode rays had become known as electrons, and their importance in understanding not just electricity but also matter itself had come to be accepted by physicists. The discovery of this first subatomic particle had a profound effect on the research conducted in the Cavendish Laboratory, and an effect too on the Cambridge Philosophical Society.

The closing decades of the nineteenth century had seen a sharp rise in the number of papers on physiology and the life sciences in the Society's *Proceedings*. Mathematics and mathematical physics had also been the subjects of many papers in the journal, but experimental physics had lagged behind. In the *Proceedings* published in 1898 (Volume IX) and 1900 (Volume X), approximately 15 per cent of papers had concerned experimental physics—many of them by Thomson himself. After 1900, as Thomson's theory of electrons became more widely accepted, the number of experimental physics papers presented to the Society jumped suddenly, and many of them were focused on studies of electrons or x-rays. Almost 30 per cent of papers published in *Proceedings* in 1902 (Volume XI) related to experimental physics, 35 per cent in 1904 (Volume XII), 40 per cent in 1910 (Volume XV), and 50 per cent in 1912 (Volume XVI). Thomson himself continued to publish extensively, but many of these papers were written by the young research students of the Cavendish.[6]

One such student was Lawrence Bragg—the son of the physicist William Bragg—who had entered Trinity College in 1909, studied mathematics and physics, and graduated in 1911[7] (see Figure 49). In summer 1912, it was reported that three German scientists—Max von Laue, Walter Friedrich, and Paul Knipping—had passed x-rays through crystals and observed an interference pattern similar to the kind seen when light passes through a grating. On holiday together in Yorkshire that summer,

**Figure 49** Lawrence Bragg, 1915.

William and Lawrence discussed this finding and debated the implications of the German experiment. Some people believed that x-rays were made up of particles, but in this experiment they behaved more like light waves. William Bragg continued to believe in the idea of x-rays as particles, though he grappled with the possibility that there was a way to reconcile their particle and wave properties. Meanwhile, Lawrence Bragg returned to Cambridge and continued to think over the problem.

Laue and the German scientists had hypothesized that x-rays travelled into the crystal and struck the atoms within it; thus, the atoms became excited and emitted secondary x-rays, which were the ones that gave rise to the interference pattern. Laue then explained the pattern by assuming that the crystal was made up of atoms in a cubic arrangement and that the incident x-rays were of a few discrete wavelengths. But the pattern of dots that was produced by the x-rays as they exited the crystal did not quite fit with Laue's assumptions.

Instead, Lawrence Bragg came up with a simpler solution: he believed that the x-rays that travelled into the crystal were the same ones that exited it in a neat pattern. He also posited that the crystal did not have a cubic structure but rather a face-centred lattice and that the x-rays occurred across a wide spectrum of wavelengths, not just a few discrete ones. Using these assumptions, Bragg was able to perfectly account for the pattern produced by the x-rays (see Figure 50). He was also able to formulate an equation (now known as the Bragg equation) which described the angles at which x-rays would exit the crystal, and linked this to the arrangement of atoms in the lattice. Lawrence Bragg was only twenty-two years old when he brought this theory to a meeting of the Cambridge Philosophical Society in November 1912; as he was so young, it was decided that his mentor J.J. Thomson would read the paper on his behalf.[8] The paper caused an instant sensation in Cambridge, for Bragg's idea not only explained the pattern produced by the x-rays; it also created a new way of understanding matter.

The pattern produced was dependent on the structure of the atoms within the crystal. The crystal acted as a three-dimensional diffraction

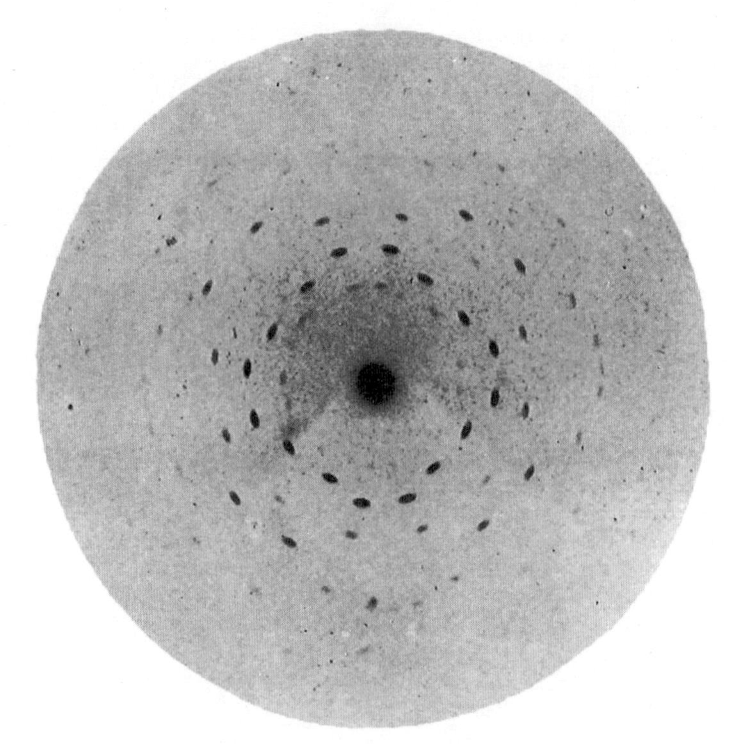

**Figure 50** A plate showing an x-ray diffraction pattern, from Lawrence Bragg's first paper on the topic, presented to the Cambridge Philosophical Society in November 1912.

grating so it was possible to know the arrangement of atoms in it by knowing the angle of incidence of the x-rays and by recording the layout of dots in the interference pattern. In trying to explain Laue's result, Bragg had inadvertently created a new way of understanding the structure of matter, and a new branch of physics: x-ray crystallography. One of the first converts to this method was Lawrence's own father, William. A couple of weeks after the younger Bragg's paper was presented to the Philosophical Society, William wrote to a friend: 'My family are quite all right. [Lawrence] is coaching and demonstrating at Cambridge, and has just brought off a rather fine bit of work in explaining the new x-ray and crystal experiment.'[9] William elaborated further in a letter sent to *Nature* the following week:

> In a paper read recently before the Cambridge Philosophical Society my son has given a theory which makes it possible to calculate the positions of the spots for all dispositions of crystal and photographic plate. It accounts also for the form of the spots and other details... It is based on the idea that any plane within the crystal which is 'rich' in atoms can be looked on as a reflecting plane; the positions of the spots can then be calculated by the reflection laws in the ordinary way.[10]

William was still not entirely convinced that his son's work conclusively proved the wave theory of x-rays, but he was certainly won over by the idea of using x-rays to understand crystals. The two began to collaborate almost immediately.

Using the theory that Lawrence had presented to the Philosophical Society, William designed a new scientific instrument called the x-ray spectrometer. With the technician C.H. Jenkinson, William Bragg constructed the first such instrument in his laboratory in Leeds in the winter of 1912–13. The Braggs and others began to use these new spectrometers to investigate the x-rays emitted from different sources and soon discovered that each element gives out a characteristic x-ray spectrum made up of particular wavelengths. In Oxford, Henry Moseley used a spectrometer to determine the atomic number (now thought of as the number of positively charged particles in the nucleus of an atom) of various elements, allowing the abstract theory behind the periodic table to be linked to the physical structures of atoms. X-ray crystallography has been one of the most significant techniques of modern science; not just physics but also chemistry and biology have been shaped by it. It has been used to show that electrons have both wave and particle properties, to understand chemical bonds and, most famously of all, to discover the double-helical structure of DNA molecules. William and Lawrence Bragg were awarded the Nobel Prize in 1915 for their work, and more Nobel Prizes have been awarded for work involving x-ray crystallography than any other scientific technique.

Lawrence Bragg was not the only researcher in the Cavendish who tried to solve one small puzzle and inadvertently created a new way of doing physics. Charles Wilson had studied at Cambridge and graduated from the Natural Sciences Tripos in 1892. His first fascination was for meteorology, and he spent as much time as possible after graduation at the observatory on Scotland's highest mountain, Ben Nevis, watching cloud formations and the sunlight as it filtered through the Scottish mists. He later recalled how

> [t]he wonderful optical phenomena shown when the sun shone on the clouds surrounding the hill-top, and especially the coloured rings surrounding the sun (coronas) or surrounding the shadow cast by the hill-top observer on mist and clouds (glories), greatly excited my interest and made me wish to imitate them in the laboratory.[11]

Wilson returned to Cambridge and the Cavendish as a research student in 1894, and gave his first-ever scientific paper to the Cambridge Philosophical Society in May 1895.[12] This paper was on a piece of apparatus Wilson had developed for studying atmospheric phenomena in a laboratory: a cloud chamber (see Figures 51 and 52). He was a skilled glass-blower and the glass chambers he created allowed water vapour to be condensed or expanded by varying temperature or pressure; this meant that he could study cloud formation in a carefully controlled environment. His first paper gave special mention to dust, and particularly to the importance of excluding it from the chamber because the presence of dust would lead to water droplets condensing around the particles.

Though Wilson's main interest was in meteorology and in the physics of water droplets, being in the Cavendish in the mid-1890s meant that he had easy access to x-ray tubes and thus he began to investigate the effects of x-rays, ultraviolet light, and other radiation on his artificial clouds. He found that introducing x-rays into his cloud systems enormously increased the number of nuclei around which droplets could condense. Next, he tried irradiating his clouds with 'the Uranium radiation discovered by Becquerel' and found again that large numbers of nuclei were formed within the water vapour. He concluded that 'the electrical properties of gases under the action of Röntgen and Uranium rays

**Figure 51** Charles T.R. Wilson with a cloud chamber, 1924.

point to the presence of free ions. It is natural to identify with these the nuclei made manifest in the gas under the same conditions by the condensation phenomena.'[13]

J.J. Thomson was struck by Wilson's results. At around the same time that Wilson was irradiating his cloud chambers with x-rays and uranium radiation, Thomson was working out the charge-to-mass ratio of the particles in cathode rays. He established that the ratio was almost 2,000 times that found in the hydrogen ion. Next, he needed to measure the charge of the particles in order to find out their mass. Thomson looked to Wilson's work, writing that, when there are charged particles in a cloud chamber, 'each of the charged particles becomes the centre round which a drop of water forms; the drops form a cloud, and thus the charged particles, however small to begin with, now become visible and can be observed'. When the chamber was exposed to radium, large numbers of

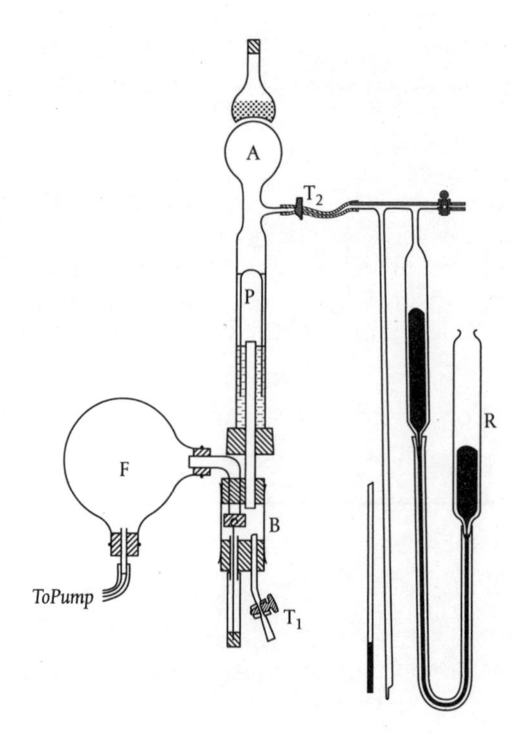


**Figure 52** Diagram of Charles Wilson's cloud chamber apparatus, from one of his early papers to the Cambridge Philosophical Society.

positively and negatively charged particles were formed within it, resulting in the formation of many water droplets. Thomson explained what this could tell him about those charged particles:

> We can use the drops to find the charge on the particles, for when we know the travel of the piston [used to expand and condense the air in the chamber], we can deduce the amount of supersaturation, and hence the amount of water deposited when the cloud forms. The water is deposited in the form of a number of small drops all of the same size; thus the number of drops will be the volume of the water deposited divided by the volume of one of the drops. Hence, if we find the volume of one of the drops, we can find the number of drops which are formed round the charged particles. If the particles are not too numerous, each will have a drop round it, and we can thus find the number of electrified particles.
>
> From the rate at which the drops slowly fall we can determine their size.

Using a formula derived by G.G. Stokes, Thomson could then use the velocity of the falling drops to estimate their radii:

> We can in this way find the volume of a drop, and may therefore, as explained above, calculate the number of drops and therefore the number of electrified particles.
>
> It is a simple matter to find by electrical methods the total quantity of electricity on these particles; and hence, as we know the number of particles, we can deduce at once the charge on each particle.[14]

Wilson's invention, designed to study atmospheric phenomena, first presented to the Cambridge Philosophical Society in a short and unassuming paper, and which cost about £5 to make, had been used to determine the charge of the first known subatomic particle.[15] Following Thomson's use of the cloud chamber, it became one of the key pieces of equipment in the Cavendish Laboratory in the early twentieth century. In the 1910s, Wilson developed the apparatus further, allowing photographs to be taken of the tracks of charged particles within the chamber.[16] By this time, many different kinds of radiation and charged particles were known: as well as x-rays and electrons, there were the rays that Henri Becquerel had discovered emanating from uranium, and the so-called alpha and beta particles that Ernest Rutherford had discovered while studying radium. Each of these different kinds of radiation or particle could be studied through the use of Wilson's cloud chamber. The images of their tracks through the chamber were used to understand what kinds of particles they were—their ionizing properties, their mass, their charge, and their place within the atom could all begin to be fathomed. The secrets of the atom were being unlocked, thanks to Wilson's love of meteorology. No other apparatus in use at the time had the power of Wilson's cloud chamber. As the physicist Patrick Blackett—a former student of the Cavendish—later observed:

> There are many decisive experiments in the history of physics which, if they had not been made when they were made, would surely have been

> made not much later by someone else. This might not have been true of Wilson's discovery of the cloud method. In spite of its essential simplicity, the road to its final achievement was long and arduous: without C.T.R. Wilson's vision and superb experimental skill, mankind might have had to wait many years before someone else found the way.[17]

Ernest Rutherford called the cloud chamber 'the most original and wonderful instrument in scientific history'.[18]

In the opening decades of the twentieth century, the Cambridge Philosophical Society became more and more closely tied to the researchers of the Cavendish Laboratory. Approximately a third of all research papers produced in the Cavendish were published by the Society, and many talks were given at Society meetings that were eventually published in London or elsewhere. The Cavendish was a unique blend of established physicists like Thomson and young research students like Bragg or Wilson—both kinds of researcher were welcomed by the Society and given opportunities to present and publish their work. Because of the link between the Society and the Laboratory, and the Society's place within the University, the Society was more likely to publish 'blue-skies' research from unknown researchers than other journals were. The slightly informal nature of the *Proceedings*—which accepted very short papers and works in progress—also facilitated these young researchers and allowed them a platform that might not otherwise have been available to them. The Society's meetings and publications provided a vital boost to the early careers of many who would go on to become stars of twentieth-century science. Bragg, just twenty-two when he gave his first paper to the Society, and Wilson, twenty-six, both went on to win Nobel Prizes for the work that they had first introduced to the world at meetings of the Cambridge Philosophical Society.

The Cavendish Laboratory seemed to have an unstoppable momentum in the opening years of the twentieth century, but its productivity came to an abrupt halt in 1914 with the outbreak of World War I. Though the laboratory remained open, many of its researchers left Cambridge to fight in the war. Undergraduate enrolments also dropped considerably from 1914 and fell off even more sharply after conscription began in 1916; before the war, there had been almost 4,000 undergraduates in Cambridge; by 1916, that figure had dropped to about 400.[19] The University went into a kind of hibernation as lectures were cancelled, practical classes cut, and students replaced by soldiers. A student of Newnham College (which, being a women's college, was not much affected by the drop in student enrolment) recalled that

> [t]he colleges housed soldiers doing courses, studying this and that, marching and counter-marching through the streets and the countryside. Working in the library at Newnham you could hear them going by, singing the haunting songs of those years, which still have the power to twist the heart.[20]

while F.J.M. Stratton—who would later become Professor of Astrophysics—remembered how

> [a]ll the open spaces of the town were occupied, the encampments extending to the fields around as far as Grantchester; horses were picketed in quiet roads such as Adams Road. The officers of the 3rd Rifle Brigade who were encamped on Midsummer Common were invited to become honorary members of the [Caius College] Senior Combination Room and took their meals in College and dined at the high table.[21]

Meanwhile, parts of Trinity College, and the cricket ground of King's and Clare Colleges (now the site of the University Library), were turned into temporary hospitals.[22]

At the Cavendish Laboratory, most of the researchers joined the Officer Training Corps. A small amount of research continued, with some of it related to the war effort, while the Laboratory's workshops were used to manufacture gauges to be used in armament factories. Few at the time saw a particularly strong link between experimental physics and war, and so resources were diverted away from the Laboratory, though some more established members of the Laboratory became government advisers (e.g. J.J. Thomson sat on the Board of Invention and Research, which was operated by the Royal Navy).[23]

In the Philosophical Society, business continued as normal—in theory. Meetings continued to be held every second Monday evening, and periodicals continued to be printed, but, with so many researchers absent from Cambridge, the output of the Society was greatly diminished during the war years. No volumes of the *Transactions* were published during the war and, though two volumes of the *Proceedings* were published, they contained only half as many articles as the pre-war volumes. The Society's social events, such as its annual dinner, and its 'at homes' were cancelled or postponed indefinitely. From 1916, the practice of serving tea at meetings was halted.[24]

One question that troubled the Society's council was what to do about the German journals that the Society received through its programme of periodical exchanges, and whether to continue sending its publications to German scientific societies. After some debate, it was resolved to continue the exchanges.[25] In 1916, the government formally requested that learned societies not publish papers concerning munitions or related topics without having them vetted by the Ministry of Munitions.[26] By the start of 1918, the Society had to apply for permission from the Board of Trade to continue its periodical exchanges with Germany.[27] Though previous wars had affected scientific communication between rival nations, the wars of the twentieth century had a more profound effect on scientists and the sciences. Science—especially chemistry and acoustics—played a significant role in World War I and, because there was increasing government funding for the sciences, there was a new sense that scientific discoveries were neither the property of the scientist behind them nor the property of the general public: there grew up the view that new discoveries (particularly in sensitive fields) should not necessarily be disseminated as widely as before.

This forced learned societies to rethink the ways in which they communicated with colleagues abroad. Following the war, some fellows of London's Royal Society went so far as to demand that Germans and Austrians be stripped of their membership, while 1919 saw the setting up of the International Research Council, which was composed of Entente and neutral countries but excluded Germany.[28] Many British scientists and scientific societies mistrusted their German counterparts and doubted the value of their work. H.H. Turner, Professor of Astronomy at Oxford, wrote that 'the basis of [scientific societies] is the good faith of the contracting parties: can we accept in scientific matters assurances which are, by some of the parties, not considered binding in other connections?'[29] The Royal Astronomical Society refused to send copies of its journal to 'enemy countries', even after the war had ended.[30] But while some were cutting ties with Germany, the Cambridge Philosophical Society continued to stock its library with German periodicals, perhaps believing that science should transcend politics.

Following the war, researchers returned to Cambridge, and the laboratories and societies enthusiastically resumed their previous work. Individuals not only had to reignite their scientific research but also had to work within the new international framework of the sciences, where using German ideas might be frowned upon. In physics, it would have been rash for researchers to cut themselves off from all German work, when so many interesting results, especially in theoretical physics, were coming from that part of the world.

One Cambridge scientist who firmly believed in using German ideas in his work was the Quaker and pacifist Arthur Stanley Eddington. Eddington had been a fellow of the Cambridge Philosophical Society since 1907; he was its vice-president from 1915 to 1917, and remained on its council into the 1920s.[31] Like many Quakers, he believed that it was important to humanize those seen as enemies, through personal contact and cooperative projects. And so it was that Eddington began to think about experimenting with the ideas of the German scientist Albert Einstein.[32]

Eddington had been studying for the Mathematical Tripos in 1905 when Einstein first published his theory of special relativity, but the theory had not been the subject of much interest in Cambridge.[33] In 1915 and 1916, Einstein put forward his theory of general relativity.[34] By this time,

Eddington was Plumian Professor of Astronomy and Director of Cambridge Observatory. At a time when many British scientists were pointedly ignoring German ideas, Eddington quickly saw the importance of general relativity, and began to think about proving Einstein's theory experimentally. Eddington was impressed by Einstein's theory and by the mathematical proofs he used to support it; but he was also pleased to learn that Einstein was a pacifist and 'interested to hear that so fine a thinker as Einstein is anti-Prussian'. Einstein too was happy to work with a British colleague and to 'throw a bridge over the abyss of misunderstanding'.[35]

Eddington realized that a solar eclipse due in 1919 would provide the perfect opportunity to test the theory that gravity is a geometric consequence of space and time, by seeing whether large objects such as stars could gravitationally bend rays of light. Together with the Astronomer Royal, Frank Dyson, Eddington began to plan an expedition to observe the solar eclipse from the island of Príncipe off the west coast of Africa, while a sister expedition was planned to Brazil. The war was still raging when Eddington and Dyson began planning the trip; Eddington had applied to be exempted from military service on the grounds of his religion, but in the end it was his unique scientific skills that meant that Eddington was not conscripted (though, by the time Eddington set off for the Gulf of Guinea, the war was over).

By April 1919, Eddington and his team were on Príncipe. They had a month to set up their equipment on the heavily wooded and 'very charming' island, but their progress was slow due to heavy rain, plagues of mosquitoes, and mischievous monkeys interfering with the apparatus. On the morning of the eclipse—29 May 1919—with telescopes and cameras in place, the recording station was hit by a fierce rainstorm and clouds filled the skies completely. But, with only half an hour to go before the five minutes of total eclipse, the clouds began to thin. As the sun was fully obscured by the moon, and darkness fell upon the island, Eddington's photographic equipment managed to record the moment. The plates then had to be developed and the painstaking business of measuring and analysing them could begin. Of the sixteen images taken in Príncipe, and the nineteen in Brazil, only a handful were of sufficient quality to be scientifically useful, but what they revealed astonished astronomers. They showed starlight from several constellations that appeared to be next to the sun. But, in reality, these stars were directly behind the sun and should not be visible from Earth. The sun's gravitational force was causing light from the other stars to bend around it. This result was extraordinary. Classical physics did not allow light rays to bend, but Einstein's new general theory of relativity saw gravity as simply a result of the curvature of space-time around a mass; if space-time itself was curved, then light rays would be bent around large objects. Eddington's observations confirmed the predictions of Einstein's theory.[36]

In November 1919, a special joint meeting of the Royal Astronomical Society and the Royal Society was convened, chaired by J.J. Thomson, then President of the Royal Society. The sole purpose of this meeting was to allow Eddington to present his results from Príncipe. That night, Eddington announced to the world that, in a test between Newton's laws and Einstein's, Einstein's had won out. Our understanding of the universe had shifted.

A few weeks later, following excited reportage of the London meeting in the national press, Eddington gave a talk to his home crowd. He appeared before the Cambridge Philosophical Society on the evening of 24 November. The meeting took place in the Cavendish Laboratory and was presided over by Charles Wilson. Eddington later wrote to his friend Einstein to describe the meeting, saying: 'I had a huge audience at the Cambridge Philosophical Society a few days ago, and hundreds were turned away unable to get near the room.'[37]

Eddington was warmly welcomed by his colleagues at the Philosophical Society. Many of its fellows remembered him as the one who had played a crucial role in maintaining periodical exchanges between the Society and the German scientific bodies during the war. The Philosophical Library was *the* primary scientific library in Cambridge at the time and those exchanges gave the researchers of Cambridge the opportunity to remain up to date with German science through the war years and for many years afterwards, when communications between the former enemies remained stilted. The Society strove to maintain scientific communication with colleagues abroad despite political barriers. For example, they sought to keep contact with 'the unhappy men of science in Russia' in the years following the Russian Revolution; as the country was ravaged by civil war, the Society resolved to send copies of their *Proceedings* and *Transactions* to Russian scientists and institutions.[38]

In the same letter to Einstein in which Eddington had described the crowds at the Philosophical Society, he also wrote that 'one feels that things have turned out very fortunately in giving this object-lesson of the solidarity of German and British science even in time of war'.[39] Eddington's observation of the eclipse was of tremendous scientific importance, but it also did much to repair the relationship between British and German scientists after World War I. Ernest Rutherford explained the feeling of the time:

> The war had just ended, and the complacency of the Victorian and Edwardian times had been shattered. The people felt that all their values and all their ideals had lost their bearings. Now, suddenly, they learnt that an astronomical prediction by a German scientist had been confirmed by expeditions...by British astronomers....An astronomical discovery, transcending worldly strife, struck a responsive chord.[40]

It was not an accident that many came to view the expedition in this way: Eddington had carefully crafted the expedition as one of international cooperation and he firmly believed that advocating peace, advancing international understanding, and forwarding the cause of science should all go hand in hand.

Late in 1919, just weeks after Eddington's talk had drawn such a large audience to the Philosophical Society, the Society celebrated its centenary. The occasion was marked with a nine-course dinner in the hall of Sidney Sussex College on 13 December—the same date that had seen the Society's first official meeting—with the dinner being interspersed with ten speeches.[41] The next day, the celebrations continued with a programme of scientific talks.[42] Following the war, the centenary was a good opportunity for the Society's fellows to think about reforms that might be made to the Society: there was talk of holding some meetings 'about a topic of general interest instead of one for original research' to attract more non-specialists and even members of the public; changing the time of some meetings; including more display objects to entice new audiences; altering the structure of the journals to increase circulation; and setting up a common room for fellows on the New Museums Site.[43] Slowly, some of these changes unfolded over the next decade, with particular emphasis on restructuring the journals and making some meetings more accessible. But efforts at further reform, and particularly making reforms that would have demanded monetary investment—such as a new common room—were stymied by financial worries.

The post-war years saw increased inflation in Britain, and the Society's finances suffered.[44] Up to the beginning of the war, the Society's outgoings had roughly matched their income, though they did not have much in reserve.[45] But, in 1920, the Society needed to appoint an emergency committee to examine their finances. The committee found that things 'were not as yet such as to cause serious anxiety' but suggested several cost-cutting measures, such as ceasing to send copies of *Proceedings* to associate members and making the annual dinner a biennial event.[46] But, by 1922, with Britain in recession, things had deteriorated and the Society's annual expenditure exceeded the annual income by £125. It was decided to raise the subscription fees for fellows and associates, to ask the scientific departments to contribute towards some of the running costs of the library, and to make an appeal to life members for assistance.[47]

The Society's two main sources of expense were its publishing endeavours and the upkeep of the library. But, though they were expensive, the periodicals and the library were central to the mission of the Society: the publications were 'a record of Cambridge research ... [and] a medium of exchange' while 'the importance of [the library could] hardly be overestimated'.[48] Though the University supported the library financially, notably through providing a home and staff for it, the Society had to bear the entire cost of binding the many periodicals which were sent to it each month. Binding costs rose from £50 in 1913 to £181 in 1920, meaning that the Society could only afford to bind the most popular journals. But the library was worth the expense, both in terms of the contents of the journals it housed and in terms of the money it saved the University. Of the 600 journals taken by the library in 1922, all but forty-five were received through the Society's programme of periodical exchange. Exchanging these journals rather than paying full market price for them meant that the University could save more than £500 each year.[49]

Over the next few years, the Society managed to alleviate its financial difficulties somewhat by its appeal to life fellows and through selling off some rare books.[50] Further relief came when Cambridge University Press wrote off £180 of debt and the Royal Society gave some money towards the costs of producing periodicals.[51] Membership subscriptions were raised slightly and, when women who had been associates were elevated to the status of fellow in 1929, their higher subscriptions caused another jump in revenue. By 1925, the Society was in a position to make some small investments and, by 1929, the deficit had been reduced to almost nothing.[52]

But perhaps the greatest contributor to the long-term financial stability of the Society was the reorganization of its journals following the war. By the turn of the century, the *Transactions* had become dedicated exclusively to mathematics and mathematical physics, while the *Proceedings* published across a much wider spectrum of subjects. The first two decades of the twentieth century saw experimental physics become the dominant subject in the *Proceedings*, but physiology and other topics were still well represented. But suddenly, in 1920, the *Proceedings* changed its focus to mathematics: of fifty-four papers published in Volume XIX, twenty-three were on pure mathematics. This change was brought about by G.H. Hardy.

Hardy was a natural mathematician, said to have been able to count into the millions when he was only two years old and amusing himself as a child in church by factorizing the hymn numbers. He came to Cambridge in 1896 and was fourth wrangler in the Mathematical Tripos in 1898; he became a fellow of Trinity College in 1900 and a fellow of the Cambridge Philosophical Society in 1901.[53] Though he had always been regarded as a fine mathematician, Hardy's career really took off in 1911 when he began to collaborate with John Littlewood, another fellow of Trinity. Together, Hardy and Littlewood would write a hundred papers on many different aspects of number theory and analysis. They would become the most important mathematicians in twentieth-century England, completely reviving analysis there.[54] Hardy saw mathematics as the most beautiful of all subjects, writing that

> [a] mathematician, like a painter or a poet, is a maker of patterns. If his patterns are more permanent than theirs, it is because they are made with *ideas*... The mathematician's patterns, like the painter's or the poet's must be *beautiful*; the ideas, like the colours or the words must fit together in a harmonious way. Beauty is the first test; there is no permanent place in the world for ugly mathematics.[55]

Hardy was only interested in the aesthetics of mathematics and he thought most practical and applied mathematics was 'rather dull'. He deliberately avoided any mathematical problems that might be considered of practical utility.[56]

In addition to working with Littlewood, Hardy also had a stunning, but tragically short-lived, collaboration with the Indian mathematician Srinivasa Ramanujan from 1913. Ramanujan, who had had no formal training in mathematics, sent Hardy some papers which so impressed the British mathematician that Hardy began making efforts to bring Ramanujan to Cambridge. Ramanujan spent several fruitful years as a fellow of Trinity College solving many supposedly 'unsolvable' problems, but sadly died young in 1920.[57]

Hardy's work with both Littlewood and Ramanujan spurred him on to greater achievements while making him more aware of the low state of British mathematics at that time. His awareness of the differences between British and European mathematics was also heightened by the

International Congress of Mathematicians, which met in Cambridge—under the auspices of the Philosophical Society—in 1912. Hardy served on the organizing committee and played a role in welcoming 400 mathematicians to Cambridge.[58] The meeting allowed him to see clearly how underdeveloped Cambridge mathematics—especially pure mathematics and analysis—was in comparison to its overseas counterparts.

In addition, 1912 was the year in which Hardy joined the council of the Cambridge Philosophical Society, acting as one of its secretaries until 1919. The secretaries had editorial responsibility for the Society's publications and thus Hardy was in a position to influence their contents. So it was that, following the war, that highly mathematical issue of the Society's *Proceedings* appeared. It included several papers by Hardy, by Ramanujan, and by Littlewood, amongst others. It also contained many papers which were written by other mathematicians but had been read by Hardy at meetings of the Society—either because the mathematician was seen as too young to present their own work or because they were absent from Cambridge. Hardy was clearly pushing his mathematical agenda hard at the Society. In a report to the Society's council in 1918, he wrote:

> [I]t is a rather distressing anomaly that Cambridge, while fully maintaining its position as the leading centre of mathematics in England, possesses no first-rate mathematical periodical. In the time of [G.G.] Stokes and [Arthur] Cayley our *Transactions* was perhaps the best such periodical in England, certainly second only to the *Phil. Trans.* of the Royal Society. Both of these papers have, so far as pure mathematics is concerned, almost ceased to count.[59]

Hardy's suggested remedy to this problem was to split the *Proceedings* into two journals: one for mathematics, astronomy, and physics, and one for biology, chemistry, and other sciences. He hoped that, by making the Society's publications more specialized, they would attract more mathematical papers and that those papers would be of a higher standard than the pre-war mathematics papers.

Though the Society's council decided not to split the *Proceedings* into two journals at that time, Hardy was given permission to experiment with his idea of creating a more specialized mathematical journal—resulting in that heavily mathematical volume of 1920. Others on the council also saw a need to reform the journals, and various suggestions were made: make *Transactions* a mathematical journal while *Proceedings* dealt with experimental subjects, make *Proceedings* exclusively biological, or discontinue *Transactions.*[60] Finally, in 1922, it was agreed that the Society should have a separate biological journal and so the journal that would come to be called *Biological reviews* was established, while the Society's *Proceedings* was reserved for mathematics and non-biological sciences.[61] *Biological reviews* began accepting papers in 1923 and the first issue was completed in 1925.[62] Meanwhile, after many years of debate, it was agreed to cease publishing the *Transactions* (which was, by this time, entirely mathematical in content) and to consolidate all mathematics in the *Proceedings.*[63] The final volume of the *Transactions,* which covered the years 1923 to 1928, appeared in 1931.

The restructuring of the Society's journals, and their division into biological and non-biological sciences, allowed them to thrive. The reputation of both journals grew as they published more specialized material. Crucially, the restructuring put both journals on a much more secure financial footing. Sales of *Biological reviews* rose quickly, and the number of periodical exchanges was increased.

Though Hardy had envisioned a purely mathematical future for the *Proceedings,* it continued to publish both mathematical and experimental physics papers (other subjects like chemistry and geology fell away as the 1920s progressed). Following the war, though struggling somewhat for lack of apparatus, the Cavendish began to regain its previous momentum, and many papers at meetings of the Philosophical Society were given by Cavendish researchers working on subatomic particles, radiation, quantum theory, and ideas linked to relativity.[64] In 1919, Ernest Rutherford was appointed Cavendish Professor. He had spent many years as a professor in Montreal and Manchester investigating radioactive decay and the different kinds of radioactive particles. In 1911, he had proposed that atoms have a nucleus, where most of the atom's mass and charge is centralized. In 1917, he discovered that the hydrogen nucleus is present in the other atoms—essentially discovering the proton. When he returned to Cambridge from Manchester in 1919, he continued Thomson's legacy of inspiring fundamental research into the nature of matter, and much of this research was first reported to meetings of the Cambridge Philosophical Society.

Rutherford himself gave several papers on alpha particles and radioactive elements, as well as presenting countless papers by his students.[65] George Paget Thomson, the son of J.J. Thomson and Rose Paget, gave a paper on anode rays—the positive rays sometimes produced in discharge tubes—and proposed that they were made up of ions.[66] James Chadwick, who had studied under Rutherford at Manchester and followed him to Cambridge in 1920, gave a paper on the beta rays produced by different kinds of radium.[67] J. Robert Oppenheimer, an American research student, presented his first paper to the Society in July 1926.[68] Paul Dirac's first paper—on the dissociation of gas molecules—was presented to the Society in March 1924 by his supervisor Ralph Fowler.[69] G.P. Thomson, Chadwick, Oppenheimer, and Dirac would all go on to become eminent physicists: Thomson would prove the wave nature of electrons in 1927 (work complementary to his father's proof that electrons are particles), Chadwick would discover the neutron in 1932, Oppenheimer would lead the Los Alamos Laboratory towards the creation of the atomic bomb, and Dirac would work out many of the key equations behind quantum mechanics.

These were just some of the students and researchers whose earliest work was given encouragement by the Cambridge Philosophical Society. Another important part of the Society's work was bringing new ideas to Cambridge, and to the anglophone scientific community. In 1923, it was proposed that the Society should publish Niels Bohr's papers on quantum theory, in translation. Bohr had studied in his native Copenhagen before coming to Cambridge as a research student in 1911, then travelling to Manchester to study under Rutherford, and finally returning to Copenhagen. During the 1910s, Bohr had developed his model of the atom, a model in which electrons orbit the nucleus in discrete energy levels. This work elegantly combined results from experimental work on electrons and other atomic particles with theoretical results from quantum physics. Bohr received many plaudits for his work, including an Honorary Fellowship of the Cambridge Philosophical Society. In the citation for this fellowship, the Society praised Bohr's contributions to mathematics, to physics, and to chemistry, calling his work 'beautiful' and 'brilliant'.[70] This appreciation for Bohr's work, coupled with the Society's dedication to promoting innovative ideas, made the Society seem like the natural place to publish a translation of Bohr's key papers into English. These appeared as a supplement to the *Proceedings* in 1924 and allowed a huge number of researchers access to Bohr's theory.[71]

Throughout the 1920s and 1930s, the Society continued to publish a variety of mathematical and physical papers in its *Proceedings*. These were written by Cambridge researchers—both established and just beginning—and also by overseas scientists who saw the *Proceedings* as a desirable place to publish. Pyotr Kapitsa, Ernest Walton, C.P. Snow, Erwin Schrödinger, Paul Erdős, and Fred Hoyle were just some of the illustrious names that appeared in the *Proceedings* in this period.

The scientific landscape of Cambridge had shifted subtly throughout the opening decades of the twentieth century as different branches of science had risen and fallen in popularity, as new results had created new avenues of research, and as assorted kinds of researchers had passed through the city's laboratories. The year 1933 saw the first of a new sort of scientist: the émigré. In April 1933, the new German National Socialist government passed a 'law for the re-establishment of the professional civil service'. The new law forbade 'non-Aryan' or 'unreliable' people from holding state jobs. As almost all university staff were state employees, this forced countless scientists from their posts. Many of these scientists, seeing no future for themselves in Germany, fled the country.[72] In Cambridge and many other university towns around the world, these refugees were welcomed into pre-existing scientific communities. Germany was a world leader in science in the 1930s, particularly in physics, and these émigrés were held in high regard in Britain. The Cambridge

Philosophical Society did its bit to help the new arrivals re-establish their work by ensuring that they had free access to the periodicals of the Philosophical Library.[73]

As the 1930s progressed, it became clear that the situation in Germany was worsening. In November 1938, a month after Germany had occupied the Sudetenland, and just days after Kristallnacht, the Cambridge Philosophical Society began making preparations for air raids 'in the event of war'. They started to move the contents of the Philosophical Library into basement storage, they arranged for a second copy of the library catalogue to be stored in another building, and they moved much of the stock of back issues of the *Proceedings, Transactions,* and *Biological reviews* into the cellars of the Cavendish Laboratory, the Zoology Department, and the Old Anatomy School.[74] When war did finally break out in the autumn of 1939, the Society resolved, as during World War I, to carry on its activities as normal—as far as it could. Publication of journals and exchanges of periodicals would continue as usual, and a handful of temporary bye-laws were added to the regulations to allow new fellows and associates to be elected in the absence of a quorum.[75] Minor inconveniences like paper shortages, press censorship, and the loss of library staff to war service could be overcome fairly easily.[76]

Cambridge was not as badly affected by the war as much of the country, though day-to-day life changed for its inhabitants. The University ceased making new appointments and froze new projects, some buildings were requisitioned for government use, and volunteers signed up for fire-watch duties, but undergraduate numbers did not drop as dramatically as during the previous war, and research was not so fiercely curtailed.[77] During the 1930s, it had come to be seen that modern physics might play a role in warfare, which is why the Cavendish was able to maintain many of its research programmes. To the same end, undergraduate teaching became much more concentrated on subjects like electronics, which were seen as potentially useful in the development of radar systems.[78]

The British government was very interested in the possibilities of radar, and work on it was heavily funded in the years leading up to the war. The scientists who were fleeing continental Europe and taking up positions in British laboratories were often expressly forbidden from working on radar projects, as radar was considered so central to the war effort. One such scientist who was banned from researching radar was Rudolph Peierls. Peierls, a German citizen with a Jewish father and a Catholic mother, had been studying in Cambridge when Hitler came to power in 1933, and was granted leave to remain in Britain. He had previously studied in several German universities—Berlin, Munich, and Leipzig—before continuing his postgraduate studies in Zurich (where he worked with Wolfgang Pauli), Odessa, and Rome. He became an expert on solid state physics, working on the application of quantum mechanics to questions about electrons in metals and semiconductors. Once granted permission to stay in Britain, Peierls spent some time in Manchester (where a fund had been set up to enable refugee scientists to continue their research) before returning to Cambridge.[79] He began working in the Mond Laboratory, which was a sister laboratory to the Cavendish, set up in 1933. He was elected an associate of the Cambridge Philosophical Society in 1935 and a fellow in 1936.[80]

Late in 1938, a series of experiments was carried out by Otto Hahn and Fritz Strassmann in Berlin in which atoms of uranium were bombarded with neutrons, resulting in the release of barium atoms and large amounts of energy. The release of barium was unexpected and so Hahn sent his results to his former colleague Lise Meitner (an Austrian Jew who had fled to Sweden) to ask her opinion on what was happening to the uranium atoms. Meitner, together with her nephew Otto Frisch, correctly interpreted and explained the results early in 1939. This, they said, was nuclear fission: the uranium atoms were splitting into two smaller atoms of approximately equal size when hit by neutrons, releasing energy and further neutrons. Within a few weeks of the discovery of nuclear fission, attention turned to the large amount of energy being released and its potential uses. It was quickly realized that there might be a possibility for a 'chain reaction' in which the neutrons resulting from one fission event could trigger further atomic fissions, releasing more energy and further neutrons, and so on. A chain reaction would be able to rapidly unleash vast quantities of energy.

Given the political situation in Europe, thoughts turned to using this energy for military purposes—a bomb unlike anything seen before. First, the question of 'critical mass' had to be worked out—just how much uranium was needed to sustain a chain reaction? The French physicist Francis Perrin estimated that the critical mass might be as much as 44 tonnes—a block of uranium about 3 metres wide.[81] This was an enormous amount and made the idea of a nuclear fission weapon seem completely unfeasible. But Rudolph Peierls was not convinced by Perrin's calculations. In the summer of 1939, Peierls decided to make his own calculation. Peierls clearly thought of this as an academic problem, rather than a practical one: he casually wrote that the problem of the critical mass of uranium 'seems of some interest' and he sent his completed paper to the Cambridge Philosophical Society for publication. The Society received the paper in June 1939, and agreed with Peierls that the problem was indeed an interesting one—they published it later that year just as war was breaking out.[82] Peierls used a series of equations to calculate a critical mass that was considerably smaller than Perrin's, but still, he thought, too large to be feasibly manufactured. Peierls later explained:

> This size was of the order of tons. It therefore appeared to me that the paper had no relevance to a nuclear weapon. The whole mass, quite apart from anything else, would have been the size of the present Windscale reactor. There was of course no chance of getting such a thing into an aeroplane, and the paper appeared to have no practical significance.[83]

By the time the war started, Peierls had moved to Birmingham to take up a chair. The Austrian Otto Frisch, who had been one of the first to study and confirm the existence of nuclear fission, had been visiting Birmingham when war broke out and was stranded there, unable to return to Copenhagen, where he had been working with Niels Bohr. Peierls invited Frisch to stay with him and the two men quickly became close collaborators and developed a warm friendship. Together, discussing the equations that

Peierls had published in the Cambridge *Proceedings*, Frisch and Peierls began to realize that there was another way to interpret them.

There are two commonly occurring types of uranium: the isotope $^{235}U$ has 92 protons and 143 neutrons in its nucleus, and is highly fissionable; the isotope $^{238}U$ has 92 protons and 146 neutrons, and was generally considered non-fissionable. There are also two kinds of neutron bombardment that could lead to fission: fast and slow. This meant that there were four possible combinations that might lead to fission:

Slow-neutron bombardment of $^{238}U$
Fast-neutron bombardment of $^{238}U$
Slow-neutron bombardment of $^{235}U$
Fast-neutron bombardment of $^{235}U$

Fission by the first of these had not been observed and it was assumed that $^{238}U$ could not be fissioned by slow neutrons. Fission did occur by the second method but was not an efficient way to produce energy. The third produced enough energy to generate electricity but was thought too slow for use in a weapon. The fourth had never been seriously considered, as $^{235}U$ makes up less than 1 per cent of the uranium found in nature.[84] But Frisch, who had been working on ways to separate the two isotopes of uranium and was confident that he could produce enough $^{235}U$ to experiment upon, began to think about the fourth option. He went to Peierls for help, and together the two men recalculated the critical mass of uranium needed for a bomb.

Peierls's formula required the 'cross section' of fast-neutron fission of $^{235}U$ to be plugged into it—the cross section represents the probability of a fission taking place when $^{235}U$ is bombarded by fast neutrons. No one had calculated this experimentally but Frisch and Peierls were confident that there was a high probability of fission; as Peierls put it, 'if a neutron hit the [$^{235}U$] nucleus something was bound to happen'.[85] Peierls and Frisch picked what they considered a plausible cross section—'just sort of playfully'—and plugged it into Peierls's equations.[86] To their astonishment, they calculated a critical mass that was significantly smaller than the several tonnes that Peierls had first calculated; they found that only about a kilogram of $^{235}U$ would be enough to make a nuclear bomb of unimaginable power. With a material as dense as uranium, that would mean a volume less than the size of a small apple.

When Peierls had published his paper in the *Proceedings*, he had no idea that it might contain the information needed to prove that making an atomic bomb was a realistic possibility but, as soon as he realized the power of the information, he and Frisch began writing a secret memorandum for the British government. Ironically, these two friends had been able to spend time working on the problem of nuclear fission because, as foreign nationals, they had been excluded from working on war-time projects such as radar. As both had worked extensively in the German-speaking world, they knew that their German counterparts were far advanced in nuclear physics and were extremely worried that Hitler might already be directing research towards the construction of a nuclear weapon. In their memorandum to the British government, Peierls and Frisch described the possibilities of such a weapon:

> The attached detailed report concerns the possibility of constructing a 'super-bomb' which utilises the energy stored in atomic nuclei as a source of energy. The energy liberated in the explosion of such a super-bomb is about the same as that produced by the explosion of 1,000 tons of dynamite. This energy is liberated in a small volume, in which it will, for an instant, produce a temperature comparable to that in the interior of the sun. The blast from such an explosion would destroy life in a wide area. The size of this area is difficult to estimate, but it will probably cover the centre of a big city.
>
> In addition, some part of the energy set free by the bomb goes to produce radioactive substances, and these will emit very powerful and dangerous radiations. The effects of these radiations is greatest immediately after the explosion, but it decays only gradually and even for days after the explosion any person entering the affected area will be killed.
>
> Some of this radioactivity will be carried along with the wind and will spread the contamination; several miles downwind this may kill people.[87]

The British government responded by setting up the MAUD Committee—peopled almost entirely by staff or former staff of the Cavendish

Laboratory—to investigate the practical implications of the theoretical ideas proposed by Peierls and Frisch. From the MAUD Committee grew the deliberately dull-sounding Tube Alloys Project to develop a British nuclear weapon. In America, another set of émigrés—the Hungarians Eugene Wigner and Leó Szilárd (who had first postulated the idea of a nuclear chain reaction) and the German Albert Einstein—was urging the government there to consider the power of nuclear physics to build weapons. Though the weapon they initially speculated about was too large to be carried in an aircraft, their lobbying eventually led to the creation of the Manhattan Project.

After national governments began to take the possibility of a nuclear weapon seriously, the number of published articles about nuclear physics fell sharply and remained low throughout the remainder of the war. Scientists were self-censoring, and government regulations required that potentially sensitive papers be checked by officials before publication. As during World War I, the number of scientific papers overall fell as scientists were redeployed to war duties or covert research operations. Volumes of the Society's *Proceedings* became thinner as the war progressed, and experimental papers gave way to pure mathematical ones and to theoretical subjects such as cosmology.

Happily, the significance of Peierls's paper on critical mass was not picked up by German scientists. In 1943, Peierls moved to America to join the Manhattan Project, bringing with him his assistant Klaus Fuchs. Fuchs, though never a member of the Cambridge Philosophical Society, had published several papers in its *Proceedings*, presumably at Peierls's suggestion.[88] Fuchs was regarded as a fine physicist, and a particularly skilled theoretician. Peierls and Fuchs worked at the Los Alamos Laboratory until the end of the war. Some years after the war ended, just as the Cold War was intensifying, Fuchs was revealed to be a Soviet spy. He had been passing information about the creation of uranium and plutonium bombs to Russian agents.[89]

Physics had changed beyond recognition in just a few decades, from an esoteric and largely academic pursuit to a state secret that could affect the lives of millions. In the early years of the twentieth century, when J.J. Thomson's comic toast of 'the electron: may it never be of any use to anybody' used to ring out at Cavendish dinners, few could imagine how soon those days of innocence would end. The Cambridge Philosophical Society was present at many of the most significant moments of early twentieth-century science: from Thomson's first inklings that there existed subatomic particles to the invention of x-ray diffraction and the cloud chamber; through two devastating world wars and the changes they wrought on academic communities, leading to heated debates about sharing scientific information across national borders; and from financial instability to increased scientific specialization. Though meetings of the Society heard reports of many important experiments and though the Society published countless new results, perhaps the Society's single most important contribution to science in this period was giving a voice to younger researchers, those who had not yet established themselves enough to be invited to the national societies and whose ideas were still works in progress. Lawrence Bragg, Charles Wilson, George Thomson, James Chadwick, Paul Dirac, Ernest Walton, and Fred Hoyle are just some of the giants of twentieth-century physics who found an early welcome in Cambridge's Philosophical Society.

# 9

# FOLLOWING THE FOOTSTEPS

*Haootia* is a tangled phantom of a beast. Its sinuous arms fold one upon another, its exposed muscles ripple this way and that: all the glorious chaos of life frozen forever in a bed of dull grey siltstone. *Haootia* is thought to be one of the world's oldest animals. It lived and died some 560 million years ago, and the impression of its twisted remains was found captured in rock on the coast of Newfoundland in 2011. Its name, meaning 'demon', comes from the Beothuk language of that island, a language that has also died out, leaving only a few unconnected words to testify to a vanished world.

*Haootia* lived in the waters off Avalonia. Not the island of Arthurian legend, but an ancient microcontinent named after the very real and rugged Avalon Peninsula at the south-eastern end of Newfoundland, not far from the Bonavista Peninsula on which *Haootia* was discovered. These exposed peninsulas, which once lay at the bottom of the sea and have since been thrust up to make dry land, abound with strange fossils, including many mysterious frond-like forms as big as half a metre and more. These fossils are evidence of thriving communities of organisms that existed millions of years ago on a continent that no longer survives. The Avalonian rocks and the fossils they bear date from a time known as the Ediacaran period, which lasted from about 635 million years ago until the start of the Cambrian period, 541 million years ago. The Cambrian period, which was first identified and named by Adam Sedgwick, is when large-scale complex animal life on Earth is commonly supposed to have begun. It is during the Cambrian that the world is said to have *exploded* with life. For the first time, so the story goes, large organisms appeared that shared many of the features that we recognize in animals today. Almost every major animal phylum can be traced to this period. The Cambrian explosion is said to be a defining moment in the evolution of life on our planet, and it is recorded in exquisite detail in fossils such as those of the famous Burgess Shale.

So what are probable animals such as *Haootia* doing in rocks that predate the Cambrian explosion by up to 40 million years? Could it be that the neat story told of all animal life appearing in a sudden dramatic burst is not quite right? That is what many scientists, including Alexander Liu, part of the team that first scientifically described *Haootia*, believe.

Liu and his colleagues published their description of *Haootia* in 2014, while Liu was a Henslow Fellow at Girton College, Cambridge.[1] The Henslow Fellowship is a research scheme set up by the Cambridge Philosophical Society in association with several Cambridge colleges in 2010 to continue the mission propounded by Sedgwick and Henslow all those years ago in 1819: 'to keep alive the spirit of inquiry'.[2] In many ways, the world of science is very different now from that in the days of Henslow and Sedgwick. Science no longer involves individuals striding out, exploring the world's frontiers, with little more than their passion and some simple apparatus. Today, science is an international enterprise, involving teamwork and collaboration, and often complex and expensive equipment, paid for by large funding bodies. Liu worked with an international team of colleagues from the University of Oxford and the Memorial University of Newfoundland; their funding came from a variety of sources, including the Canadian and British governments, as well as private philanthropic and charitable institutions; their techniques included things, like high-resolution photography, that would have been unimaginable a few generations before. Still, there is much of Liu's work that Sedgwick and Henslow would have recognized: the scrambling about on exposed rocks in all weathers, the difficulties of trying to tell a trace of ancient life from a hundred other marks on a surface, and the thrill of discovery.

Liu, who was a doctoral student at Oxford University when the fossil was first discovered, tells me of how he and fellow students had walked over the outcrop containing *Haootia* several times before their supervisor, the Oxford palaeobiologist Martin Brasier, spotted the strange rippling form embedded in the surface and drew their attention to it (see Figure 53). On a dull day, on dull, dry grey rock, the marks are easy to miss. But pour a little water over them, add a little oblique light, and something begins to reveal itself. To the untrained eye, *Haootia* is a confused jumble. But to Liu, Brasier, and their colleagues, it was immediately apparent that they were looking at something remarkable. In particular, those wonderfully preserved twisted strands arranged into little bundles are strikingly like muscle fibres. If their interpretation is right, Liu et al. have discovered not just one of the oldest known examples of an animal, but the oldest known example of a muscle.[3] Reconstructed, the creature resembles nothing so

**Figure 53** *Haootia quadriformis*, at ~560 million years old considered one of the earliest known examples of an animal, and perhaps the earliest known example of muscle fibres in the fossil record.

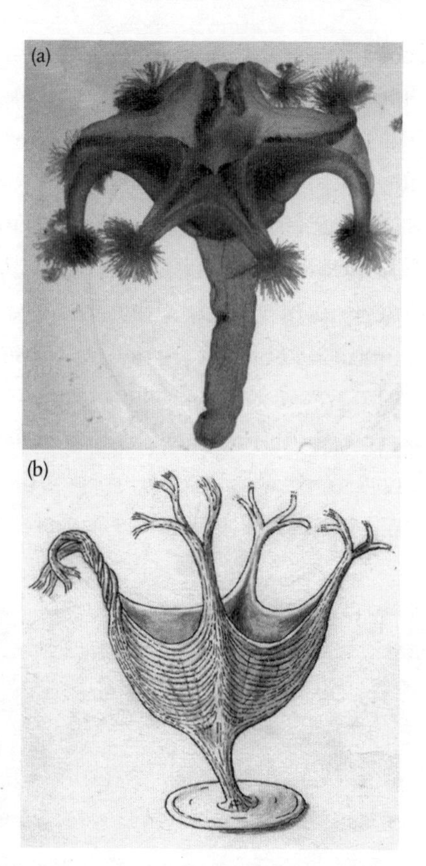

**Figure 54** Artist's impression of *Haootia*—it is believed the creature would have resembled a stalked jellyfish.

much as a modern stalked jellyfish which anchors itself to the sea floor, while waving its many tentacles through the water to feed itself (see Figure 54).

*Haootia* is just one of many kinds of fossils known from the Ediacaran time. The first Ediacaran fossils were discovered by Reginald Sprigg in 1946 in the Ediacara Hills on the north side of the Flinders Range, north of Adelaide, when he was looking for minerals. In the decades since their discovery, similar fossils have been found in ancient rocks across the world. But, in the nineteenth century, Charles Darwin had puzzled over the apparent lack of fossils in rocks earlier than the Cambrian. They should be there, according to his theory, but no one had ever found any. Therefore, he must have been elated to hear that the palaeontologist John William Salter had discovered fossils in rocks in Shropshire, England, which were thought 'azoic'—that is, rocks that were so old that it was not believed they could contain traces of life. Salter had once worked alongside Adam Sedgwick on the Woodwardian Cabinet before joining the young Geological Survey in 1846. That dramatic diagram showing how plant life had evolved on Earth (Figure 19) was by Salter, published in one of the many papers he wrote for the *Transactions* and *Proceedings of the Cambridge Philosophical Society*. In the 1850s, he was working in Shropshire, in the moorland plateau of the Long Mynd, when he saw what he thought might be 'traces of marine worms' as well as the remains of some kind of crustacean in an outcrop.[4] It was an extraordinary find, and Darwin cited the fossils in *On the origin of species*.[5] The veracity of Salter's discovery was debated, and the markings have since been reinterpreted as the result of natural mechanical processes and microbial action, but certainly his work encouraged the search for large-scale life in the Precambrian.[6]

Today, Liu, and another Henslow Fellow, Emily Mitchell, who works on the ecology of these ancient communities, are supported by the Cambridge Philosophical Society. One cannot help but feel that Adam Sedgwick and John Stevens Henslow would very much approve of the research that is now being sponsored by the Society they set up two hundred years ago—the Society that was first conceived as they chipped fossils from the rocks on the Isle of Wight, and tried to imagine the world as it had once been, millions of years earlier. Just as Sedgwick and Henslow tried to untangle the strata, Liu has been involved in untangling the limbs of *Haootia*, and Mitchell seeks to untangle the webs of interaction between the Ediacaran sea-creatures. Their work joins that of scientists worldwide to give us new ideas about early evolution, and a growing understanding of the history of life on Earth.

The Henslow Fellowship Scheme was inaugurated in 2010 to promote research across the sciences, and to facilitate the communication of the results of scientific research.[7] As with so much of the Society's work over the centuries, the fellowships benefit researchers in the early stages of their careers, allowing blue-skies thinking across a range of disciplines. Henslow Fellows have done everything from mapping the brain to developing the next generation of optical communication links, from trying to understand the processes of ageing to seeking to explain the properties of matter using quantum mechanics, and from tracking molecular changes in acute head injury patients to measuring the interactions of molecules within cells.[8] Like many of their forebears in the Society, they seek order amongst the seeming chaos of the natural world. But the ways these researchers operate, as part of vast international efforts using expensive technology, would be almost unrecognizable to the early fellows of the Society.

The sciences underwent dramatic changes through the twentieth century. Science was no longer the pursuit of solitary gentlemen in private studies or workrooms. With the creation of research laboratories like the Cavendish in the late nineteenth century, scientific research had come to be seen as a group activity, with many minds working together to solve a single problem. The Cavendish and similar laboratories were just the start of this trend towards ever larger undertakings. The twentieth century saw a trajectory towards increasingly sophisticated and costly projects, with bigger and bigger teams behind them. The Manhattan Project is commonly seen as the culmination of this trend towards so-called big science: a state-backed project with a single utilitarian aim—the creation of a nuclear weapon. It is the best-known example of big science and, though it seems extraordinary, in fact many aspects of it have been assimilated into ordinary, everyday science. Science worldwide has been deeply affected by increased funding and space, and its place in society has changed as it has come to be associated with authority and utility.

Cambridge, as much as anywhere, has felt these changes. It is now one of the leading centres for the sciences, not just in the UK but in the world. The University has gone from grudgingly awarding just a handful of science degrees each year in the 1850s to training thousands of undergraduates and graduate students, as well as facilitating countless postdoctoral and senior researchers today. Cambridge science is no longer bound within the walls of the University or colleges; the so-called silicon fen which has grown up around Cambridge employs tens of thousands of people in high-tech jobs. Money pours in from national research councils, private donors, and specialist charities. Where once Cambridge had one scientific society, it now has countless discussion groups, seminars, clubs, and associations. Where once Cambridge had one scientific journal, there are now hundreds of journals produced by the University Press or edited by its academics. The Society's library, its museum, and its reading room—once singular things within the city—now have dozens of modern counterparts.

The development of science in Cambridge is closely linked to the story of the Cambridge Philosophical Society. As the Society rose, the University changed. At the same time, the Society was itself subject to outside influences: greater global events work their way through the Society's minute books, showing how events trickle down from the national to the local level. Just as during World War I the Society strove to maintain ties to European colleagues, during the Cold War, the fellows (many of them left-leaning) tried to continue their relationship with Russian scientists, even arranging for the *Mathematical proceedings* to be printed in the USSR.[9] The Society's records tell of war and peace, booms and busts, the arrival of more and more women in Cambridge, the internationalization of science, and the advent of computing (much of the work on the latter being done by the pioneering computer scientist Maurice Wilkes, Treasurer of the Society from 1946 to 1958) (see Figure 55).

Changes in the way science has been communicated can also be traced through the Society's minute books. The earliest library grew from a collection of individually donated books to being the largest systematic science library in Cambridge, keeping itself at the centre of the city's scientific life by developing its enormous network of periodical exchanges (see Figure 56). In the 1960s, the library, known simply as the Philosophical Library to many generations, became the Scientific Periodicals Library—a more modern-sounding name that reflected the

**Figure 55** Maurice Wilkes with EDSAC (Electronic Delay Storage Automatic Calculator)—one of the earliest electronic-digital stored-programme computers. Wilkes was Treasurer of the Cambridge Philosophical Society from 1946 to 1958.

fast-paced periodical-based scientific publishing of the time. In the 1970s, the library was wholly assimilated into the University Library; this allowed the University to finally realize its ambition of having its own comprehensive scientific library, while also providing greater stability for the library's staff as they linked up with their peers across Cambridge's many libraries. In 2001, the mathematics section was separated from the main body of the library and rehoused in a new Mathematics Library—a sign of increasing specialization in the sciences. In 2004, what remained of the Scientific Periodicals Library became the Central Science Library. And, in 2015, as a redevelopment project began on the New Museums

**Figure 56** The Philosophical Library, New Museums Site. The Library moved to the New Museums Site in 1865 and remained there (under a succession of different names) until 2015.

Site, the periodical exchange programme, which had run for almost two hundred years, was halted, and the Library was closed, its collections disbanded.[10] This final step speaks to the power of electronic communication in the sciences today: though print journals are still produced, the closure of the physical library will hardly have hampered researchers' ability to access the journals they need. Though some might mourn the passing of the old library with its lovely wooden galleries and wrought iron spiral staircases, science moves on, and the Society with it. The story of the library could be the story of science, or the story of communication—changing and evolving, without a fixed endpoint.

There have been other changes in the Society too. Its meetings, still held every second Monday evening during term, as they have been since 1819, have become more accessible. The public are invited in to hear about the latest research going on in the laboratories of Cambridge, but now the meetings take the form of polished lectures: they are more likely to be the culmination of many years' work rather than hastily prepared works in progress. This change traces its roots back to 1919, when Ernest Rutherford, then the Society's vice-president, suggested holding an occasional meeting that focused on discussion and exhibitions as a way to make science more accessible.[11] This was also the time when the Society's journals were becoming more specialized, and perhaps Rutherford's suggestion was a reaction to that: a desire to ensure that the public could still engage with new science. From then on, the Society has always ensured that at least some of its meetings appeal to non-specialists. By the closing decades of the twentieth century, the majority of the Society's meetings were dedicated to disseminating scientific knowledge to the public rather than to sharing knowledge between peers.

As the meetings have become more general, the Society's publications have become more specialized. *Mathematical proceedings* (renamed from the original *Proceedings* in 1975) and *Biological reviews* both continue to be high-impact journals in their areas, and both have become increasingly technical as the subjects they represent have branched and sub-branched into ever more discrete fields. There is no longer any formal link between the talks given at meetings and the content of the Society's publications—a reflection of how science communication has changed.

One unforeseen consequence of the success of the journals is that it has, finally, given the Society the financial stability it lacked in earlier years. Since the late 1980s, the revenue from the journals has increased significantly. This has not just allowed the Society to maintain itself; it has enabled it to move away from the extremely cautious financial policy it adopted in the mid-twentieth century, towards a more tactical investment policy. These investments fund the Society's work and secure its future. As well as the Henslow Fellowships, the Society supports final-year doctoral students as they complete their theses, provides travel grants for students and early career researchers, and funds mathematicians who wish to study in Cambridge, as well as schoolteachers who want to encourage their pupils into research. The programme of honorary fellowships, which had lapsed in the 1970s, has been reinstituted. The Society still awards the William Hopkins Prize (for research in physical or geological sciences), to which has been added, since the 1960s, the William Bate Hardy Prize for biological sciences. Membership has grown into the thousands.

In material terms, almost nothing is left of the early Society. Gone are its house, its museum, and its library. A handful of artefacts remain: the astronomical clock which adorned the old reading room; the charter and the seal; and some of the beautifully bound rare books which once formed the heart of the library. But the Society has always been much more than a physical place or a collection of interesting things. It has been a community—a place where ideas can be shared, talent can be nurtured, and minds can meet. At its heart, it does the same thing today that it has always done: it facilitates science for its own sake.

No one could have predicted where the young Darwin's observations would lead when members of the Society first listened to his letters being read aloud in the flickering gaslight of the meeting room in All Saints' Passage, no one could have guessed that Lawrence Bragg's short paper on how x-rays diffract through crystals would be the key to unlocking the secrets of DNA by two later fellows of the Society—Francis Crick and James Watson, and no one could have foreseen that Charles Wilson's obsession with clouds would reveal the workings of the subatomic world. The Society was, and is, a place to think and a place in which thoughts can be shared, and its support for young researchers, those whose ideas are still nebulous, has ever paid dividends for science. For it is often from a ferment of uncertainty that the most interesting concepts will grow.

The Cambridge Philosophical Society was the parent of modern science in the city, a city that is now one of the most sophisticated centres for science in the modern world. The Society was the first dedicated space for science in a place that once saw the sciences as, at best, a pleasant distraction for its students—young men who would probably go on to be country clerics or minor aristocrats. Its fellows agitated for reform of the curriculum, for the creation of science degrees, and for scientific libraries, laboratories, and museums, at a time when such things seemed irrelevant to many; and today Cambridge abounds with them. The Society persuaded the University to take science seriously. Its vision has been realized: it has been so successful in creating the idea of Cambridge as a place for science that the many scientific spaces within Cambridge now greatly overshadow the original institution. The spirit of inquiry which drove Adam Sedgwick and John Stevens Henslow as they clambered about the rocks of the Isle of Wight two hundred years ago is now reproduced in many ways and a hundredfold across the city. The Philosophical Society has become just a small part of the vast landscape of Cambridge science—and that is the true mark of its success.

But this is not just a story about that fenland town which has grown to be a giant in the world of science. It is a story which goes far beyond Cambridge; it is a kind of microcosm which can be used to explain the rise of science nationally and internationally, to understand what it is, why we do it, and where it can lead us.

# ENDNOTES

## Front Matter

1. For information on the history of ancient natural philosophy, see Stephen Gaukroger *The emergence of a scientific culture* Oxford: Clarendon Press 2006 chapter 6. On natural philosophy as an essentially religious activity see Andrew Cunningham 'How the *Principia* got its name; or, taking natural philosophy seriously' *History of science* XXIX (1991) pp. 377–92, pp. 380–1; for a critique of Cunningham, and a discussion of the relationship between natural philosophy and science, see Edward Grant *A history of natural philosophy* Cambridge: Cambridge University Press 2007 chapter 10.
2. On the bilateral influence of natural philosophy and the exact sciences, see Edward Grant *A history of natural philosophy* Cambridge: Cambridge University Press 2007 chapter 10.
3. S.J.M.M. Alberti 'Natural history and the philosophical societies of late Victorian Yorkshire' *Archives of natural history* XXX (2003) pp. 342–58, p. 343; this article also has a good discussion of provincial philosophical societies in the late eighteenth and early nineteenth centuries. On the moral aspects of natural philosophy, see James A. Secord *Visions of science* Oxford: Oxford University Press 2014 p. 81 and Steven Shapin *The scientific life* Chicago: University of Chicago Press 2008 pp. 24–6. On the moral aspect of philosophical societies, see Roy Porter 'Science, provincial culture and public opinion in Enlightenment England' in Peter Borsay (editor) *The eighteenth-century town* London and New York: Longman Group 1990 pp. 243–67, p. 257.
4. For a discussion of the most popular topics of discussion in provincial philosophical societies, the books most borrowed from their libraries, and their teaching activities, see S.J.M.M. Alberti 'Natural history and the philosophical societies of late Victorian Yorkshire' *Archives of natural history* XXX (2003) pp. 342–58, pp. 346–50.
5. The Society's archive was catalogued by Joan Bullock-Anderson. Most of the archive is held in the Whipple Library, Department of the History and Philosophy of Science, University of Cambridge, with a small number of objects remaining in the Society's offices. The catalogue is available online via the Janus database.
6. There have been two previous histories of the Society. The first was written in 1890 by John Willis Clark and published in the Society's *Proceedings* in 1892. Clark was the son of Mary Willis and William Clark. William Clark had once been Professor of Anatomy and President of the Cambridge Philosophical Society and he and Mary had worked together to build up the University's anatomy museum (see Chapter 3 of this book on William Clark's involvement with the Society); John's

uncle Robert Willis was a former Jacksonian Professor of Natural Philosophy and another former president of the Society (see Chapter 2 of this book for more on Willis). John Willis Clark, whose interests veered more towards architectural history, was himself President of the Society when he wrote his brief history, which focused only on its first few decades: John Willis Clark 'The foundation and early years of the Society' *Proceedings of the Cambridge Philosophical Society* VII (1892) pp. i–l.

The second history was written by A. Rupert Hall in 1969 to mark the Society's 150th anniversary. This very short book almost entirely omits twentieth-century history, as Hall did not wish to discuss the work of persons still living: A. Rupert Hall *The Cambridge Philosophical Society: a history 1819–1969* Cambridge: Cambridge Philosophical Society 1969.

The Cambridge Philosophical Society is strangely absent from many histories of Cambridge science, and it is particularly notable that Susan Cannon's *Science in culture* with its meticulous dissection of 'the Cambridge network' almost entirely neglects the Society; see Susan F. Cannon *Science in culture: the early Victorian period* New York: Dawson and Science History Publications 1978 and Walter F. Cannon 'Scientists and broad churchmen: an early Victorian intellectual network' *Journal of British studies* IV (1964) pp. 65–88.

## Chapter 1

1. Adam Sedgwick 'On the geology of the Isle of Wight' *The annals of philosophy* III (1822) pp. 340–2.
2. S. Max Walters 'John Stevens Henslow (1796–1861)' *Oxford dictionary of national biography* Retrieved 2 March 2018 from http://www.oxforddnb.com/view/10.1093/ref:odnb/9780198614128.001.0001/odnb-9780198614128-e-12990.
3. Robert Fox 'John Dawson (bap. 1735–1820)' *Oxford dictionary of national biography* Retrieved 2 March 2018 from http://www.oxforddnb.com/view/10.1093/ref:odnb/9780198614128.001.0001/odnb-9780198614128-e-7350. Dawson taught eleven (or perhaps twelve) boys who would go on to be 'senior wrangler' (i.e. the highest-ranked student in mathematics in Cambridge) between 1781 and 1807.
4. John Willis Clark and Thomas McKenny Hughes *The life and letters of Adam Sedgwick, volume I* Cambridge: Cambridge University Press 2009 p. 71.
5. R.M. Beverley *A letter to His Royal Highness the Duke of Gloucester* London 1833 p. 14. The author of the pamphlet from which these quotations were taken was Robert Mackenzie Beverley, a Yorkshire man who matriculated at Trinity College in 1816. This rather hysterical pamphlet was filled with wild allegations about moral lapses in the University. While some allegations may have been unfounded, his descriptions of undergraduate 'wine-parties' were probably based in reality.
6. George Gordon, Lord Byron 'Thoughts suggested by a college examination' *Hours of idleness* Newark 1807 pp. 113–14.
7. William Ainger to Adam Sedgwick, 11 April 1807, quoted in John Willis Clark and Thomas McKenny Hughes *The life and letters of Adam Sedgwick, volume I* Cambridge: Cambridge University Press 2009 p. 87.

8. John Willis Clark and Thomas McKenny Hughes *The life and letters of Adam Sedgwick, volume I* Cambridge: Cambridge University Press 2009 pp. 86–7.
9. Colin Speakman *Adam Sedgwick: geologist and dalesman* Broad Oak: Broad Oak Press 1982 p. 49.
10. Adam Sedgwick to William Ainger, 19 February 1808, quoted in John Willis Clark and Thomas McKenny Hughes *The life and letters of Adam Sedgwick, volume I* Cambridge: Cambridge University Press 2009 pp. 96–7.
11. Adam Sedgwick to Fanny Hicks, 12 August 1854, quoted in John Willis Clark and Thomas McKenny Hughes *The life and letters of Adam Sedgwick, volume I* Cambridge: Cambridge University Press 2009 p. 202. Sedgwick had met this woman in Derbyshire in 1818. Another letter to his niece reveals that the lady in question went on to marry a goldsmith in Glasgow and had at least one son.
12. John Willis Clark and Thomas McKenny Hughes *The life and letters of Adam Sedgwick, volume I* Cambridge: Cambridge University Press 2009 pp. 142–7.
13. J.M. Levine 'John Woodward (1665/1668–1728)' *Oxford dictionary of national biography* Retrieved 2 March 2018 from http://www.oxforddnb.com/view/10.1093/ref:odnb/9780198614128.001.0001/odnb-9780198614128-e-29946. Woodward's cabinets of fossils can be seen today in the Sedgwick Museum of Earth Sciences.
14. Roy Porter 'John Woodward: "a droll sort of philosopher"' *Geological magazine* CXVI (September 1979) pp. 335–343, p. 342.
15. John D. Pickles 'John Hailstone (1759–1847)' *Oxford dictionary of national biography* Retrieved 2 March 2018 from http://www.oxforddnb.com/view/10.1093/ref:odnb/9780198614128.001.0001/odnb-9780198614128-e-11874.
16. Adam Sedgwick to William Ainger, 19 March 1818, quoted in John Willis Clark and Thomas McKenny Hughes *The life and letters of Adam Sedgwick, volume I* Cambridge: Cambridge University Press 2009 p. 153.
17. John Willis Clark and Thomas McKenny Hughes *The life and letters of Adam Sedgwick, volume I* Cambridge: Cambridge University Press 2009 pp. 155–8.
18. J.A. Secord 'Adam Sedgwick (1785–1873)' *Oxford dictionary of national biography* Retrieved 2 March 2018 from http://www.oxforddnb.com/view/10.1093/ref:odnb/9780198614128.001.0001/odnb-9780198614128-e-25011; Colin Speakman *Adam Sedgwick: geologist and dalesman* Broad Oak: Broad Oak Press 1982.
19. John Willis Clark and Thomas McKenny Hughes *The life and letters of Adam Sedgwick, volume I* Cambridge: Cambridge University Press 2009 pp. 153–4; Colin Speakman *Adam Sedgwick: geologist and dalesman* Broad Oak: Broad Oak Press 1982 p. 55.
20. Adam Sedgwick to William Ainger, 23 October 1818, quoted in John Willis Clark and Thomas McKenny Hughes *The life and letters of Adam Sedgwick, volume I* Cambridge: Cambridge University Press 2009 pp. 200–1.
21. S.M. Walters and E.A. Stow *Darwin's mentor: John Stevens Henslow, 1796–1861* Cambridge: Cambridge University Press 2001 pp. 22–3.
22. On the state of the geology museum, see *Report of the inspectors*, 7 May 1819, quoted in John Willis Clark and Thomas McKenny Hughes *The life and letters of Adam Sedgwick, volume I* Cambridge: Cambridge University Press 2009 p. 205.

23. Leonard Jenyns *Memoir of the Rev. John Stevens Henslow, M.A., F.L.S., F.G.S., F.C.P.S.: late rector of Hitcham and Professor of Botany in the University of Cambridge* London: John Van Voorst 1862 pp. 13, 17; John Willis Clark and Thomas McKenny Hughes *The life and letters of Adam Sedgwick, volume I* Cambridge: Cambridge University Press 2009 p. 204.
24. Jack Morrell and Arnold Thackray *Gentlemen of science: early years of the British Association for the Advancement of Science* Oxford: Clarendon Press 1981 pp. 12–13.
25. John Willis Clark 'The foundation and early years of the Society' *Proceedings of the Cambridge Philosophical Society* VII (1892) pp. i–l, p. xlviii; George Cornelius Gorham *Memoirs of John Martyn, F.R.S., and of Thomas Martyn, B.D., F.R.S., F.L.S., Professors of Botany in the University of Cambridge* London: Hatchard 1830 pp. 165–7.
26. Leonard Jenyns *Memoir of the Reverend John Stevens Henslow, M.A., F.L.S., F.G.S., F.C.P.S.: late rector of Hitcham and Professor of Botany in the University of Cambridge* London: John Van Voorst 1862 p. 17.
27. Leonard Jenyns *Memoir of the Reverend John Stevens Henslow, M.A., F.L.S., F.G.S., F.C.P.S.: late rector of Hitcham and Professor of Botany in the University of Cambridge* London: John Van Voorst 1862 p. 17.
28. Colin Speakman *Adam Sedgwick: geologist and dalesman* Broad Oak: Broad Oak Press 1982 p. 62.
29. Adam Sedgwick to William Ainger, 14 August 1819, quoted in John Willis Clark and Thomas McKenny Hughes *The life and letters of Adam Sedgwick, volume I* Cambridge: Cambridge University Press 2009 pp. 211–12.
30. John Stevens Henslow to Adam Sedgwick, 31 March 1820, quoted in John Willis Clark and Thomas McKenny Hughes *The life and letters of Adam Sedgwick, volume I* Cambridge: Cambridge University Press 2009 p. 214; Anonymous 'Replies to queries' *Yn lioar Manninagh* I (1889) pp. 23–4; Joseph Train *An historical and statistical account of the Isle of Man, from the earliest times to the present date: with a view of its ancient laws, peculiar customs, and popular superstitions, volume I* Douglas: Mary A. Quiggin 1845 pp. 7–8.
31. William Otter *The life and remains of the Reverend Edward Daniel Clarke, LL.D., Professor of Mineralogy in the University of Cambridge* London: J.F. Dove 1824 pp. 54–5.
32. William Otter *The life and remains of the Reverend Edward Daniel Clarke, LL.D., Professor of Mineralogy in the University of Cambridge* London: J.F. Dove 1824 p. 55.
33. William Otter *The life and remains of the Reverend Edward Daniel Clarke, LL.D., Professor of Mineralogy in the University of Cambridge* London: J.F. Dove 1824 pp. 503–6.
34. See http://webapps.fitzmuseum.cam.ac.uk/explorer/index.php?oid=65755.
35. William Otter *The life and remains of the Reverend Edward Daniel Clarke, LL.D., Professor of Mineralogy in the University of Cambridge* London: J.F. Dove 1824 p. 555.
36. Edward Daniel Clarke to John Marten Cripps, 12 February 1807 and 18 February 1807, quoted in William Otter *The life and remains of the Reverend Edward Daniel Clarke, LL.D., Professor of Mineralogy in the University of Cambridge* London: J.F. Dove 1824 p. 557.
37. John Martin Frederick Wright *Alma mater, or, seven years at the University of Cambridge, volume II* Cambridge: Cambridge University Press 2010 pp. 30–1; Brian Dolan

*Exploring European frontiers: British travellers in the age of Enlightenment* Basingstoke: Macmillan 2000 pp. 167–70.

38. Leonard Jenyns *Memoir of the Reverend John Stevens Henslow, M.A., F.L.S., F.G.S., F.C.P.S.: late rector of Hitcham and Professor of Botany in the University of Cambridge* London: John Van Voorst 1862 p. 17.
39. Cambridge Philosophical Society Archives, 3/1/1, minutes of general meeting, 30 October 1819.
40. John Willis Clark 'The foundation and early years of the Society' *Proceedings of the Cambridge Philosophical Society* VII (1892) pp. i–l, pp. iii–iv.
41. J.P.C. Roach (editor) 'The University of Cambridge: The Schools and University Library' in *A history of the county of Cambridge and the Isle of Ely, volume III: the City and University of Cambridge* London 1959 pp. 312–21. Retrieved 2 March 2018 from British History Online http://www.british-history.ac.uk/vch/cambs/vol3/pp312-321.
42. Peter Searby *A history of the University of Cambridge, volume III: 1750–1870* Cambridge: Cambridge University Press 1997 p. 208.
43. Anonymous 'Preface' *Transactions of the Cambridge Philosophical Society* I (1822) pp. iii–viii, pp. iv–vi.
44. John Willis Clark 'The foundation and early years of the Society' *Proceedings of the Cambridge Philosophical Society* VII (1892) pp. i–l, p. iv.
45. Mark W. Weatherall 'John Haviland (1785–1851)' *Oxford dictionary of national biography* Retrieved 2 March 2018 from http://www.oxforddnb.com/view/10.1093/ref:odnb/9780198614128.001.0001/odnb-9780198614128-e-12636.
46. Anonymous 'Preface' *Transactions of the Cambridge Philosophical Society* I (1822) pp. iii–viii, p. vi; Anonymous *Cambridge chronicle* 5 November 1819.
47. John Gascoigne 'The universities and the scientific revolution: the case of Newton and Restoration Cambridge' *History of science* XXIII (1985) pp. 391–434, p. 392.
48. Christopher N.L. Brooke *A history of the University of Cambridge, volume IV: 1870–1990* Cambridge: Cambridge University Press 1993 p. 167; Mark Weatherall *Gentlemen, scientists and doctors: medicine at Cambridge 1800–1940* Woodbridge: Boydell Press 2000 pp. 30–1.
49. Charles Babbage *Reflections on the decline of science in England, and on some of its causes* London: B. Fellowes 1830 p. 3.
50. David Brewster, 1830, quoted in Jack Morrell and Arnold Thackray *Gentlemen of science: early years of the British Association for the Advancement of Science* Oxford: Clarendon Press 1981 p. 52.
51. Andrew Warwick *Masters of theory: Cambridge and the rise of mathematical physics* Chicago: University of Chicago Press 2003 p. 186.
52. Andrew Warwick *Masters of theory: Cambridge and the rise of mathematical physics* Chicago: University of Chicago Press 2003 pp. 56–8.
53. David B. Wilson 'The educational matrix: physics education at early-Victorian Cambridge, Edinburgh and Glasgow Universities' in P.M. Harman (editor) *Wranglers and physicists: studies on Cambridge physics in the nineteenth century* Manchester: Manchester University Press 1985 pp. 12–48, pp. 12–13.

54. Andrew Warwick *Masters of theory: Cambridge and the rise of mathematical physics* Chicago: University of Chicago Press 2003 pp. 51, 59–60.
55. I. Grattan-Guinness 'Mathematics and mathematical physics from Cambridge, 1815–1840: a survey of the achievements and of the French influences' in P.M. Harman (editor) *Wranglers and physicists: studies on Cambridge physics in the nineteenth century* Manchester: Manchester University Press 1985 pp. 84–111, pp. 85–7.
56. R.M. Beverley *A letter to His Royal Highness the Duke of Gloucester* London 1833 pp. 38–9.
57. Doron Swade 'Charles Babbage (1791–1871)' *Oxford dictionary of national biography* Retrieved 2 March 2018 from http://www.oxforddnb.com/view/10.1093/ref:odnb/9780198614128.001.0001/odnb-9780198614128-e-962.
58. Philip C. Enros 'The Analytical Society (1812–1813): precursor of the renewal of Cambridge mathematics' *Historia mathematica* 10 (1983), pp. 24–47, p. 31.
59. Philip C. Enros 'The Analytical Society (1812–1813): precursor of the renewal of Cambridge mathematics' *Historia mathematica* 10 (1983) pp. 24–47.
60. John Herschel to Charles Babbage, 12 January 1814, quoted in Philip C. Enros 'The Analytical Society (1812–1813): precursor of the renewal of Cambridge mathematics' *Historia mathematica* 10 (1983) pp. 24–47, p. 34.
61. John Whittaker to Edward Bromhead, 16 February 1813 and 20 March 1813; Charles Babbage to Edward Bromhead, 30 November 1813; and John Herschel to Edward Bromhead, 19 November 1813; quoted in Philip C. Enros 'The Analytical Society (1812–1813): precursor of the renewal of Cambridge mathematics' *Historia mathematica* 10 (1983) pp. 24–47, pp. 35, 37, 41.
62. Sylvestre Lacroix (translated by Charles Babbage, John Herschel, and George Peacock) *An elementary treatise on the differential and integral calculus* Cambridge: Cambridge University Press 1816.
63. Harvey W. Becher 'George Peacock (1791–1858)' *Oxford dictionary of national biography* Retrieved 2 March 2018 from http://www.oxforddnb.com/view/10.1093/ref:odnb/9780198614128.001.0001/odnb-9780198614128-e-21673; Philip C. Enros 'The Analytical Society (1812–1813): precursor of the renewal of Cambridge mathematics' *Historia mathematica* X (1983) pp. 24–47.
64. Adam Sedgwick to John Herschel, 26 February 1820, quoted in John Willis Clark and Thomas McKenny Hughes *The life and letters of Adam Sedgwick, volume I* Cambridge: Cambridge University Press 2009 p. 208.
65. Jephson later fled the country after accusations of committing an 'unnatural crime'. Anonymous *The crimes of the clergy; or, the pillars of priestcraft shaken* London 1823 pp. 239–40; William Cobbett 'The parson and the boy' *Cobbett's weekly register* XLVII (2 August 1823) pp. 256–319.
66. Cambridge Philosophical Society Archives, 3/1/1, minutes of general meeting, 8 November 1819.
67. Anonymous 'Preface' *Transactions of the Cambridge Philosophical Society* I (1822) pp. iii–viii, p. vii.
68. Charles Babbage *Reflections on the decline of science in England, and on some of its causes* London: B. Fellowes 1830 pp. 186–7.

69. Anita McConnell 'William Farish (1759–1837)' *Oxford dictionary of national biography* Retrieved 2 March 2018 from http://www.oxforddnb.com/view/10.1093/ref:odnb/9780198614128.001.0001/odnb-9780198614128-e-9162.
70. Thomas Hamilton (revised by John D. Haigh) 'Samuel Lee (1783–1852)' *Oxford dictionary of national biography* Retrieved 2 March 2018 from http://www.oxforddnb.com/view/10.1093/ref:odnb/9780198614128.001.0001/odnb-9780198614128-e-16309.
71. Thompson Cooper (revised by Julia Tompson) 'Bewick Bridge (1767–1833)' *Oxford dictionary of national biography* Retrieved 2 March 2018 from http://www.oxforddnb.com/view/10.1093/ref:odnb/9780198614128.001.0001/odnb-9780198614128-e-3386.
72. A.M. Clerke (revised by Anita McConnell) 'Thomas Catton (1758–1838)' *Oxford dictionary of national biography* Retrieved 2 March 2018 from http://www.oxforddnb.com/view/10.1093/ref:odnb/9780198614128.001.0001/odnb-9780198614128-e-4903.
73. Thompson Cooper (revised by M.C. Curthoys) 'Thomas Turton (1780–1864)' *Oxford dictionary of national biography* Retrieved 2 March 2018 from http://www.oxforddnb.com/view/10.1093/ref:odnb/9780198614128.001.0001/odnb-9780198614128-e-27895.
74. Pamela Tudor-Craig 'Thomas Kerrich (1748–1828)' *Oxford dictionary of national biography* Retrieved 2 March 2018 from http://www.oxforddnb.com/view/10.1093/ref:odnb/9780198614128.001.0001/odnb-9780198614128-e-15471.
75. Harvey W. Becher 'Robert Woodhouse (1773–1827)' *Oxford dictionary of national biography* Retrieved 2 March 2018 from http://www.oxforddnb.com/view/10.1093/ref:odnb/9780198614128.001.0001/odnb-9780198614128-e-29926.
76. Christopher F. Lindsey 'James Cumming (1777–1861)' *Oxford dictionary of national biography* Retrieved 2 March 2018 from http://www.oxforddnb.com/view/10.1093/ref:odnb/9780198614128.001.0001/odnb-9780198614128-e-6896.
77. Jonathan Smith and Christopher Stray (editors) *Teaching and learning in nineteenth-century Cambridge* Woodbridge: Boydell Press 2001.
78. Negley Harte and John North *The world of University College London 1828–1978* London: University College London 1978.
79. Hugh James Rose *The tendency of prevalent opinions about knowledge considered* Cambridge 1826 pp. 236–7. More and Mede were seventeenth-century Cambridge academics; More was a Platonist, and Mede was a biblical scholar and Professor of Greek.
80. David A. Valone 'Hugh James Rose's Anglican critique of Cambridge: science, antirationalism, and Coleridgean idealism in late Georgian England' *Albion: a quarterly journal concerned with British studies* XXXIII (2001) pp. 218–42.
81. Charles Babbage *Reflections on the decline of science in England, and on some of its causes* London: B. Fellowes 1830 pp. 1–30.
82. Robert E. Schofield 'History of scientific societies: needs and opportunities for research' *History of science* II (1963) pp. 70–83, p. 70; Jack Morrell and Arnold Thackray *Gentlemen of science: early years of the British Association for the Advancement of Science* Oxford: Clarendon Press 1981 p. 14; David Brewster 'Decline of science in England' *Quarterly review* XLIII (1830) p. 327.

## Chapter 2

1. Leonard Jenyns *Memoir of the Reverend John Stevens Henslow, M.A., F.L.S., F.G.S., F.C.P.S.: late rector of Hitcham and Professor of Botany in the University of Cambridge* London: John Van Voorst 1862; S. Max Walters 'John Stevens Henslow (1796–1861)' *Oxford dictionary of national biography* Retrieved 2 March 2018 from http://www.oxforddnb.com/view/10.1093/ref:odnb/9780198614128.001.0001/odnb-9780198614128-e-12990; S.M. Walters and E.A. Stow *Darwin's mentor: John Stevens Henslow, 1796–1861* Cambridge: Cambridge University Press 2001.
2. Mr Dryden 'Quarterly night' 2 October 1816, quoted in Walter Jerrold *Michael Faraday: man of science* London: S.W. Partridge & Co. 1892 p. 58. For more on the history of the City Philosophical Society, see Frank James 'Michael Faraday, the City Philosophical Society and the Society of Arts' *Royal Society of Arts Journal* CXL (1992) pp. 192–9.
3. J.S. Rowlinson *Sir James Dewar, 1842–1923: a ruthless chemist* London: Routledge 2012 p. 21. Slaughterhouse Lane is now called Corn Exchange Street; it had previously been known as Fairyard Lane. Bird Bolt Lane is now called Downing Street. The Beast Market was located on what now remains of St Tibb's Row (also known as St Tibbs Row, or Tib Row, as it appears on some maps, once extended all the way to Petty Cury); see J.P.C. Roach (editor) *A history of the county of Cambridge and the Isle of Ely, volume III: the City and University of Cambridge* London: Victoria County History 1959 pp. 101–8.
4. John Gascoigne *Cambridge in the age of the Enlightenment* Cambridge: Cambridge University Press 1989 pp. 277, 292.
5. Anita McConnell 'William Farish (1759–1837)' *Oxford dictionary of national biography* Retrieved 2 March 2018 from http://www.oxforddnb.com/view/10.1093/ref:odnb/9780198614128.001.0001/odnb-9780198614128-e-9162; William Farish *A plan of a course of lectures on arts and manufactures, more particularly such as relate to chemistry* Cambridge: J. Burgess 1796.
6. Edward Daniel Clarke 'Observations upon the ores which contain cadmium, and upon the discovery of this metal in the Derbyshire silicates and other ores of zinc' *Annals of philosophy* XV (1820) pp. 272–6.
7. For examples of post-meeting displays in later years, see Anonymous 'Review of a meeting of the Cambridge Philosophical Society, 18 June 1835' *The London, Edinburgh, and Dublin philosophical magazine and journal of science: series 3* VII (1835) p. 71; Anonymous 'Review of a meeting of the Cambridge Philosophical Society, 30 November 1840' *The annals and magazine of natural history* VI (1841) p. 379.
8. Cambridge Philosophical Society Archives, 3/1/1, minutes of general meeting, 21 February 1820; Anonymous 'Review of meeting of Cambridge Philosophical Society' *The Edinburgh philosophical journal* III (1820) pp. 184–5; Anonymous 'Review of meeting of Cambridge Philosophical Society' *The London literary gazette and journal of belles lettres* 1820 p. 172.
9. Adam Sedgwick to John Herschel, 26 February 1820, quoted in John Willis Clark and Thomas McKenny Hughes *The life and letters of Adam Sedgwick, volume I* Cambridge: Cambridge University Press 2009 p. 209.

10. Anonymous 'Review of meeting of Cambridge Philosophical Society' *The Edinburgh philosophical journal* III (1820) pp. 184–5; Anonymous 'Review of meeting of Cambridge Philosophical Society' *The London literary gazette and journal of belles lettres* 1820 p. 172.
11. Cambridge Philosophical Society Archives, 3/1/1, minutes of general meeting, 6 March 1820, 20 March 1820, 17 April 1820; William Whewell 'On the position of the apsides of orbits of great eccentricity' *Transactions of the Cambridge Philosophical Society* I (1822) pp. 179–94.
12. Richard Yeo 'William Whewell (1794–1866)' *Oxford dictionary of national biography* Retrieved 15 March 2018 from http://www.oxforddnb.com/view/10.1093/ref:odnb/9780198614128.001.0001/odnb-9780198614128-e-29200.
13. Cambridge Philosophical Society Archives, 2/1/1, minutes of Council meeting, 13 March 1820, 23 March 1820.
14. See advertisements for Smith's auctions in, for example, the 5 November 1819 issue of the *Cambridge chronicle.*
15. Cambridge Philosophical Society Archives, 2/1/1, minutes of Council meeting, 22 May 1820.
16. Cambridge Philosophical Society Archives, 2/1/1, minutes of Council meeting, 22 May 1820, 24 October 1820, 27 January 1824; 2/1/2, minutes of Council meeting, 24 October 1836.
17. Cambridge Philosophical Society Archives, 2/1/1, minutes of Council meeting, 22 May 1820, 24 October 1820; John Herschel 'On certain remarkable instances of deviation from Newton's scale in the tints developed by crystals, with one axis of double refraction, on exposure to polarised light' *Transactions of the Cambridge Philosophical Society* I (1822) pp. 21–42.
18. Cambridge Philosophical Society Archives, 3/1/1, minutes of general meeting, 1 May 1820, 13 November 1820; 2/1/1, minutes of Council meeting, 1 May 1820.
19. Carla Yanni *Nature's museums: Victorian science and the architecture of display* London: Athlone Press 2005.
20. Cambridge Philosophical Society Archives, 2/1/1, minutes of Council meeting, 4 December 1819; 3/1/1, minutes of general meeting, 13 December 1819.
21. Leonard Jenyns *Chapters in my life* Cambridge: Cambridge University Press 2011 p. 23.
22. Leonard Jenyns 'Cambridge Philosophical Society museum' in J.J. Smith (editor) *The Cambridge portfolio* London: J.W. Parker 1840 pp. 127–9, p. 127. For lists of donations, see Anonymous 'Donations to the museum' *Transactions of the Cambridge Philosophical Society* I (1822) pp. 463–4; Anonymous 'Donations to the museum' *Transactions of the Cambridge Philosophical Society* II (1827) pp. 450–1; Anonymous 'Donations to the museum' *Transactions of the Cambridge Philosophical Society* III (1830) pp. 447–8.
23. Haswell later renamed himself Halswell.
24. William Buckland *Reliquiae diluvianae, or, observations on the organic remains contained in caves, fissures and diluvial gravel, and on other geological phenomena* London: John Murray 1823. The Kirkdale fossils in the Cambridge Philosophical Society's

museum were donated by Rev. Francis Wrangham, a noted author, poet, and abolitionist who had studied in Cambridge in the 1780s.

25. M.E. Bury and J.D. Pickles (editors) *Romilly's Cambridge Diary, 1842–1847* Cambridge: Cambridgeshire Records Society 1994 p. 101.
26. Leonard Jenyns 'Cambridge Philosophical Society museum' in J.J. Smith (editor) *The Cambridge portfolio* London: J.W. Parker 1840 pp. 127–9, p. 128; Cambridge Philosophical Society Archives, 2/1/1, minutes of Council meeting, 24 November 1821, 21 July 1823; 3/1/1, minutes of general meeting, 19 March 1821.
27. A. Rupert Hall *The Cambridge Philosophical Society: a history 1819–1969* Cambridge: Cambridge Philosophical Society 1969 p. 9. Cambridge Philosophical Society Archives, 6/2/1, membership records.
28. Cambridge Philosophical Society Archives, 2/1/2, minutes of Council meeting, 14 November 1831, undated April 1832, 4 May 1832, 9 May 1832.
29. Cambridge Philosophical Society Archives, 1/2/2, lease of the Society's house.
30. The name is occasionally spelled 'Humphrey'. For more details on his buildings, see Anonymous *An inventory of the historical monuments in the City of Cambridge* London: Her Majesty's Stationery Office 1959 pp. 209–44. Christopher N.L. Brooke *A history of the University of Cambridge, volume IV: 1870–1990* Cambridge: Cambridge University Press 1993 pp. 252, 304.
31. Cambridge Philosophical Society Archives, 7/2/1, specification for the Philosophical Society's house in All Saints' Passage; 2/1/2, minutes of Council meeting, 5 June 1832, 12 June 1833.
32. Cambridge Philosophical Society Archives, 2/1/2, minutes of Council meeting, undated April 1832, 5 June 1832.
33. Leonard Jenyns 'Cambridge Philosophical Society museum' in J.J. Smith (editor) *The Cambridge portfolio* London: J.W. Parker 1840 pp. 127–9, p. 127.
34. Cambridge Philosophical Society Archives, 2/1/1, minutes of Council meeting, 18 March 1822.
35. John Stevens Henslow to Leonard Jenyns, 8 April 1823, quoted in S.M Walters and E.A. Stow *Darwin's mentor: John Stevens Henslow, 1796–1861* Cambridge: Cambridge University Press 2001 p. 40.
36. Charles Darwin to William Darwin Fox, 8 October 1830, Darwin Correspondence Project letter #86.
37. John Willis Clark 'The foundation and early years of the Society' *Proceedings of the Cambridge Philosophical Society* VII (1892) pp. i–l, p. xii.
38. David McKitterick *Cambridge University Library: a history: the eighteenth and nineteenth centuries* Cambridge: Cambridge University Press 2009.
39. Cambridge Philosophical Society Archives, 2/1/1, minutes of Council meeting, 24 November 1821, undated May 1822, 23 May 1831, 28 November 1831.
40. Cambridge Philosophical Society Archives, 2/1/2, minutes of Council meeting, 25 July 1833, 9 December 1833, 17 March 1834; 2/1/3, minutes of Council meeting, 16 November 1840.
41. Anonymous 'Donations to the library' *Transactions of the Cambridge Philosophical Society* I (1822) pp. 461–3; Anonymous 'Donations to the library' *Transactions of the*

*Cambridge Philosophical Society* II (1827) pp. 446–9; Anonymous 'Donations to the library' *Transactions of the Cambridge Philosophical Society* III (1830) pp. 445–7.

42. Cambridge Philosophical Society Archives, 7/2/1, specification for the Philosophical Society's house in All Saints' Passage, pp. 8–11. On the architecture of provincial philosophical societies, see Sophie Forgan 'Context, image and function: a preliminary enquiry into the architecture of scientific societies' *British journal for the history of science* XIX (1986) pp. 89–113.
43. Cambridge Philosophical Society Archives, 3/1/1, minutes of general meeting, 15 May 1820.
44. William Whewell 'On the results of observations made with a new anemometer' *Transactions of the Cambridge Philosophical Society* VI (1838) pp. 301–15; Anonymous 'Anemometers of Messrs. Whewell and Osler' *Arcana of science and art* (1838) pp. 279–81.
45. Cambridge Philosophical Society Archives, 2/1/2, minutes of Council meeting, 24 October 1836.
46. Cambridge Philosophical Society Archives, 2/1/3, minutes of Council meeting, 6 November 1839; Christopher Webster and John Elliott (editors) *'A church as it should be': the Cambridge Camden Society and its influence* Stamford: Shaun Tyas 2000.
47. Cambridge Philosophical Society Archives, 2/1/2, minutes of Council meeting, undated April 1832. Charters had been awarded to the Royal Institution (1800), the Linnean Society (1802), the Horticultural Society (1809), the Geological Society (1825), the Zoological Society (1829), and the Royal Astronomical Society (1831). Retrieved 19 July 2016 from http://privycouncil.independent.gov.uk/royal-charters/chartered-bodies/.
48. Adam Sedgwick to Prince William Frederick, 21 May 1832, Cambridge Philosophical Society Archives 1/2/1a. This letter was found in 2016 in the basement of a building once occupied by the Privy Council; it was rescued from the recycling bin by Patricia Mulcahy.
49. Cambridge Philosophical Society Archives, Royal Charter, 1832.
50. Adam Sedgwick, quoted in John Willis Clark and Thomas McKenny Hughes *The life and letters of Adam Sedgwick, volume I* Cambridge: Cambridge University Press 2009 p. 397.
51. Cambridge Philosophical Society Archives, 2/1/2, minutes of Council meeting, 23 October 1832, 29 October 1832; 1/2/2, Charter and bye laws of the Cambridge Philosophical Society.
52. M.E. Bury and J.D. Pickles (editors) *Romilly's Cambridge Diary, 1842–1847* Cambridge: Cambridgeshire Records Society 1994 p. 21. Connop Thirlwall was a fellow of Trinity College. In the 1830s, he was involved in disputes about allowing greater religious freedoms in Cambridge; Thirlwall supported the idea of allowing dissenters to study at Cambridge and lost his fellowship over this stance. His main opponent was another fellow of the Cambridge Philosophical Society—Thomas Turton, who had sat on the Society's first Council. J.W. Clark (revised by H.G.C. Matthew) 'Connop Thirlwall (1797–1875)' *Oxford dictionary of national biography* Retrieved 15 March 2018 from http://www.oxforddnb.com/view/10.1093/

ref:odnb/9780198614128.001.0001/odnb-9780198614128-e-27185. Lodge may have been John Lodge, the University Librarian and a member of the Society's council.

53. Adam Sedgwick to Roderick Impey Murchison, 7 November 1832, quoted in John Willis Clark and Thomas McKenny Hughes *The life and letters of Adam Sedgwick, volume I* Cambridge: Cambridge University Press 2009 p. 397.
54. Cambridge Philosophical Society Archives, 2/1/2, minutes of Council meeting, 29 October 1832, 14 November 1832; Nicolas Carlisle *A memoir of the life of William Wyon* London: W. Nichol 1837 p. 206.
55. David McKitterick *A history of Cambridge University Press, volume II* Cambridge: Cambridge University Press 1998 chapter 1; Aileen Fyfe *Steam-powered knowledge: William Chambers and the business of publishing, 1820–1860* Chicago: University of Chicago Press 2012; Adrian Johns 'Miscellaneous methods: authors, societies and journals in early modern England' *British journal for the history of science* XXXIII (2000) pp. 159–86.
56. Cambridge Philosophical Society Archives, 2/1/1, minutes of Council meeting, 1 May 1820, 5 June 1820. For a history of the Geological Society of London, see C.L.E. Lewis and S.J. Knell (editors) *The making of the Geological Society of London* London: Geological Society 2009. For more information on the style and content of the *Transactions*, see M.J.S. Rudwick 'The early Geological Society in its international context' in C.L.E. Lewis and S.J. Knell (editors) *The making of the Geological Society of London* London: Geological Society 2009 pp. 145–54, pp. 149–50. See also *Transactions of the Geological Society* I (1811); Anonymous 'Review of *Transactions of the Geological Society*, volume I' *The Edinburgh review, or critical journal* XIX (1812) pp. 207–29; and Charles Babbage *Reflections on the decline of science in England, and on some of its causes* London: B. Fellowes 1830 p. 45.
57. Cambridge Philosophical Society Archives, 2/1/1, minutes of Council meeting, 4 December 1820; Anonymous 'Preface' *Transactions of the Cambridge Philosophical Society* I (1822) pp. iii–viii, p. viii; Anonymous 'Regulations' *Transactions of the Cambridge Philosophical Society* I (1822) pp. xvii–xxiii, p. xx. On the financial difficulties of publishing journals, see Aileen Fyfe 'Journals, learned societies and money: *Philosophical Transactions*, ca. 1750–1900' *Notes and records of the Royal Society* LXIX (2015) pp. 277–99. On the new landscape of literacy, printing, and scientific publications, see Jack Morrell and Arnold Thackray *Gentlemen of science: early years of the British Association for the Advancement of Science* Oxford: Clarendon Press 1981 pp. 18–19; David M. Night 'Scientists and their publics: popularisation of science in the nineteenth century' in Mary Jo Nye (editor) *The Cambridge history of science, volume V: the modern physical and mathematical sciences* Cambridge: Cambridge University Press 2003 pp. 72–90, p. 75; David McKitterick *A history of Cambridge University Press, volume II* Cambridge: Cambridge University Press 1998 p. 25; Cambridge Philosophical Society Archives, 2/1/2, minutes of Council meeting, 18 March 1837.
58. David McKitterick *A history of Cambridge University Press, volume II* Cambridge: Cambridge University Press 1998 p. 6.

59. Cambridge Philosophical Society Archives, 2/1/1, minutes of Council meeting, 5 June 1820, 24 January 1821, 15 December 1821, 10 May 1822. For more on John Smith, see David McKitterick *A history of Cambridge University Press, volume II* Cambridge: Cambridge University Press 1998 chapter 15. For more on John Murray, see William Zachs, Peter Isaac, Angus Fraser, and William Lister 'Murray family (per. 1768–1967)' *Oxford dictionary of national biography* Retrieved 15 March 2018 from http://www.oxforddnb.com/view/10.1093/ref:odnb/9780198614128.001.0001/odnb-9780198614128-e-64907. Mr Lowry may have been Wilson Lowry or his son, Joseph Wilson Lowry, both engravers active at this time. Mr Bowtell may have been John Bowtell, the nephew of a very well-regarded Cambridge bookbinder also called John Bowtell.
60. Aileen Fyfe 'Peer review: not as old as you might think' *Times higher education* 25 June 2015 Retrieved 8 October 2018 from http://www.timeshighereducation.com/features/peer-review-not-old-you-might-think; Alex Csiszar 'Peer review: troubled from the start' *Nature* DXXXII (2016) pp. 306–8; Noah Moxham and Aileen Fyfe 'The Royal Society and the prehistory of peer review, 1665–1965' *The historical journal* (2017) doi:10.1017/S0018246X17000334. Before the publication of the first volume of the Society's *Transactions*, Charles Babbage's paper on the notation used in calculus was examined by the newest Council members—George Peacock and William Whewell; Adam Sedgwick's first papers on Cornish geology were reviewed by Edward Clarke and George Peacock; Edward Clark's mineralogical papers were read by William Farish and John Haviland; and William Whewell's paper on orbits was reviewed by Thomas Turton and Richard Gwatkin. In the early days, most papers passed the review process, but a few were rejected—like that by one Mr Emmett, rejected after a report by John Haviland and Adam Sedgwick. Though any fellow of the Society could be called upon to review an article, in reality, most reviewing was done by members of the Council. Cambridge Philosophical Society Archives, 2/1/1, minutes of Council meeting, 22 May 1820.
61. Cambridge Philosophical Society Archives, 2/1/1, minutes of Council meeting, 22 May 1820; Anonymous 'Preface' *Transactions of the Cambridge Philosophical Society* I (1822) pp. iii–viii, p. xx.
62. Anonymous 'Preface' *Transactions of the Cambridge Philosophical Society* I (1822) pp. iii–viii.
63. Adam Sedgwick to William Ainger, 13 April 1821, quoted in John Willis Clark and Thomas McKenny Hughes *The life and letters of Adam Sedgwick, volume I* Cambridge: Cambridge University Press 2009 p. 225.
64. Anonymous 'Review of *Transactions of the Cambridge Philosophical Society, volume I, part II*' *The British critic: a new review* XVIII (1822) pp. 386, 395.
65. Anonymous 'Review of *Transactions of the Cambridge Philosophical Society, volume I, part I*' *The Cambridge quarterly review and academical register* I (1824) pp. 163, 166, 171, 181, 182.
66. Cambridge Philosophical Society Archives, 2/1/1, minutes of Council meeting, 2 April 1821, 10 May 1822.

67. Cambridge Philosophical Society Archives, 2/1/1, minutes of Council meeting, undated May 1822; 2/1/2, minutes of Council meeting, 16 November 1829; Anonymous 'Donations to the library' *Transactions of the Cambridge Philosophical Society* I (1822) pp. 461–3; Anonymous 'Donations to the library' *Transactions of the Cambridge Philosophical Society* II (1827) pp. 446–9; Anonymous 'Donations to the library' *Transactions of the Cambridge Philosophical Society* III (1830) pp. 445–7.
68. John Willis Clark and Thomas McKenny Hughes *The life and letters of Adam Sedgwick, volume I* Cambridge: Cambridge University Press 2009 p. 396.
69. Anonymous 'Preface' *Transactions of the Cambridge Philosophical Society* I (1822) pp. iii–viii, p. v.
70. Roderick Impey Murchison to Adam Sedgwick, 1828, quoted in Crosbie Smith 'Geologists and mathematicians: the rise of physical geology' in P.M. Harman (editor) *Wranglers and physicists: studies on Cambridge physics in the nineteenth century* Manchester: Manchester University Press 1985 pp. 49–83, p. 52.
71. James A Secord *Controversy in Victorian geology: the Cambrian–Silurian dispute* Princeton: Princeton University Press 1986 pp. 61–6; Crosbie Smith 'Geologists and mathematicians: the rise of physical geology' in P.M. Harman (editor) *Wranglers and physicists: studies on Cambridge physics in the nineteenth century* Manchester: Manchester University Press 1985 pp. 49–83, p. 52.
72. Adam Sedgwick 'On the physical structure of those formations which are immediately associated with the primitive ridge of Devonshire and Cornwall' *Transactions of the Cambridge Philosophical Society* I (1822) pp. 89–146.
73. Crosbie Smith 'Geologists and mathematicians: the rise of physical geology' in P.M. Harman (editor) *Wranglers and physicists: studies on Cambridge physics in the nineteenth century* Manchester: Manchester University Press 1985 pp. 49–83, p. 55.
74. William Whewell 'Mathematical exposition of some of the leading doctrines in Mr Ricardo's "Principles of political economy and taxation"' *Transactions of the Cambridge Philosophical Society* IV (1833) pp. 155–98, p. 156.
75. Anonymous 'Review of a meeting of the Cambridge Philosophical Society, 22 February 1836' *The London, Edinburgh, and Dublin philosophical magazine and journal of science: series 3* VIII (1836) p. 429; Isaac Todhunter (editor) *William Whewell, Master of Trinity College, Cambridge: an account of his writings, volume I* London: Macmillan 1876 chapter 6.
76. Henry J.H. Bond 'A statistical report of Addenbrooke's Hospital, for the year 1836' *Transactions of the Cambridge Philosophical Society* VI (1838) pp. 361–78; Henry J.H. Bond 'A statistical report of Addenbrooke's Hospital, for the year 1837' *Transactions of the Cambridge Philosophical Society* VI (1838) pp. 565–75.
77. Terence D. Murphy 'Medical knowledge and statistical methods in early nineteenth-century France' *Medical history* XXV (1981) 301–19; Andrea A. Rusnock *Vital accounts: quantifying health and population in eighteenth-century England and France* Cambridge: Cambridge University Press 2002.
78. Adam Sedgwick, quoted in John Willis Clark and Thomas McKenny Hughes *The life and letters of Adam Sedgwick, volume I* Cambridge: Cambridge University Press 2009 p. 208.

79. Adam Sedgwick to William Wordsworth, May 1842, quoted in John Willis Clark and Thomas McKenny Hughes *The life and letters of Adam Sedgwick, volume I* Cambridge: Cambridge University Press 2009 p. 248.
80. Adam Sedgwick, 1853, quoted in John Willis Clark and Thomas McKenny Hughes *The life and letters of Adam Sedgwick, volume I* Cambridge: Cambridge University Press 2009 p. 249.
81. William Whewell, 1824, quoted in John Willis Clark and Thomas McKenny Hughes *The life and letters of Adam Sedgwick, volume I* Cambridge: Cambridge University Press 2009 p. 247.
82. Adam Sedgwick, 1853, quoted in John Willis Clark and Thomas McKenny Hughes *The life and letters of Adam Sedgwick, volume I* Cambridge: Cambridge University Press 2009 p. 249.
83. William Wordsworth *The excursion* London 1814 pp. 103–4.
84. Adam Sedgwick to William Wordsworth, 1842, quoted in Crosbie Smith 'Geologists and mathematicians: the rise of physical geology' in P.M. Harman (editor) *Wranglers and physicists: studies on Cambridge physics in the nineteenth century* Manchester: Manchester University Press 1985 pp. 49–83, p. 83.
85. Charles Babbage *Reflections on the decline of science in England, and on some of its causes* London: B. Fellowes 1830 pp. 10, 30–1, 36–7.
86. Adam Sedgwick to William Ainger, 19 February 1825, quoted in John Willis Clark and Thomas McKenny Hughes *The life and letters of Adam Sedgwick, volume I* Cambridge: Cambridge University Press 2009 p. 263.
87. John Herschel to Charles Babbage, 18 December 1815, quoted in Günther Buttmann *The shadow of the telescope: a biography of John Herschel* Guildford: Lutterworth Press 1974 p. 18.
88. Michael J. Crowe 'John Frederick William Herschel (1792–1871)' *Oxford dictionary of national biography* Retrieved 15 March 2018 from http://www.oxforddnb.com/view/10.1093/ref:odnb/9780198614128.001.0001/odnb-9780198614128-e-13101.
89. George Peacock to John Herschel, 3 December 1816, quoted in Roger Hutchins *British university observatories, 1772–1939* Aldershot: Ashgate 2008 p. 31.
90. Adam Sedgwick to John Herschel, 14 November 1820, quoted in John Willis Clark and Thomas McKenny Hughes *The life and letters of Adam Sedgwick, volume I* Cambridge: Cambridge University Press 2009 p. 208.
91. Roger Hutchins *British university observatories, 1772–1939* Aldershot: Ashgate 2008 pp. 30–7.
92. George Biddell Airy 'On the use of silvered glass for the mirrors of reflecting telescopes' *Transactions of the Cambridge Philosophical Society* II (1827) pp. 105–18.
93. Allan Chapman 'George Biddell Airy (1801–1892)' *Oxford dictionary of national biography* Retrieved 15 March 2018 from http://www.oxforddnb.com/view/10.1093/ref:odnb/9780198614128.001.0001/odnb-9780198614128-e-251; Susan F. Cannon *Science in culture: the early Victorian period* New York: Dawson and Science History Publications 1978 p. 8.
94. S.M Walters and E.A. Stow *Darwin's mentor: John Stevens Henslow, 1796–1861* Cambridge: Cambridge University Press 2001 p. 42.

95. Roger Hutchins *British university observatories, 1772–1939* Aldershot: Ashgate 2008 pp. 32–3.
96. Richard Dunn and Rebecca Higgitt *Ships, clocks, and stars: the quest for longitude* Glasgow: Collins and London: Royal Museums Greenwich 2014.
97. Jack Morrell and Arnold Thackray *Gentlemen of science: early years of the British Association for the Advancement of Science* Oxford: Clarendon Press 1981 p. 42.
98. Charles Babbage *Reflections on the decline of science in England, and on some of its causes* London: B. Fellowes 1830 p. 148.
99. John Willis Clark and Thomas McKenny Hughes *The life and letters of Adam Sedgwick, volume I* Cambridge: Cambridge University Press 2009 p. 320.
100. Jack Morrell and Arnold Thackray *Gentlemen of science: early years of the British Association for the Advancement of Science* Oxford: Clarendon Press 1981 p. 48; Anonymous [David Brewster] 'Decline of science in England' *Quarterly review* XLIII (1830) pp. 305–42.
101. Charles Babbage *Reflections on the decline of science in England, and on some of its causes* London: B. Fellowes 1830.
102. Jack Morrell and Arnold Thackray *Gentlemen of science: early years of the British Association for the Advancement of Science* Oxford: Clarendon Press 1981 p. 49.
103. David Brewster to John Philips, 23 February 1831, quoted in Jack Morrell and Arnold Thackray *Gentlemen of science: early years of the British Association for the Advancement of Science* Oxford: Clarendon Press 1981 p. 59.
104. A.D. Orange *Philosophers and provincials: the Yorkshire Philosophical Society from 1822 to 1844* York: Yorkshire Philosophical Society 1973 pp. 7–10. For the foundation story of another philosophical society in Yorkshire, see E. Kitson Clark *The history of 100 years of life of the Leeds Philosophical and Literary Society* Leeds: Jowett and Sowry 1924 pp. 5–25.
105. Jack Morrell and Arnold Thackray *Gentlemen of science: early years of the British Association for the Advancement of Science* Oxford: Clarendon Press 1981 pp. 38, 58; Jack Morrell 'William Vernon Harcourt (1789–1871)' *Oxford dictionary of national biography* Retrieved 15 March 2018 from http://www.oxforddnb.com/view/10.1093/ref:odnb/9780198614128.001.0001/odnb-9780198614128-e-12249.
106. Jack Morrell and Arnold Thackray *Gentlemen of science: early years of the British Association for the Advancement of Science* Oxford: Clarendon Press 1981 pp. 58–88.
107. William Buckland to Adam Sedgwick, 19 April 1832, quoted in John Willis Clark and Thomas McKenny Hughes *The life and letters of Adam Sedgwick, volume I* Cambridge: Cambridge University Press 2009 p. 390.
108. Susan F. Cannon *Science in culture: the early Victorian period* New York: Dawson and Science History Publications 1978 p. 214.
109. Jack Morrell and Arnold Thackray *Gentlemen of science: early years of the British Association for the Advancement of Science* Oxford: Clarendon Press 1981 pp. 168–71, 216; Anonymous *Report of the third meeting of the British Association for the Advancement of Science, held at Cambridge in 1833* London: John Murray 1834. The other members of the organizing committee (apart from Sedgwick et al.) were George Peacock, Joseph Cape, Edward John Ash, William Hodgson, Charles Currie, James

Bowstead, Frederick Hildyard, William Henry Hanson, Henry Philpott, Charles Thomas Whitley, Charles Cardale Babington, William Lewes Pugh Garnons, James William Lucas Heaviside, John Maurice Herbert, John Matthews Robinson, John Thompson, and Thomas Thorp. Charles Dickens *The Mudfog Papers* Richmond: Alma Classics 2014.

110. Simon Schaffer 'Scientific discoveries and the end of natural philosophy' *Social studies of science* XVI (1986) pp. 387–420 p. 409. For more on Coleridge's views of natural philosophy and science, see Samuel Taylor Coleridge 'General introduction, or, a preliminary treatise on method' in Edward Smedley, Hugh James Rose, and Henry John Rose (editors) *Encyclopædia metropolitana, volume I* London: J.J. Griffin 1849, pp. (1)–(27). For more on the role of the British Association in shaping the idea of 'science', see Richard Yeo *Defining science* Cambridge: Cambridge University Press 1993 pp. 24–35. On the narrowing of scientific terminology in the nineteenth century, see Lorraine Daston and H. Otto Sibum 'Scientific personae and their histories' *Science in context* XVI (2003) pp. 1–8; David Cahan (editor) *From natural philosophy to the sciences* Chicago: Chicago University Press 2003 chapter 1; and Jed Z. Buchwald and Sungook Hong 'Physics' in David Cahan (editor) *From natural philosophy to the sciences* Chicago: Chicago University Press 2003 pp. 163–95.
111. William Whewell 'Review of *On the connexion of the physical sciences*' *The quarterly review* LI (1834) pp. 54–68, pp. 59–60; James A. Secord *Visions of science: books and readers at the dawn of the Victorian age* Oxford: Oxford University Press 2014 pp. 104–6.
112. For a discussion of the so-called Cambridge network, see Susan F. Cannon *Science in culture: the early Victorian period* New York: Dawson and Science History Publications 1978, and Walter F. Cannon 'Scientists and broad churchmen: an early Victorian intellectual network' *Journal of British studies* IV (1964) pp. 65–88.
113. Adam Sedgwick to William Ainger, 16 August 1825, quoted in John Willis Clark and Thomas McKenny Hughes *The life and letters of Adam Sedgwick, volume I* Cambridge: Cambridge University Press 2009 p. 267.
114. John Willis Clark and Thomas McKenny Hughes *The life and letters of Adam Sedgwick, volume I* Cambridge: Cambridge University Press 2009 p. 208; Cambridge Philosophical Society Archives, 2/1/3, minutes of Council meeting, 30 October 1839.

## Chapter 3

1. Charles Lyell to Caroline Lyell, 26 February 1829, quoted in K.M. Lyell (editor) *Life, letters and journals of Sir Charles Lyell, Bart* Cambridge: Cambridge University Press 2010 pp. 251–2.
2. John Herschel (edited by David S. Evans, Terence J. Deeming, Betty Hall Evans, and Stephen Goldfarb) *Herschel at the Cape: diaries and correspondence of Sir John Herschel, 1834–1838* Austin: University of Texas Press 1969 pp. 47, 72.
3. Michael J. Crowe 'John Frederick William Herschel (1792–1871)' *Oxford dictionary of national biography* Retrieved 15 March 2018 from http://www.oxforddnb.com/view/10.1093/ref:odnb/9780198614128.001.0001/odnb-9780198614128-e-13101.

4. John Herschel to Charles Lyell, 12 June 1837, quoted in Walter F. Cannon 'The impact of uniformitarianism: two letters from John Herschel to Charles Lyell, 1836–1837' *Proceedings of the American Philosophical Society* CV (1961) pp. 301–14, p. 314.
5. John Herschel (edited by David S. Evans, Terence J. Deeming, Betty Hall Evans, and Stephen Goldfarb) *Herschel at the Cape: diaries and correspondence of Sir John Herschel, 1834–1838* Austin: University of Texas Press 1969 pp. 144, 152, 192, 194–5, 200.
6. Anonymous 'Cambridge Philosophical Society, 16 November 1835' *The London, Edinburgh, and Dublin philosophical magazine and journal of science: series 3* VIII (1836) pp. 78–80, pp. 78–9.
7. Cambridge Philosophical Society Archives, 3/1/2, minutes of general meeting, 16 November 1835.
8. George Peacock to John Stevens Henslow, first letter before 24 August 1831, second letter undated, quoted in Nora Barlow (editor) *Darwin and Henslow: the growth of an idea, letters 1831–1860* London: Murray 1967 pp. 28–9.
9. Charles Darwin to Susan Darwin, 4 September 1831, Darwin Correspondence Project letter #115; John Stevens Henslow to Charles Darwin, 24 August 1831, quoted in Nora Barlow (editor) *Darwin and Henslow: the growth of an idea, letters 1831–1860* London: Murray 1967 pp. 29–30.
10. Charles Darwin to John Stevens Henslow, 15 November 1831, Darwin Correspondence Project letter #147.
11. Charles Darwin *Narrative of the surveying voyages of His Majesty's ships* Adventure *and* Beagle, *volume III: journal and remarks, 1832–1836* London: Henry Colburn 1839 pp. 484–5.
12. Charles Darwin *Extracts from letters addressed to Professor Henslow* Cambridge: Cambridge Philosophical Society 1835 pp. 1–31; Cambridge Philosophical Society Archives, 3/1/2, minutes of general meeting, 16 November 1835; Anonymous 'Cambridge Philosophical Society, 16 November 1835' *The London, Edinburgh, and Dublin philosophical magazine and journal of science: series 3* VIII (1836) pp. 78–80, p. 79.
13. Charles Darwin (edited by Nora Barlow) *The works of Charles Darwin, volume I: diary of the voyage of H.M.S.* Beagle New York: New York University Press 1986 p. 315.
14. Charles Darwin to Catherine Darwin, 3 June 1836, Darwin Correspondence Project letter #302.
15. Charles Darwin to John Stevens Henslow, 9 July 1836, Darwin Correspondence Project letter #304. Herschel also recorded the dinner with Darwin in his diary: John Herschel (edited by David S. Evans, Terence J. Deeming, Betty Hall Evans, and Stephen Goldfarb) *Herschel at the Cape: diaries and correspondence of Sir John Herschel, 1834–1838* Austin: University of Texas Press 1969 p. 242.
16. Caroline Darwin to Charles Darwin, 29 December 1835, Darwin Correspondence Project letter #291; Anonymous 'Geological Society of London' *The Athenaeum* Issue 421, 21 November 1835 p. 876.
17. John Stevens Henslow to William Lonsdale, 28 November 1835, quoted in Susan F. Cannon *Science in culture: the early Victorian period* New York: Dawson and Science History Publications 1978 p. 56.

18. Charles Darwin *Extracts from letters addressed to Professor Henslow* Cambridge: Cambridge Philosophical Society 1835 pp. 23–31; Cambridge Philosophical Society Archives, 3/1/2, minutes of general meeting, 14 December 1835; Anonymous 'Cambridge Philosophical Society, 14 December 1835' *The London, Edinburgh, and Dublin philosophical magazine and journal of science: series 3* VIII (1836) p. 80.
19. See the article by the French naturalist M. Gay in Charles Darwin *Extracts from letters addressed to Professor Henslow* Cambridge: Cambridge Philosophical Society 1835 pp. 30–1.
20. Charles Darwin to John Stevens Henslow, 18 May 1832, Darwin Correspondence Project letter #171.
21. Charles Darwin to John Stevens Henslow, 11 April 1833, Darwin Correspondence Project letter #204.
22. Charles Darwin *Extracts from letters addressed to Professor Henslow* Cambridge: Cambridge Philosophical Society 1835 p. 24.
23. Cambridge Philosophical Society Archives, 2/1/2, minutes of Council meeting, 30 November 1835.
24. Charles Darwin *Extracts from letters addressed to Professor Henslow* Cambridge: Cambridge Philosophical Society 1835 preface, dated 1 December 1835.
25. Charles Darwin to Catherine Darwin, 3 June 1836, Darwin Correspondence Project letter #302.
26. Caroline Darwin to Charles Darwin, 29 December 1835, Darwin Correspondence Project letter #291.
27. Charles Darwin to John Stevens Henslow, 30 October 1836, quoted in Nora Barlow (editor) *Darwin and Henslow: the growth of an idea, letters 1831–1860* London: Murray 1967 pp. 118–23.
28. Charles Darwin to John Stevens Henslow, 9 September 1831, quoted in Nora Barlow (editor) *Darwin and Henslow: the growth of an idea, letters 1831–1860* London: Murray 1967 p. 40.
29. Adam Sedgwick to Samuel Butler, 1835, quoted in quoted in John Willis Clark and Thomas McKenny Hughes *The life and letters of Adam Sedgwick, volume I* Cambridge: Cambridge University Press 2009 p. 280; Charles Darwin to John Stevens Henslow, 30 October 1836, quoted in Nora Barlow (editor) *Darwin and Henslow: the growth of an idea, letters 1831–1860* London: Murray 1967 p. 119. The Zoological Museum referred to is the Museum of the Zoological Society of London.
30. Leonard Jenyns 'Cambridge Philosophical Society museum' in J.J. Smith (editor) *The Cambridge portfolio* London: J.W. Parker 1840 pp. 127–9, p. 128.
31. Cambridge Philosophical Society Archives, 3/1/2, minutes of general meeting, 27 February 1837; Anonymous 'Cambridge Philosophical Society, 27 February 1837' *The London, Edinburgh, and Dublin philosophical magazine and journal of science: series 3* X (1837) p. 316.
32. Charles Darwin to Caroline Darwin, 27 February 1837, Darwin Correspondence Project letter #346. Adam Sedgwick was a prebendary of Norwich Cathedral.

33. Cambridge Philosophical Society Archives, 3/1/2, minutes of general meeting, 12 March 1838; Charles Darwin to John Stevens Henslow, 26 March 1838, Darwin Correspondence Project letter #406.
34. The best-known pre-Darwinian theory of transmutation of species was proposed by Jean-Baptiste Lamarck; see R.W. Burkhardt *The spirit of system: Lamarck and evolutionary biology* Cambridge, Mass., and London: Harvard University Press 1995.
35. James A. Secord *Victorian sensation* Chicago: University of Chicago Press 2000 p. 2.
36. Book reviews: Anonymous 'Review of *Vestiges of creation*' *The Examiner* 9 November 1844 pp. 707–9; Anonymous 'Review of *Vestiges of creation*' *The Spectator* 9 November 1844 p. 1072.
37. James A. Secord *Victorian sensation* Chicago: University of Chicago Press 2000 pp. 1–2, 168–9.
38. Anonymous [Robert Chambers] *Vestiges of the natural history of creation* London: John Churchill 1844 pp. 12–15, 44–54, 58, 63, 76–80, 94–104, 111.
39. Anonymous [Robert Chambers] *Vestiges of the natural history of creation* London: John Churchill 1844 pp. 168, 185.
40. Anonymous [Robert Chambers] *Vestiges of the natural history of creation* London: John Churchill 1844 pp. 204–5.
41. Anonymous [Robert Chambers] *Vestiges of the natural history of creation* London: John Churchill 1844 pp. 212–13.
42. James A. Secord *Victorian sensation* Chicago: University of Chicago Press 2000 p. 131.
43. James A. Secord *Vestiges of the natural history of creation and other evolutionary writings* Chicago and London: University of Chicago Press 1994 introduction.
44. Anonymous [Adam Sedgwick] 'Review of *Vestiges of the natural history of creation*' *Edinburgh review* CLXV (1845) pp. 1–85, pp. 3, 74.
45. William Whewell to Adam Sedgwick, 6 June 1845, quoted in John Willis Clark and Thomas McKenny Hughes *The life and letters of Adam Sedgwick, volume I* Cambridge: Cambridge University Press 2009 p. 98.
46. James A. Secord *Victorian sensation* Chicago: University of Chicago Press 2000 pp. 227–8; William Whewell *Indications of the Creator* London: John W. Parker 1845.
47. William Whewell to R. Jones, 18 July 1845, quoted in James A. Secord *Victorian sensation* Chicago: University of Chicago Press 2000 p. 228.
48. Cambridge Philosophical Society Archives, 3/1/2, minutes of general meeting, 28 April 1845; James A. Secord *Victorian sensation* Chicago: University of Chicago Press 2000 p. 250.
49. William Clark 'A case of human monstrosity, with a commentary' *Transactions of the Cambridge Philosophical Society* IV (1833) pp. 219–56, p. 220.
50. William Clark, quoted in Anonymous [Adam Sedgwick] 'Review of *Vestiges of the natural history of creation*' *Edinburgh review* CLXV (1845) pp. 1–85, pp. 78–9. Clark did not publish the paper he delivered to the Society in 1845, but it can be reconstructed from this review which was published a few months later by Sedgwick in the *Edinburgh review* and which quoted Clark extensively.

51. Anonymous [Adam Sedgwick] 'Review of *Vestiges of the natural history of creation*' *Edinburgh review* CLXV (1845) pp. 1–85, pp. 74, 80–1.
52. See Darwin's comments in Charles Darwin to Caroline Darwin, 27 February 1837, Darwin Correspondence Project letter #346 about discussion at a meeting of the Society; and there are several instances of discussion mentioned in *Proceedings of the Cambridge Philosophical Society* I (1843–1863).
53. William Whewell *Indications of the Creator* London: John W. Parker 1845 p. 26.
54. J.W. Clark (revised by Michael Bevan) 'William Clark (1788–1869)' *Oxford dictionary of national biography* Retrieved 30 May 2018 from http://www.oxforddnb.com/view/10.1093/ref:odnb/9780198614128.001.0001/odnb-9780198614128-e-5478; James A. Secord *Victorian sensation* Chicago: University of Chicago Press 2000 pp. 249–50.
55. James A. Secord *Victorian sensation* Chicago: University of Chicago Press 2000 pp. 231–3.
56. For a list of the members of the Association's council, see Anonymous *Report of the fifteenth meeting of the British Association for the Advancement of Science, held at Cambridge in June 1845* London: John Murray 1846 p. viii; for details of the planning of the Association's meeting in Cambridge, see Cambridge Philosophical Society Archives, 2/1/3, minutes of Council meeting, 12 May 1845, 18 June 1845, 10 November 1845.
57. Michael J. Crowe 'John Frederick William Herschel (1792–1871)' *Oxford dictionary of national biography* Retrieved 15 March 2018 from http://www.oxforddnb.com/view/10.1093/ref:odnb/9780198614128.001.0001/odnb-9780198614128-e-13101.
58. Walter F. Cannon 'The impact of uniformitarianism: two letters from John Herschel to Charles Lyell, 1836–1837' *Proceedings of the American Philosophical Society* CV (1961) pp. 301–14.
59. James A. Secord *Victorian sensation* Chicago: University of Chicago Press 2000 pp. 406–9; John Herschel 'Address' *Report of the fifteenth meeting of the British Association for the Advancement of Science, held at Cambridge in June 1845* London: John Murray 1846 pp. xxvii–xliv.
60. M.E. Bury and J.D. Pickles (editors) *Romilly's Cambridge Diary, 1842–1847* Cambridge: Cambridgeshire Records Society 1994 p. 129.
61. John Herschel 'Address' *Report of the fifteenth meeting of the British Association for the Advancement of Science, held at Cambridge in June 1845* London: John Murray 1846 pp. xxvii–xliv, pp. xlii–xliii.
62. John Herschel 'Address' *Report of the fifteenth meeting of the British Association for the Advancement of Science, held at Cambridge in June 1845* London: John Murray 1846 pp. xxvii–xliv, p. xxviii.
63. John Herschel 'Address' *Report of the fifteenth meeting of the British Association for the Advancement of Science, held at Cambridge in June 1845* London: John Murray 1846 pp. xxvii–xliv, p. xxviii.
64. *Prospectus* quoted in Morse Peckham 'Dr. Lardner's *Cabinet Cyclopaedia*' *The papers of the Bibliographical Society of America* XLV (1951) pp. 37–58, p. 41.

65. Raymond D. Tumbleson '"Reason and religion": the science of Anglicanism' *Journal of the history of ideas* LVII (1996) pp. 131–56, pp. 132–3. For a case study of Professor William Buckland's approach to combining geology and religion, see Mott T. Green 'Genesis and geology revisited: the order of nature and the nature of order in nineteenth-century Britain' in David C. Lindberg and Ronald L. Numbers (editors) *When science and Christianity meet* Chicago and London: University of Chicago Press 2003 pp. 139–60.
66. John Herschel *A preliminary discourse on the study of natural philosophy* London: Longman, Rees, Orme, Brown and Green 1831 p. 7.
67. William Whewell *Astronomy and general physics considered with reference to natural theology* London: William Pickering 1833 pp. 356–7.
68. Walter F. Cannon 'John Herschel and the idea of science' *Journal of the history of ideas* XXII (1961) pp. 215–39, pp. 215–19.
69. John Herschel *A preliminary discourse on the study of natural philosophy* London: Longman, Rees, Orme, Brown and Green 1831 p. 76.
70. Adam Sedgwick to Charles Darwin, 24 November 1859, Darwin Correspondence Project letter #2548.
71. John Herschel *A preliminary discourse on the study of natural philosophy* London: Longman, Rees, Orme, Brown and Green 1831 pp. 144–5.
72. Adam Sedgwick *A discourse on the studies of the University of Cambridge* Cambridge: University of Cambridge Press 1833 pp. 9–13. See also Crosbie Smith 'Geologists and mathematicians: the rise of physical geology' in P.M. Harman (editor) *Wranglers and physicists: studies on Cambridge physics in the nineteenth century* Manchester: Manchester University Press 1985 pp. 49–83, p. 61; Eric Ashby and Mary Anderson 'Introduction' in Adam Sedgwick *A discourse on the studies of the University of Cambridge* Leicester: Leicester University Press 1969 pp. (7)–(17).
73. Adam Sedgwick *A discourse on the studies of the University of Cambridge* Fifth edition London: Parker 1850 p. xi.
74. Adam Sedgwick *A discourse on the studies of the University of Cambridge* Fifth edition London: Parker 1850 p. cccxxiii. The sequel to *Vestiges* was Anonymous *Explanations: a sequel to 'Vestiges of the natural history of creation'* London: Churchill 1845. The sequel to *Vestiges* was published in 1845 as a short response to some of the criticisms faced by it: Anonymous *Explanations, a sequel* London: John Churchill 1845.
75. Adam Sedgwick *A discourse on the studies of the University of Cambridge* Cambridge: University of Cambridge Press 1833 p. 23.
76. Cambridge Philosophical Society Archives, 3/1/2, minutes of general meeting, 7 May 1849; 3/1/3, minutes of general meeting, 5 March 1855.
77. Charles Darwin to John Stevens Henslow, 11 November 1859, Darwin Correspondence Project letter #2522.
78. Charles Darwin to Adam Sedgwick, 11 November 1859, Darwin Correspondence Project letter #2525; Charles Darwin *On the origin of species by means of natural selection* London: John Murray 1859.

79. Adam Sedgwick to Charles Darwin, 24 November 1859, Darwin Correspondence Project letter #2548.
80. Adam Sedgwick to Charles Darwin, 24 November 1859, Darwin Correspondence Project letter #2548.
81. John Stevens Henslow to Charles Darwin, 5 May 1860, Darwin Correspondence Project letter #2783
82. John Stevens Henslow to Charles Darwin, 5 May 1860, Darwin Correspondence Project letter #2783.
83. Cambridge Philosophical Society Archives, 3/1/3, minutes of general meeting, 7 May 1860.
84. Anonymous *Cambridge Herald and Huntingdonshire Gazette* 19 May 1860.
85. John Stevens Henslow to Joseph Dalton Hooker, 10 May 1860, Darwin Correspondence Project letter #2794. Dr Clark is William Clark, the anatomy professor who had chaired the meetings at which Darwin's letters from South America were read to the Society and who had spoken against *Vestiges*.
86. Charles Darwin to John Stevens Henslow, 14 May 1860, Darwin Correspondence Project letter #2801.
87. Charles Darwin to Joseph Dalton Hooker, 15 May 1860, Darwin Correspondence Project letter #2802.
88. J.A. Secord 'Adam Sedgwick (1785–1873)' *Oxford dictionary of national biography* Retrieved 2 March 2018 from http://www.oxforddnb.com/view/10.1093/ref:odnb/9780198614128.001.0001/odnb-9780198614128-e-25011; Roy Porter 'The natural science tripos and the "Cambridge school of geology", 1850–1914' *History of universities* II (1982) pp. 193–216, p. 206.
89. Charles Darwin to John Stevens Henslow, 2 April 1860, Darwin Correspondence Project letter #2742.
90. J.W. Salter 'On the succession of plant life upon the earth' *Proceedings of the Cambridge Philosophical Society* II (1876) pp. 125–8.
91. J.W. Salter 'Diagram of the relations of the univalve to the bivalve, and of this to the brachiopod' *Transactions of the Cambridge Philosophical Society* XI (1871) p. 485–8, pp. 488.
92. Cambridge Philosophical Society Archives, 3/1/3, minutes of general meeting, 31 March 1862, 27 February 1865. For more on Newton's views on Darwinism, see Jonathan Smith 'Alfred Newton: the scientific naturalist who wasn't' in Bernard Lightman and Michael S. Reidy (editors) *The age of scientific naturalism* Pittsburgh: University of Pittsburgh Press 2016 pp. 137–56.
93. Cambridge Philosophical Society Archives, 2/1/5, minutes of Council meeting, 1 May 1882.
94. Cambridge Philosophical Society Archives, 2/1/6, minutes of Council meeting, 27 April 1896.
95. A.C. Seward (editor) *Darwin and modern science* Cambridge: Cambridge Philosophical Society and Cambridge University Press 1910.

## Chapter 4

1. Anonymous 'Fatal accident on the London and Brighton Railway four lives lost' *The Times* 4 October 1841; Anonymous 'Report on the Inquest' *The Times* 5 October 1841 Retrieved 14 September 2016 from Sussex History Forum: http://sussexhistoryforum.co.uk/index.php?topic=3615.0;wap2.
2. Dionysius Lardner (editor) *The museum of science and art, volume I* London: Walton and Maberly 1854 p. 162.
3. For a report of the inquest, see Anonymous 'Report on the Inquest' *The Times* 5 October 1841. The official report was published by the Board of Trade, 15 October 1841 Retrieved 14 September 2016 from Railway Archives http://www.railways-archive.co.uk/documents/BoT_HaywardsHeath1841.pdf. For an example of public and press interest, see Walter Fletcher 'Letter to the editor' *The Spectator* 30 October 1841 p. 12.
4. Joseph Power 'An enquiry into the causes which led to the fatal accident on the Brighton Railway (Oct. 2 1841), in which is developed a principle of motion of the greatest importance in guarding against the disastrous effects of collision under whatever circumstances it may occur' *Transactions of the Cambridge Philosophical Society* VII (1842) pp. 301–17; David McKitterick 'Joseph Power (1798–1868)' *Oxford dictionary of national biography* Retrieved 31 May 2018 from http://www.oxforddnb.com/view/10.1093/ref:odnb/9780198614128.001.0001/odnb-9780198614128-e-22666.
5. Alexandrina Buchanan *Robert Willis (1800–1875) and the foundation of architectural history* Cambridge: Cambridge University Press 2013 p. 259.
6. Joseph Power 'An enquiry into the causes which led to the fatal accident on the Brighton Railway (Oct. 2 1841), in which is developed a principle of motion of the greatest importance in guarding against the disastrous effects of collision under whatever circumstances it may occur' *Transactions of the Cambridge Philosophical Society* VII (1842) pp. 301–17, p. 317. Power's detailed knowledge of railway accidents did not put him off investing in the proposed Cambridge and Oxford Railway of 1845: Anonymous *The railway chronicle* 26 April 1845 p. 488.
7. Cambridge Philosophical Society Archives, 3/1/2, minutes of general meeting, 29 November 1841.
8. Robert Willis to George Gabriel Stokes, 21 July 1849, quoted in George Gabriel Stokes and Joseph Larmor (editor) *Memoir and scientific correspondence of the late George Gabriel Stokes, Bart, volume I* Cambridge: Cambridge University Press 2010 pp. 124–5; George Gabriel Stokes 'Discussion of a differential equation relating to the breaking of railway bridges' *Transactions of the Cambridge Philosophical Society* VIII (1849) pp. 707–35, p. 707.
9. Andrew Warwick *Masters of theory: Cambridge and the rise of mathematical physics* Chicago: University of Chicago Press 2003 pp. 89–90.
10. Crosbie Smith 'William Hopkins (1793–1866)' *Oxford dictionary of national biography* Retrieved 31 May 2018 from http://www.oxforddnb.com/view/10.1093/ref:odnb/9780198614128.001.0001/odnb-9780198614128-e-13756.

11. Cambridge Philosophical Society Archives, 3/1/2, minutes of general meeting, 25 April 1842, 29 March 1843, 14 April 1845. David B. Wilson 'George Gabriel Stokes (1819–1903)' *Oxford dictionary of national biography* Retrieved 31 May 2018 from http://www.oxforddnb.com/view/10.1093/ref:odnb/9780198614128.001.0001/odnb-9780198614128-e-36313.
12. Cambridge Philosophical Society Archives, 3/1/2, minutes of general meeting, 12 May 1845, 6 December 1847, 21 May 1849, 10 December 1849, 7 March 1853, 25 May 1868.
13. George Gabriel Stokes 'Discussion of a differential equation relating to the breaking of railway bridges' *Transactions of the Cambridge Philosophical Society* VIII (1849) pp. 707–35, p. 707.
14. J.L.A. Simmons *Report to the Commissioners of Railways, by Mr. Walker and Captain Simmons, R.E., on the fatal accident on the 24th day of May 1847, by the falling of the bridge over the River Dee, on the Chester and Holyhead Railway* London 1849.
15. Cambridge Philosophical Society Archives, 3/1/2, minutes of general meeting, 21 May 1849, 10 December 1849.
16. George Gabriel Stokes and Joseph Larmor (editor) *Memoir and scientific correspondence of the late George Gabriel Stokes, Bart, volume I* Cambridge: Cambridge University Press 2010 p. 8.
17. Cambridge Philosophical Society Archives, 3/1/2, minutes of general meeting, 26 April 1852, 1 May 1854, 11 May 1857. W. Towler Kingsley 'Application of the microscope to photography' *Journal of the Society of Arts, London* I (1853) pp. 289–92; W. Towler Kingsley 'Application of photography to the microscope' *Proceedings of the Cambridge Philosophical Society* I (1863) pp. 117–19.
18. Anonymous letter in P.M. Harman (editor) *The scientific letters and papers of James Clerk Maxwell, volume I: 1846–1862* Cambridge: Cambridge University Press 1990 p. 314; Roger Taylor and Larry John Schaaf *Impressed by light: British photographs from paper negatives, 1840–1860* New Haven: Yale University Press 2007 p. 337–8.
19. Cambridge Philosophical Society Archives, 3/1/2, minutes of general meeting, 23 February 1846, 9 March 1846; Anonymous 'Cambridge Philosophical Society' *The London, Edinburgh, and Dublin philosophical magazine and journal of science: series 3* XXIX (1846) pp. 65–6; John Timbs *The year-book of facts in science and art* London 1847 pp. 194–5, 242.
20. Roger Hutchins 'John Couch Adams (1819–1892)' *Oxford dictionary of national biography* Retrieved 31 May 2018 from http://www.oxforddnb.com/view/10.1093/ref:odnb/9780198614128.001.0001/odnb-9780198614128-e-123.
21. Walter White *The journals of Walter White* Cambridge: Cambridge University Press 2012 p. 72.
22. George Biddell Airy *Account of the Northumberland equatoreal and dome attached to the Cambridge Observatory* Cambridge: Cambridge University Press 1844.
23. Anonymous *Inverness courier* 28 October 1846; Anonymous 'Professor Airey's [sic] statement regarding Leverrier's planet' *The British quarterly review* VI (1847) pp. 1–40, pp. 34, 36; Anonymous *Royal Cornwall gazette* 17 September 1847.

24. Cambridge Philosophical Society Archives, 3/1/2 minutes of general meeting, 3 May 1847; 2/1/3 minutes of Council meeting, 13 May 1847; Susan F. Cannon *Science in culture: the early Victorian period* New York: Dawson and Science History Publications 1978 pp. 191–2.
25. Cambridge Philosophical Society Archives, 3/1/2, minutes of general meeting, 9 November 1846.
26. J.P.T. Bury (editor) *Romilly's Cambridge Diary, 1842–1847* Cambridge: Cambridgeshire Records Society 1984 p. 180. Professor Miller was William Hallowes Miller, who had been Professor of Mineralogy since 1832 and was an expert on crystallography.
27. John Herschel to Richard Sheepshanks, 17 December 1846, quoted in Robert W. Smith 'The Cambridge network in action: the discovery of Neptune' *Isis* 80 (1989) pp. 395–422, pp. 415–16.
28. Roger Hutchins 'John Couch Adams (1819–1892)' *Oxford dictionary of national biography* Retrieved 31 May 2018 from http://www.oxforddnb.com/view/10.1093/ref:odnb/9780198614128.001.0001/odnb-9780198614128-e-123; Sophia Elizabeth De Morgan *Memoir of Augustus De Morgan* Cambridge: Cambridge University Press 2010 pp. 127–30.
29. Robert W. Smith 'The Cambridge network in action: the discovery of Neptune' *Isis* 80 (1989) pp. 395–422.
30. Anonymous 'A choral ode addressed to HRH Prince Albert' *Cambridge chronicle* 10 July 1847. The editor helpfully explains that 'mother' refers to 'alma mater'. 'Granta' is another name for the River Cam.
31. Adam Sedgwick *A discourse on the studies of the University* Fifth edition London: Parker 1850 p. cccxxxiii. On the same occasion, John Herschel presented the first bound copy of his observations made at the Cape of Good Hope to the royal couple.
32. The Adams Prize is still awarded today, but the winner's essay is no longer (necessarily) published by the Society. The first prize was awarded to R. Peirson in 1850 for 'The theory of the long inequality of Uranus and Neptune' *Transactions of the Cambridge Philosophical Society* IX (1856) Appendix I pp. i–lxvii. The judges were H.W. Cookson (Vice-Chancellor of the University and later President of the Cambridge Philosophical Society), John Herschel, James Challis, and Matthew O'Brien.
33. The quotations are taken from an article William Peverill Turnbull wrote for the Parallelepiped Club and which was cited in A. Rupert Hall *The Cambridge Philosophical Society: a history 1819–1969* Cambridge: Cambridge Philosophical Society 1969 p. 17. For information on Turnbull, see John Venn and J. A. Venn (editors) *Alumni Cantabrigienses, part II: 1752–1900, volume vi* Cambridge: Cambridge University Press 1954 p. 245 and John Couch Adams 'On the meteoric shower of November, 1866' *Proceedings of the Cambridge Philosophical Society* II (1866) p. 60.
34. See the graph in A. Rupert Hall *The Cambridge Philosophical Society: a history 1819–1969* Cambridge: Cambridge Philosophical Society 1969 p. 71. There had been a

slight dip in undergraduate numbers in the 1840s, and attendance at scientific lectures had also fallen. The general fall in numbers has been attributed to the 'hungry forties'—a phenomenon caused by a combination of strict Corn Laws that imposed taxes on imported grain, and British crop failures in the 1840s; it meant that people had less disposable income to spend on education. Around the same time, the Royal College of Surgeons began to accept members who did not have a degree from Cambridge or Oxford—this too affected enrolments in Cambridge. The fall in numbers attending scientific lectures in Cambridge has been attributed to the creation of a new divinity examination in 1842 that was essential for those wishing to be ordained (still the majority of students at this period); the exam meant that students had less free time to devote to subjects like natural philosophy, which was not examined. But, by the 1850s, the numbers enrolling at the University and attending lectures was rising again; see Harvey W. Becher 'Voluntary science in nineteenth-century Cambridge University to the 1850s' *British journal for the history of science* 19 (1986) pp. 80–1 and James A. Secord *Victorian sensation* 2000 p. 224.

35. Nora Barlow (editor) *Darwin and Henslow: the growth of an idea, letters 1831–1860* London: Murray 1967 p. 5.
36. Cambridge Philosophical Society Archives, 2/1/3, minutes of Council meeting, 11 May 1846; S.M. Walters and E.A. Stow *Darwin's mentor: John Stevens Henslow, 1796–1861* Cambridge: Cambridge University Press 2001 pp. 128–53.
37. Lucilla Burn *The Fitzwilliam Museum: a history* London: Philip Wilson 2016.
38. Adam Sedgwick *Discourse on the studies of the university* Fifth edition London: Parker 1850 pp. cccxxxv–cccxxxvi.
39. Adam Sedgwick *Discourse on the studies of the university* Fifth edition London: Parker 1850 pp. cccxxxi–cccxxxii.
40. Peter Searby *A history of the University of Cambridge, volume II: 1750–1870* Cambridge: Cambridge University Press 1997 pp. 507–13.
41. Robert Peel to Prince Albert, 27 October 1847, quoted in Theodore Martin *The life of His Royal Highness the Prince Consort, volume II* Cambridge: Cambridge University Press 2013 pp. 117–18.
42. Peter Searby *A history of the University of Cambridge, volume II: 1750–1870* Cambridge: Cambridge University Press 1997 pp. 507–30.
43. Compare this to Oxford where, until the 1880s, the undergraduate curriculum focused primarily on the classics. Science became a degree subject in Oxford in 1886. M.G. Brock and M.C. Curthoys (editors) *The history of the University of Oxford,* volume VI: nineteenth-century Oxford, part 1 Oxford: Clarendon Press 1997 p. 355.
44. Fenton J. Hort to his mother, 8 March 1851, quoted in Peter Searby *A history of the University of Cambridge, volume II: 1750–1870* Cambridge: Cambridge University Press 1997 p. 203.
45. Peter Searby *A history of the University of Cambridge, volume II: 1750–1870* Cambridge: Cambridge University Press 1997 pp. 203–8.

46. In reply to an address which the Senate ordered to be presented to him on the 75th anniversary of his matriculation: Anonymous 'Address to the President' *The Eagle XLII* (1922) pp. 161–70, p. 163; W.C.D. Dampier (revised by Frank A.J.L. James) 'George Liveing (1827–1924)' *Oxford dictionary of national biography* Retrieved 31 May 2018 from http://www.oxforddnb.com/view/10.1093/ref:odnb/9780198614128.001.0001/odnb-9780198614128-e-34559; Peter Searby *A history of the University of Cambridge, volume II: 1750–1870* Cambridge: Cambridge University Press 1997 pp. 223–4.
47. Anonymous *The Cambridge Ray Club. Instituted March 11, 1837. A short account of the Club, with its laws, and a list of members* Cambridge 1857; Anna Maria Babington (editor) *Memorials, journal and botanical correspondence of Charles Cardale Babington* Cambridge: Cambridge University Press 2013 pp. liii, 70, 97, 111, 122; William C. Lubenow *'Only connect': learned societies in nineteenth-century Britain* Woodbridge: Boydell Press 2015 pp. 48–9.
48. For information on the natural sciences and medical sciences societies, see Mark Weatherall *Gentlemen, scientists, and doctors: medicine at Cambridge 1800–1940* Woodbridge: Boydell Press 2000 pp. 104–5. The Antiquarian Society was also founded by Babington. The Camden Society was ostensibly devoted to the study of church architecture, but involved itself heavily in the politics of the Church of England; see James F. White *The Cambridge movement: the ecclesiologists and the Gothic revival* Cambridge: Cambridge University Press 1962. For information on the Eranus Society, see William C. Lubenow *'Only connect': learned societies in nineteenth-century Britain* Woodbridge: Boydell Press 2015 pp. 44–8; for information on the Grote Society, see John Gibbins *John Grote, Cambridge University and the development of Victorian thought* Exeter: Imprint Academic 2007; for information on the Cambridge Apostles, see William C. Lubenow *The Cambridge Apostles, 1820–1914* Cambridge: Cambridge University Press 1998.
49. Augustus De Morgan to John Herschel, 28 May 1845, quoted in Tony Crilly 'The *Cambridge mathematical journal* and its descendants: the linchpin of a research community in the early and mid-Victorian age' *Historia mathematica* XXXI (2004) pp. 455–97, p. 470.
50. Cambridge Philosophical Society Archives, 3/1/2, minutes of general meeting, 2 May 1836, 16 May 1836, 18 March 1839, 2 March 1840, 10 November 1840, 25 May 1846; 3/1/3, minutes of general meeting, 8 May 1848, 6 December 1858, 9 November 1868.
51. Adam Sedgwick *A discourse on the studies of the University* Fifth edition London: Parker 1850 p. cccxxxii.
52. George Biddell Airy 'On an inequality of long period in the motions of the Earth and Venus' *Philosophical transactions of the Royal Society* CXXI (1832) pp. 67–124; Adam Sedgwick *A discourse on the studies of the University* Fifth edition London: Parker 1850 p. cccxxxix.
53. Cambridge Philosophical Society Archives, 3/1/2, minutes of general meeting, 16 May 1836, 16 March 1840, 14 December 1840, 29 November 1841, 27 November

1843, 28 October 1844. Leslie Stephen (revised by I Grattan-Guinness) 'Augustus De Morgan (1806–1871)' *Oxford dictionary of national biography* Retrieved 31 May 2018 from http://www.oxforddnb.com/view/10.1093/ref:odnb/9780198614128.001.0001/odnb-9780198614128-e-7470.

54. Ronald M. Birse 'Philip Kelland (1808–1879)' *Oxford dictionary of national biography* Retrieved 31 May 2018 from http://www.oxforddnb.com/view/10.1093/ref:odnb/9780198614128.001.0001/odnb-9780198614128-e-15284; John C. Thackray 'David Ansted (1814–1880)' *Oxford dictionary of national biography* Retrieved 31 May 2018 from http://www.oxforddnb.com/view/10.1093/ref:odnb/9780198614128.001.0001/odnb-9780198614128-e-577. For information on Matthew O'Brien, see Alex D.D. Craik *Mr Hopkins' men: Cambridge reform and British mathematics in the 19th century* London: Springer 2007 p. 142. Cambridge Philosophical Society Archives, 3/1/2, minutes of general meeting, 16 and 30 March 1840, 4 and 18 May 1840; Anonymous 'University news: Cambridge' *The British magazine and monthly register* XVII (1840) p. 707; Matthew O'Brien 'On the symbolical equation of vibratory motion of an elastic medium, whether crystallized or uncrystallized' *Transactions of the Cambridge Philosophical Society* VIII (1849) pp. 508–23 (O'Brien had five papers in this volume of the *Transactions*).
55. Richard Owen 'Description of an extinct lacertian reptile, *Rhynchosaurus articeps*, (Owen) of which the bones and footprints characterize the Upper New Red Sandstone at Grinsill, near Shrewsbury' *Transactions of the Cambridge Philosophical Society* VII (1842) pp. 355–70; Geoffrey Tresise and Michael J. King 'History of ichnology: the misconceived footprints of rhynchosaurs' *Ichnos* XIX (2012) pp. 228–37.
56. All articles are by William Whewell: 'On the fundamental antithesis of philosophy' *Transactions of the Cambridge Philosophical Society* VIII (1849) pp. 170–82; 'Second memoir on the fundamental antithesis of philosophy' *Transactions of the Cambridge Philosophical Society* VIII (1849) pp. 614–20; 'On the intrinsic equation of a curve' *Transactions of the Cambridge Philosophical Society* VIII (1849) pp. 659–71; 'On Hegel's criticism of Newton's *Principia*' *Transactions of the Cambridge Philosophical Society* VIII (1849) pp. 696–706; 'Criticism of Aristotle's account of induction' *Transactions of the Cambridge Philosophical Society* IX (1856) part I pp. 63–72; 'Mathematical exposition of some doctrines of political economy' *Transactions of the Cambridge Philosophical Society* IX (1856) part I pp. 128–49; 'Second memoir on the intrinsic equation of a curve' *Transactions of the Cambridge Philosophical Society* IX (1856) part I pp. 150–6; 'Mathematical exposition of certain doctrines of political economy, third memoir' *Transactions of the Cambridge Philosophical Society* IX (1856) part II pp. 1–7; 'Of the transformation of hypotheses in the history of science' *Transactions of the Cambridge Philosophical Society* IX (1856) part II pp. 139–46; 'On Plato's survey of the sciences' *Transactions of the Cambridge Philosophical Society* IX (1856) part IV pp. 582–9; 'On Plato's notion of dialectic' *Transactions of the Cambridge Philosophical Society* IX (1856) part IV pp. 590–7; 'Of the intellectual powers according to Plato' *Transactions of the Cambridge Philosophical Society* IX (1856) part IV pp. 598–604.

57. John Herschel 'Address' *Report of the fifteenth meeting of the British Association for the Advancement of Science, held at Cambridge in June 1845* London 1846 p. xxviii.
58. Anonymous [William Whewell] 'Review of *Transactions of the Cambridge Philosophical Society, volume the third*' *The British critic* IX (1831) pp. 71–90, pp. 71, 89.
59. Cambridge Philosophical Society Archives, 2/1/3, minutes of Council meeting, 14 March 1842, 14 November 1842, undated May 1843, 1 July 1843, 15 February 1845, 1 November 1847, 18 November 1851, 24 November 1851; 2/1/4, minutes of Council meeting, 30 May 1859, 8 February 1869.
60. Cambridge Philosophical Society Archives, 2/1/3, minutes of Council meeting, 1 July 1843; *Proceedings of the Cambridge Philosophical Society* (1843–63).
61. Cambridge Philosophical Society Archives, 2/1/3, minutes of Council meeting, 16 October 1841, 5 February 1844, 24 November 1851, 16 February 1852; 2/1/4, minutes of Council meeting, 12 December 1853, 10 December 1855, 19 May 1856, 16 October 1869, 22 November 1869.
62. Cambridge Philosophical Society Archives, 2/1/4, minutes of Council meeting, 11 December 1854, 6 July 1858, 20 April 1863.

## Chapter 5

1. Augustus De Morgan 'On some points of the integral calculus' *Proceedings of the Cambridge Philosophical Society* I (1863) pp. 106–9.
2. Willie Sugg *A history of Cambridgeshire cricket, 1700–1890* 2008 Retrieved 4 August 2016 from http://www.cambscrickethistory.co.uk/new%20writing.shtml. On the yeoman bedell and other university posts in the nineteenth century, see Anonymous *A new guide to the University and Town of Cambridge* Cambridge 1831 pp. 14–15; for information of the kinds of task Crouch would have performed as bedell, see Charles Henry Cooper *Annals of Cambridge, volume V* Cambridge: Warwick and Co. 1850–1856 pp. 63–4; on the salaries paid to bedells, see H.P. Stokes *The esquire bedells of the University of Cambridge from the 13th century to the 20th century* Cambridge: Cambridge Antiquarian Society 1911 p. 128.
3. Anonymous *The Gazette* Issue 21180, 11 February 1851 p. 357.
4. Cambridge Philosophical Society Archives, 2/1/3, minutes of Council meeting, 10 March 1851.
5. Cambridge Philosophical Society Archives, 2/1/3, minutes of Council meeting, 5 February 1844, 10 March 1851, 14 March 1851.
6. Cambridge Philosophical Society Archives, 2/1/3, minutes of Council meeting, 15 March 1852.
7. Cambridge Philosophical Society Archives, 2/1/3, minutes of Council meeting, 7 November 1842, 5 February 1844, 18 June 1845.
8. Cambridge Philosophical Society Archives, 2/1/3, minutes of Council meeting, 10 March 1851.
9. Anonymous *Cambridge independent press* 2 August 1851; H.P. Stokes *The esquire bedells of the University of Cambridge from the 13th century to the 20th century* Cambridge: Cambridge Antiquarian Society 1911 pp. 124, 129.

10. John Willis Clark 'The foundation and early years of the Society' *Proceedings of the Cambridge Philosophical Society* VII (1892) pp. i–l, pp. xiii, xvi.
11. Cambridge Philosophical Society Archives, 2/1/3, minutes of Council meeting, 29 May 1852, 6 September 1852, 19 October 1852, 26 October 1852; 7/2/9, papers relating to the dispute between CPS and the churchwardens of the Holy Sepulchre over the poor tax.
12. Cambridge Philosophical Society Archives, 2/1/4, minutes of Council meeting, 13 November 1854, 27 November 1854.
13. Cambridge Philosophical Society Archives, 2/1/4, minutes of Council meeting, 11 December 1854.
14. Cambridge Philosophical Society Archives, 2/1/4, minutes of Council meeting, 7 April 1856.
15. Cambridge Philosophical Society Archives, 3/1/3, minutes of general meeting, 14 April 1856.
16. Cambridge Philosophical Society Archives, 2/1/4, minutes of Council meeting, 28 June 1856, 20 July 1856, 3 November 1856, 8 December 1856, 9 February 1857.
17. Cambridge Philosophical Society Archives, 2/1/4, minutes of Council meeting, 25 May 1857.
18. Cambridge Philosophical Society Archives, 3/1/3, minutes of general meeting, 28 April 1856, 6 November 1856, 26 October 1863; 2/1/4, minutes of Council meeting, 20 July 1856, 3 November 1856.
19. Cambridge Philosophical Society Archives, 2/1/4, minutes of Council meeting, 30 May 1859.
20. Cambridge Philosophical Society Archives, 2/1/3, minutes of Council meeting, 24 November 1851; 2/1/4, minutes of Council meeting, 8 February 1869.
21. Cambridge Philosophical Society Archives, 2/1/4, minutes of Council meeting, 12 December 1853, 2 November 1863, 7 December 1863.
22. Cambridge Philosophical Society Archives, 2/1/4, minutes of Council meeting, 13 February 1865, 27 February 1865.
23. Cambridge Philosophical Society Archives, 2/1/3, minutes of Council meeting, 23 November 1850; 2/1/4, minutes of Council meeting, 21 May 1855, 1 May 1865; 3/1/3, minutes of general meeting, 21 May 1855. Jenyns's specimens are still held in Cambridge University's Zoology Museum, but have not been kept together as he wished.
24. Charles Darwin to Leonard Jenyns, 9 May 1842, Darwin Correspondence Project letter #629; Cambridge Philosophical Society Archives, 2/1/4, minutes of Council meeting, 1 May 1865.
25. Cambridge Philosophical Society Archives, 2/1/4, minutes of Council meeting, 13 March 1865.
26. Cambridge Philosophical Society Archives, 2/1/4, minutes of Council meeting, 24 May 1865, 8 June 1865; 3/1/3, minutes of general meeting, 15 May 1865.
27. Cambridge Philosophical Society Archives, 2/1/4, minutes of Council meeting, 29 May 1865, 16 October 1865, 30 October 1865, 27 November 1865; 3/1/3, minutes of general meeting, 5 February 1866.

28. A.R. Catton 'On the synthesis of formic acid' *Proceedings of the Cambridge Philosophical Society* I (1863) p. 235; Arthur Cayley 'A new theorem on the equilibrium of four forces acting on a solid body' *Proceedings of the Cambridge Philosophical Society* I (1863) p. 235.
29. Peter Searby *A history of the University of Cambridge, volume III: 1750–1870* Cambridge: Cambridge University Press 1997 p. 251; Roy MacLeod and Russell Moseley 'Breaking the circle of the sciences: the Natural Sciences Tripos and the "examination revolution"' in Roy MacLeod (editor) *Days of judgement: science, examinations and the organization of knowledge in late Victorian England* Driffield: Nafferton 1982 pp. 189–212, pp. 192, 195.
30. Christopher N.L. Brooke *A history of the University of Cambridge, volume IV: 1870–1990* Cambridge: Cambridge University Press 1993 pp. 157, 295.
31. Malcolm Longair *Maxwell's enduring legacy: a scientific history of the Cavendish Laboratory* Cambridge: Cambridge University Press 2016 p. 43.
32. Christopher N.L. Brooke *A history of the University of Cambridge, volume IV: 1870–1990* Cambridge: Cambridge University Press 1993 pp. 152, 173–4; Malcolm Longair *Maxwell's enduring legacy: a scientific history of the Cavendish Laboratory* Cambridge: Cambridge University Press 2016 pp. 43–4; Cambridge Philosophical Society Archives, 2/1/4, minutes of Council meeting, 10 November 1862; 3/1/3, minutes of general meeting, 10 November 1862. For more on the Devonshire Commission, see Bernard Lightman 'Huxley and the Devonshire Commission' in Gowan Dawson and Bernard Lightman (editors) *Victorian scientific naturalism* Chicago: University of Chicago Press 2014 pp. 101–30.
33. Sedley Taylor 'Physical science at Cambridge' *Nature* II (12 May 1870) p. 28; *Cambridge University reporter* 4 November 1873 pp. 68–9.
34. Anonymous 'A voice from Cambridge' *Nature* VIII (8 May 1873) p. 21.
35. Anonymous 'A voice from Cambridge II' *Nature* VIII (15 May 1873) p. 41.
36. Ruth Barton 'Lockyer's columns of controversy in *Nature*' Retrieved 9 October 2018 from http://www.nature.com/nature/history/full/nature06260.html.
37. T.G. Bonney 'Science at Cambridge' *Nature* VIII (29 May 1873) p. 83.
38. Malcolm Longair *Maxwell's enduring legacy: a scientific history of the Cavendish Laboratory* Cambridge: Cambridge University Press 2016 p. 49; Crosbie Smith 'William Thomson (1824–1907)' *Oxford dictionary of national biography* Retrieved 31 May 2018 from http://www.oxforddnb.com/view/10.1093/ref:odnb/9780198614128.001.0001/odnb-9780198614128-e-36507.
39. Cambridge Philosophical Society Archives, 3/1/3, minutes of general meeting, 13 March 1854; James Clerk Maxwell 'On the transformation of surfaces by bending' *Transactions of the Cambridge Philosophical Society* IX (1856) pp. 445–70.
40. Cambridge Philosophical Society Archives, 2/1/4, minutes of Council meeting, 14 March 1859; James Clerk Maxwell *On the stability of the motion of Saturn's rings* Cambridge: Macmillan 1859; P.M. Harman 'James Clerk Maxwell (1831–1879)' *Oxford dictionary of national biography* Retrieved 31 May 2018 from http://www.oxforddnb.com/view/10.1093/ref:odnb/9780198614128.001.0001/odnb-9780198614128-e-5624.

41. James Clerk Maxwell 'Introductory lecture on experimental physics' in W.D. Niven (editor) *The scientific papers of James Clerk Maxwell, volume II* Cambridge: Cambridge University Press 1890, pp. 241–55, pp. 241–2; Dennis Moralee *A hundred years and more of Cambridge physics* Cambridge: Cambridge University Physics Society 1995 pp. 8–20.
42. Cambridge Philosophical Society Archives, 3/1/3, minutes of general meeting, 3 February 1873; James Clerk Maxwell 'On the proof of the equations of motion of a connected system' *Proceedings of the Cambridge Philosophical Society* II (1876) pp. 292–4; James Clerk Maxwell 'On a problem in the calculus of variations in which the solution is discontinuous' *Proceedings of the Cambridge Philosophical Society* II (1876) pp. 294–5.
43. Roy M. MacLeod 'The support of Victorian science: the endowment of research movement in Great Britain, 1868–1900' *Minerva* IX (1971) pp. 197–230, p. 209.
44. Dennis Moralee *A hundred years and more of Cambridge physics* Cambridge: Cambridge University Physics Society 1995 pp. 8–20.
45. W.M. Hicks, quoted in *A history of the Cavendish Laboratory, 1871–1910* London: Longman, Green and Co. 1910 p. 19.
46. James Clerk Maxwell, quoted in Dong-Won Kim *Leadership and creativity: a history of the Cavendish Laboratory, 1871–1919* Dordrecht: Kluwer 2002 p. 14.
47. James Clerk Maxwell 'On Faraday's lines of force' *Proceedings of the Cambridge Philosophical Society* I (1863) pp. 160–6; James Clerk Maxwell 'On Faraday's lines of force' *Transactions of the Cambridge Philosophical Society* X (1864) pp. 27–83; James Clerk Maxwell 'On physical lines of force, part I' *The London, Edinburgh, and Dublin philosophical magazine and journal of science: series 4* XXI (1861) pp. 161–75; James Clerk Maxwell 'A dynamical theory of the electromagnetic field' *Philosophical transactions of the Royal Society* CLV (1865) pp. 459–512. See also Crosbie Smith 'Force energy and thermodynamics' in Mary Jo Nye (editor) *The Cambridge history of science, volume V: the modern physical and mathematical sciences* Cambridge: Cambridge University Press 2003 pp. 289–310, pp. 305–6.
48. Albert Einstein 'Considerations concerning the fundaments of theoretical physics' Science XCI (1940) pp. 487–92.
49. *Proceedings of the Cambridge Philosophical Society* II (1876) pp. 242, 289, 292, 294, 302, 318, 338, 365, 372, 407, 427.
50. Cambridge Philosophical Society Archives, 2/1/4, minutes of Council meeting, 3 June 1861; *Proceedings of the Cambridge Philosophical Society* II (1876) p. 63.
51. Kostas Gavroglu 'John William Strutt, third Baron Rayleigh (1842–1919)' *Oxford dictionary of national biography* Retrieved 31 May 2018 from http://www.oxforddnb.com/view/10.1093/ref:odnb/9780198614128.001.0001/odnb-9780198614128-e-36359.
52. Henniker Heaton, 1887, quoted in Simon Schaffer 'Rayleigh and the establishment of electrical standards' *European journal of physics* XV (1994) pp. 277–85, p. 278; Bruce Hunt 'Doing science in a global empire: cable telegraphy and electrical physics in Victorian Britain' in Bernard Lightman (editor) *Victorian science in context* Chicago: University of Chicago Press 1997 pp. 312–33.

53. James Clerk Maxwell, quoted in Simon Schaffer 'Rayleigh and the establishment of electrical standards' *European journal of physics* XV (1994) pp. 277–85, p. 278.
54. *Proceedings of the Cambridge Philosophical Society* III (1880) p. 339; William Strutt, Lord Rayleigh 'On the minimum aberration of a single lens for parallel rays' *Proceedings of the Cambridge Philosophical Society* III (1880) pp. 373–5.
55. William Strutt, Lord Rayleigh 'On a new arrangement for sensitive flames' *Proceedings of the Cambridge Philosophical Society* IV (1883) pp. 17–18; William Strutt, Lord Rayleigh 'The use of telescopes on dark nights' *Proceedings of the Cambridge Philosophical Society* IV (1883) pp. 197–8; William Strutt, Lord Rayleigh 'On a new form of gas battery' *Proceedings of the Cambridge Philosophical Society* IV (1883) p. 198; William Strutt, Lord Rayleigh 'On the mean radius of coils of insulated wire' *Proceedings of the Cambridge Philosophical Society* IV (1883) pp. 321–4; Lord Rayleigh 'On the invisibility of small objects in a bad light' *Proceedings of the Cambridge Philosophical Society* IV (1883) p. 324.
56. Isobel Falconer 'Joseph John Thomson (1856–1940)' *Oxford dictionary of national biography* Retrieved 31 May 2018 from http://www.oxforddnb.com/view/10.1093/ref:odnb/9780198614128.001.0001/odnb-9780198614128-e-36506.
57. J.J. Thomson 'Some experiments on the electric discharge in a uniform electric field, with some theoretical considerations about the passage of electricity through gases' *Proceedings of the Cambridge Philosophical Society* V (1886) pp. 391–409, pp. 391–3.
58. Helge Kragh 'The vortex atom: a Victorian theory of everything' *Centaurus* XLVI (2002) pp. 32–114; Malcolm Longair *Maxwell's enduring legacy: a scientific history of the Cavendish Laboratory* Cambridge: Cambridge University Press 2016 pp. 118–20.
59. J.J. Thomson 'Note on the rotation of the plane of polarisation of light by a moving medium' *Proceedings of the Cambridge Philosophical Society* V (1886) pp. 250–4; J.J. Thomson and H.F. Newell 'Experiments on the magnetisation of iron rods' *Proceedings of the Cambridge Philosophical Society* VI (1889) pp. 84–90; J.J. Thomson and J. Monckman 'The effect of surface tension on chemical action' *Proceedings of the Cambridge Philosophical Society* VI (1889) pp. 264–9; J.J. Thomson 'The application of the theory of transmission of alternating currents along a wire to the telephone' *Proceedings of the Cambridge Philosophical Society* VI (1889) pp. 321–5; J.J. Thomson 'On the effect of pressure and temperature on the electric strength of gases' *Proceedings of the Cambridge Philosophical Society* VI (1889) pp. 325–33; J.J. Thomson 'On the absorption of energy by the secondary of a transformer' *Proceedings of the Cambridge Philosophical Society* VII (1892) p. 249; J.J. Thomson 'A method of comparing the conductivities of badly conducting substances for rapidly alternating currents' *Proceedings of the Cambridge Philosophical Society* VIII (1895) pp. 258–69.
60. Paula Gould 'Women and the culture of university physics in late nineteenth-century Cambridge' *British journal for the history of science* XXX (1997) pp. 127–49, p. 143; Malcolm Longair *Maxwell's enduring legacy: a scientific history of the Cavendish Laboratory* Cambridge: Cambridge University Press 2016 p. 101.

61. Cambridge Philosophical Society Archives, 2/1/6, minutes of Council meeting, 24 October 1887.
62. Thomas Archer Hirst, quoted in J. Vernon Jensen 'The X Club: fraternity of Victorian scientists' *British journal for the history of science* V (1970) pp. 63–72, p. 63. See also Roy M. MacLeod 'The X Club: a social network of science in late-Victorian England' *Notes and records of the Royal Society of London* XXIV (1970) pp. 305–22; Ruth Barton '"Huxley, Lubbock, and half a dozen others": professionals and gentlemen in the formation of the X Club, 1851–1864' *Isis* LXXXIX (1998) pp. 410–44; Roy M. MacLeod 'The support of Victorian science: the endowment of research movement in Great Britain, 1868–1900' *Minerva* IX (1971) pp. 197–230, pp. 200–2.
63. Roy M. MacLeod 'The support of Victorian science: the endowment of research movement in Great Britain, 1868–1900' *Minerva* IX (1971) pp. 197–230, p. 207.
64. Anonymous 'Our national industries' *Nature* VI (6 June 1872) p. 97.
65. Adrian Desmond 'Thomas Henry Huxley (1825–1895)' *Oxford dictionary of national biography* Retrieved 31 May 2018 from http://www.oxforddnb.com/view/10.1093/ref:odnb/9780198614128.001.0001/odnb-9780198614128-e-14320.
66. Roy M. MacLeod 'The support of Victorian science: the endowment of research movement in Great Britain, 1868–1900' *Minerva* IX (1971) pp. 197–230, p. 198.
67. Roy M. MacLeod 'The support of Victorian science: the endowment of research movement in Great Britain, 1868–1900' *Minerva* IX (1971) pp. 197–230, pp. 227–8.
68. Ruth Barton '"Huxley, Lubbock, and half a dozen others": professionals and gentlemen in the formation of the X Club, 1851–1864' *Isis* LXXXIX (1998) pp. 410–44, p. 410.
69. Roy M. MacLeod 'The support of Victorian science: the endowment of research movement in Great Britain, 1868–1900' *Minerva* IX (1971) pp. 197–230, p. 198.
70. Christopher N.L. Brooke *A history of the University of Cambridge, volume IV: 1870–1990* Cambridge: Cambridge University Press 1993 pp. 99–101.
71. Roy M. MacLeod 'The support of Victorian science: the endowment of research movement in Great Britain, 1868–1900' *Minerva* IX (1971) pp. 197–230, p. 210.
72. Gerald L. Geison *Michael Foster and the Cambridge school of physiology: the scientific enterprise in late Victorian society* Princeton: Princeton University Press 1978 p. 81.
73. Samuel J.M.M. Alberti 'Natural history and the philosophical societies of late Victorian Yorkshire' *Archives of natural history* XXX (2003) pp. 342–58, pp. 351–2.
74. Cambridge Philosophical Society Archives, 2/1/4, minutes of Council meeting, 22 February 1869.
75. Cambridge Philosophical Society Archives, 2/1/5, minutes of Council meeting, 25 October 1880.
76. Cambridge Philosophical Society Archives, 2/1/5, minutes of Council meeting, 7 February 1881.
77. For a full set of library regulations, see 'Report of the Council of the Cambridge Philosophical Society' *Proceedings of the Cambridge Philosophical Society* IV (1883) pp. 101–6. For a discussion of the library's 1882 catalogue, see Norma C. Neudoerffer

'The function of a nineteenth-century catalogue belonging to the Cambridge Philosophical Library' *Transactions of the Cambridge Bibliographical Society* IV (1967) pp. 293–301.

78. Cambridge Philosophical Society Archives, 2/1/6, minutes of Council meeting, 1 March 1886, 1 February 1904; J.W. Clark 'President's address' *Proceedings of the Cambridge Philosophical Society* VII (1892) pp. 2–4, p. 3.

## Chapter 6

1. Adrian Desmond *Archetypes and ancestors: palaeontology in Victorian London, 1850–1875* London: Blond and Briggs 1982 pp. 121–32.
2. Alice Johnson 'On the development of the pelvic girdle and skeleton of the hind limb in the chick' *Proceedings of the Cambridge Philosophical Society* IV (1883) pp. 328–31.
3. William Whewell quoted in Mary Fairfax Somerville *Personal recollections from early life to old age of Mary Somerville* London: John Murray 1874 pp. 170–2.
4. George Peacock quoted in Mary Fairfax Somerville *Personal recollections from early life to old age of Mary Somerville* London: John Murray 1874 p. 172.
5. Adam Sedgwick quoted in Charles Lyell *Life, letters and journals of Sir Charles Lyell, Bart, volume I* London: John Murray 1881 p. 368; Mary R.S. Creese 'Mary Somerville (1780–1872)' *Oxford dictionary of national biography* Retrieved 1 June 2018 from http://www.oxforddnb.com/view/10.1093/ref:odnb/9780198614128.001.0001/odnb-9780198614128-e-26024; Mary Somerville *The mechanism of the heavens* London: John Murray 1831.
6. William Buckland to Roderick Murchison, 5 April 1832, quoted in Jack Morrell and Arnold Thackray *Gentlemen of science: early years of the British Association for the Advancement of Science* Oxford: Clarendon Press 1981 p. 150. Albemarle is a reference to the fashionable Royal Institution on London's Albemarle Street.
7. Jack Morrell and Arnold Thackray *Gentlemen of science: early years of the British Association for the Advancement of Science* Oxford: Clarendon Press 1981 pp. 149, 152. Jermyn met her future husband—a geologist who had come to exhibit some coprolites—at the 1833 British Association for the Advancement of Science meeting, and she remained friends with Henslow, Sedgwick, and Whewell. Raleigh Trevelyan 'Paulina Jermyn Trevelyan, Lady Trevelyan (1816–1866)' *Oxford Dictionary of national biography* Retrieved 27 March 2018 from http://www.oxforddnb.com/view/10.1093/ref:odnb/9780198614128.001.0001/odnb-9780198614128-e-45577.
8. Elizabeth Chambers Patterson *Mary Somerville and the cultivation of science, 1815–1840* The Hague: Nijhoff 1983 p. 164.
9. Kathryn A. Neeley *Mary Somerville: science, illumination, and the female mind* Cambridge: Cambridge University Press 2001 p. 12; Mary Somerville 'On the magnetizing power of the more refrangible solar rays' *Philosophical transactions of the Royal Society* CXVI (1826) pp. 132–9.
10. Adam Sedgwick to Dr Somerville, April 1834, quoted in John Willis Clark and Thomas McKenny Hughes *The life and letters of Adam Sedgwick, volume I* Cambridge: Cambridge University Press 2009 p. 388.

11. Cambridge Philosophical Society Archives, 2/1/5, minutes of Council meeting, 18 November 1872, 25 November 1872, 17 February 1873; 3/1/3, minutes of general meeting, 3 February 1873.
12. Christopher N.L. Brooke *A history of the University of Cambridge, volume IV: 1870–1990* Cambridge: Cambridge University Press 1993 pp. 324–9.
13. Terrie M. Romano 'Michael Foster (1836–1907)' *Oxford dictionary of national biography* Retrieved 1 June 2018 from http://www.oxforddnb.com/view/10.1093/ref:odnb/9780198614128.001.0001/odnb-9780198614128-e-33218.
14. Gerald L. Geison *Michael Foster and the Cambridge school of physiology: the scientific enterprise in late Victorian society* Princeton: Princeton University Press 1978 pp. 3–4, 43, 100, 104; Christopher N.L. Brooke *A history of the University of Cambridge, volume IV: 1870–1990* Cambridge: Cambridge University Press 1993 pp. 164–5.
15. Humphry Davy Rolleston *The Cambridge medical school* Cambridge: Cambridge University Press 1932 p. 80.
16. Gerald L. Geison *Michael Foster and the Cambridge school of physiology: the scientific enterprise in late Victorian society* 1978 pp. 116–18, 120–2, 130–9.
17. George Adami, quoted in Gerald L. Geison *Michael Foster and the Cambridge school of physiology: the scientific enterprise in late Victorian society* 1978 p. 162.
18. G.M. Humphry *Reporter* 19 October 1870 p. 26. George Murray Humphry, who showed the German visitor around Cambridge, was Professor of Human Anatomy. His account of the visit appeared in the first issue of Cambridge University's *Reporter*.
19. Gerald L. Geison *Michael Foster and the Cambridge school of physiology: the scientific enterprise in late Victorian society* Princeton: Princeton University Press 1978 pp. 163–70.
20. Gerald L. Geison *Michael Foster and the Cambridge school of physiology: the scientific enterprise in late Victorian society* Princeton: Princeton University Press 1978 p. 125.
21. Anonymous 'Science teaching at Cambridge in the seventies' *British medical journal* II (9 October 1920) p. 572. It has been suggested that the author of this letter may have been Margaret Emily Pope—see Marsha L. Richmond '"A lab of one's own": the Balfour biological laboratory for women at Cambridge University, 1884–1914' in Sally Gregory Kohlstedt (editor) *History of women in the sciences* Chicago: Chicago University Press 1999 pp. 235–68, p. 240 n. 10.
22. Adam Sedgwick to Kate Malcolm, 7 February 1848, quoted in S.M Walters and E.A. Stow *Darwin's mentor: John Stevens Henslow, 1796–1861* Cambridge: Cambridge University Press 2001 p. 24.
23. Roy MacLeod (editor) *Days of judgement: science, examinations and the organization of knowledge in late Victorian England* Driffield: Nafferton 1982 p. 11; Paula Gould 'Women and the culture of university physics in late nineteenth-century Cambridge' *British journal for the history of science* XXX (1997) pp. 127–49, p. 132.
24. Katherina Rowold (editor) *Gender and science: late nineteenth-century debates on the female mind and body* Bristol: Thoemmes Press 1996, introduction.
25. Emily Davies 'The influence of university degrees on the education of women' *Victoria magazine* I (1863) pp. 260–71.

26. Anonymous 'The woman of the future' *Punch* 10 May 1884 p. 225.
27. William Whewell, quoted in Kathryn A. Neeley *Mary Somerville: science, illumination, and the female mind* Cambridge: Cambridge University Press 2001 p. 1.
28. Anonymous, quoted in Emily Davies 'The influence of university degrees on the education of women' *Victoria magazine* I (1863) pp. 260–71.
29. Miss Clough, Principle of Newnham College, quoted in Paula Gould 'Women and the culture of university physics in late nineteenth-century Cambridge' *British journal for the history of science* XXX (1997) pp. 127–49, pp. 141–3.
30. Paula Gould 'Women and the culture of university physics in late nineteenth-century Cambridge' *British journal for the history of science* XXX (1997) pp. 127–49, p. 145.
31. J.J. Thomson to Richard Threlfall, 1887, quoted in Paula Gould 'Women and the culture of university physics in late nineteenth-century Cambridge' *British journal for the history of science* XXX (1997) pp. 127–49, p. 139.
32. J.J. Thomson to H.F. Reid, 4 November 1886, quoted in Paula Gould 'Women and the culture of university physics in late nineteenth-century Cambridge' *British journal for the history of science* XXX (1997) pp. 127–49, p. 127.
33. Paula Gould 'Women and the culture of university physics in late nineteenth-century Cambridge' *British journal for the history of science* XXX (1997) pp. 127–49, pp. 137, 143.
34. Gerald L. Geison *Michael Foster and the Cambridge school of physiology: the scientific enterprise in late Victorian society* Princeton: Princeton University Press 1978 pp. 171–2.
35. Paula Gould 'Women and the culture of university physics in late nineteenth-century Cambridge' *British journal for the history of science* XXX (1997) pp. 127–49, p. 146; Prudence Waterhouse *A Victorian Monument: the buildings of Girton College* Cambridge: Girton College 1990 pp. 11–14.
36. Marsha L. Richmond '"A lab of one's own": the Balfour biological laboratory for women at Cambridge University, 1884–1914' in Sally Gregory Kohlstedt (editor) *History of women in the sciences* Chicago: Chicago University Press 1999 pp. 235–68, pp. 241–2.
37. Marsha L. Richmond '"A lab of one's own": the Balfour biological laboratory for women at Cambridge University, 1884–1914' in Sally Gregory Kohlstedt (editor) *History of women in the sciences* Chicago: Chicago University Press 1999 pp. 235–68, p. 243.
38. Alice Johnson 'On the development of the pelvic girdle and skeleton of the hind limb in the chick' *Proceedings of the Cambridge Philosophical Society* IV (1883) pp. 328–31, p. 328.
39. Marsha L. Richmond '"A lab of one's own": the Balfour biological laboratory for women at Cambridge University, 1884–1914' in Sally Gregory Kohlstedt (editor) *History of women in the sciences* Chicago: Chicago University Press 1999 pp. 235–68, pp. 247–9.
40. Marsha L. Richmond 'Adam Sedgwick (1854–1913)' *Oxford dictionary of national biography* Retrieved 1 June 2018 from http://www.oxforddnb.com/view/10.1093/ref:odnb/9780198614128.001.0001/odnb-9780198614128-e-36003.

41. Marsha L. Richmond '"A lab of one's own": the Balfour biological laboratory for women at Cambridge University, 1884–1914' in Sally Gregory Kohlstedt (editor) *History of women in the sciences* Chicago: Chicago University Press 1999 pp. 235–68, p. 244 n. 22.
42. Marsha L. Richmond '"A lab of one's own": the Balfour biological laboratory for women at Cambridge University, 1884–1914' in Sally Gregory Kohlstedt (editor) *History of women in the sciences* Chicago: Chicago University Press 1999 pp. 235–68, p. 247.
43. Edith Rebecca Saunders 'Mrs G.P. Bidder (Marion Greenwood)' *Newnham College letter* (1932) p. 65.
44. Marsha L. Richmond '"A lab of one's own": the Balfour biological laboratory for women at Cambridge University, 1884–1914' in Sally Gregory Kohlstedt (editor) *History of women in the sciences* Chicago: Chicago University Press 1999 pp. 235–68, p. 262.
45. Marsha L. Richmond '"A lab of one's own": the Balfour biological laboratory for women at Cambridge University, 1884–1914' in Sally Gregory Kohlstedt (editor) *History of women in the sciences* Chicago: Chicago University Press 1999 pp. 235–68, pp. 250–1, 261–6.
46. F. Eves 'On some experiments on the liver ferment' *Proceedings of the Cambridge Philosophical Society* V (1886) pp. 182–3; A. Bateson and F. Darwin 'On the change in shape in turgescent pith' *Proceedings of the Cambridge Philosophical Society* VI (1889) pp. 358–9; W. Bateson and A. Bateson 'On variations in the floral symmetry of certain flowers having irregular corollas' *Proceedings of the Cambridge Philosophical Society* VII (1892) p. 96; F. Darwin and D.F.M. Pertz 'On the effect of water currents on the assimilation of aquatic plants' *Proceedings of the Cambridge Philosophical Society* IX (1898) pp. 76–90; F. Darwin and D.F.M. Pertz 'On the injection of the intercellular spaces occurring in the leaves of *Elodea* during recovery from plasmolysis' *Proceedings of the Cambridge Philosophical Society* IX (1898) pp. 272–3; W. Bateson 'Notes on hybrid Cinerarias produced by Mr Lynch and Miss Pertz' *Proceedings of the Cambridge Philosophical Society* IX (1898) pp. 308–9; W. Bateson and D.F.M. Pertz 'Notes on the inheritance of variation in the corolla of Veronica Buxbaumii' *Proceedings of the Cambridge Philosophical Society* X (1900) pp. 78–93; D.F.M. Pertz and F. Darwin 'Experiments on the periodic movement of plants' *Proceedings of the Cambridge Philosophical Society* X (1900) p. 259; R. Alcock 'The digestive processes of *Ammocœtes*' *Proceedings of the Cambridge Philosophical Society* VII (1892) pp. 252–6; E. Dale 'On certain outgrowths (intumescences) on the green parts of *Hibiscus vitifolius* Linn.' *Proceedings of the Cambridge Philosophical Society* X (1900) pp. 192–210.
47. H.G. Klaassen 'On the effect of temperature on the conductivity of solutions of sulphuric acid' *Proceedings of the Cambridge Philosophical Society* VII (1892) pp. 137–41; F. Martin 'Expansion produced by electric discharge' *Proceedings of the Cambridge Philosophical Society* IX (1898) pp. 11–17.
48. Cambridge Philosophical Society Archives, 3/1/3, minutes of general meeting, 26 January 1891, 9 November 1891.

49. Physiology became more prominent in Cambridge just as Cambridge's medical school was beginning to expand. In 1870, there was only a small medical school in Cambridge; by the early 1880s, there were up to 90 medical students each year—making Cambridge the second-largest training centre in Britain after St Bartholomew's in London. Cambridge did not become an important centre for clinical training until the twentieth century, and the Medical Sciences Tripos was not created until 1966. For more information, see Christopher N.L. Brooke *A history of the University of Cambridge, volume IV: 1870–1990* Cambridge: Cambridge University Press 1993 pp. 166–71.
50. Cambridge Philosophical Society Archives, 2/1/5, minutes of Council meeting, 18 November 1872.
51. Cambridge Philosophical Society Archives, 3/1/4, minutes of general meetings (multiple entries).
52. *Proceedings of the Cambridge Philosophical Society* III (1880); *Proceedings of the Cambridge Philosophical Society* IV (1883).
53. Gerald L. Geison *Michael Foster and the Cambridge school of physiology: the scientific enterprise in late Victorian society* Princeton: Princeton University Press 1978 pp. 186–8.
54. E.M. Tansey 'George Eliot's support for physiology: the George Henry Lewes Trust 1879–1939' *Notes and records of the Royal Society of London* XLIV (1990) pp. 221–40; Gerald L. Geison *Michael Foster and the Cambridge school of physiology: the scientific enterprise in late Victorian society* Princeton: Princeton University Press 1978 pp. 177–8.
55. Roy MacLeod and Russell Moseley 'Breaking the circle of the sciences: the Natural Sciences Tripos and the "examination revolution"' in Roy MacLeod (editor) *Days of judgement: science, examinations and the organization of knowledge in late Victorian England* Driffield: Nafferton 1982 pp. 202–3.
56. Roy Porter 'The natural science tripos and the "Cambridge school of geology", 1850–1914' *History of universities* II (1982) pp. 193–216, p. 204.
57. Christopher N.L. Brooke *A history of the University of Cambridge, volume IV: 1870–1990* Cambridge: Cambridge University Press 1993 pp. 295–6.
58. Roy Porter 'The natural science tripos and the "Cambridge school of geology", 1850–1914' *History of universities* II (1982) pp. 193–216, pp. 207–8.
59. T.R. Glover *Cambridge retrospect* Cambridge: Cambridge University Press 1943 pp. 110–11; Sydney C. Roberts (revised by Herbert H. Huxley) 'Terrot Glover (1869–1943)' *Oxford dictionary of national biography* Retrieved 1 June 2018 from http://www.oxforddnb.com/view/10.1093/ref:odnb/9780198614128.001.0001/odnb-9780198614128-e-33427.
60. T.R. Glover quoted in Peter Linehan (editor) *St John's College, Cambridge: a history* Woodbridge: Boydell Press 2011 p. 494.
61. Christopher N.L. Brooke *A history of the University of Cambridge, volume IV: 1870–1990* Cambridge: Cambridge University Press 1993 p. 178; Elisabeth Leedham Green 'The arrival of research degrees in Cambridge' Darwin College Research

Report 2011. Retrieved 12 October 2018 from http://www.darwin.cam.ac.uk/drupal7/sites/default/files/Documents/publications/dcrr010.pdf.

62. Cambridge Philosophical Society Archives, 2/1/5, minutes of Council meeting, 27 January 1896; Fellows and Associates proposal forms (Oppenheimer *c.*1924); A. Rupert Hall *The Cambridge Philosophical Society: a history 1819–1969* Cambridge: Cambridge Philosophical Society 1969 p. 63.
63. John Willis Clark 'President's address' *Proceedings of the Cambridge Philosophical Society* VII (1892) pp. 2–3.
64. Marsha L. Richmond '"A lab of one's own": the Balfour biological laboratory for women at Cambridge University, 1884–1914' in Sally Gregory Kohlstedt (editor) *History of women in the sciences* Chicago: Chicago University Press 1999 pp. 235–68, pp. 253–4.
65. Paula Gould 'Women and the culture of university physics in late nineteenth-century Cambridge' *British journal for the history of science* XXX (1997) pp. 127–49, pp. 144–5.
66. Robert Olby 'William Bateson (1861–1926)' *Oxford dictionary of national biography* Retrieved 1 June 2018 from http://www.oxforddnb.com/view/10.1093/ref:odnb/9780198614128.001.0001/odnb-9780198614128-e-30641; Cambridge Philosophical Society Archives, 3/1/4, minutes of general meeting, 11 February 1884; Marsha L. Richmond 'Women in the early history of genetics: William Bateson and the Newnham College Mendelians, 1900–1910' *Isis* XCII (2001) pp. 55–90; *Proceedings of the Cambridge Philosophical Society* V (1886); *Proceedings of the Cambridge Philosophical Society* VI (1889).
67. Marsha L. Richmond 'Women in the early history of genetics: William Bateson and the Newnham College Mendelians, 1900–1910' *Isis* XCII (2001) pp. 55–90, p. 56; Mary R.S. Creese *Ladies in the laboratory? American and British women in science 1800–1900* London: Scarecrow 1998 pp. 41–6; Janet Browne 'Dorothea Pertz (1859–1939)' *Oxford dictionary of national biography* Retrieved 1 June 2018 from http://www.oxforddnb.com/view/10.1093/ref:odnb/9780198614128.001.0001/odnb-9780198614128-e-58481; Mary R.S. Creese (revised by V.M. Quirke) 'Edith Rebecca Saunders (1865–1945)' *Oxford dictionary of national biography* Retrieved 1 June 2018 from http://www.oxforddnb.com/view/10.1093/ref:odnb/9780198614128.001.0001/odnb-9780198614128-e-37936/version/0.
68. Elizabeth Crawford *The women's suffrage movement: a reference guide 1866–1928* London: University College London Press 1999 pp. 90–1.
69. A. Bateson and F. Darwin 'On the change in shape in turgescent pith' *Proceedings of the Cambridge Philosophical Society* VI (1889) pp. 358–9.
70. W. Bateson and A. Bateson 'On variations in the floral symmetry of certain flowers having irregular corollas' *Proceedings of the Cambridge Philosophical Society* VII (1892) p. 96.
71. Mary R.S. Creese *Ladies in the laboratory? American and British women in science 1800–1900* London: Scarecrow 1998 p. 42.

72. Marsha L. Richmond 'Women in the early history of genetics: William Bateson and the Newnham College Mendelians, 1900–1910' *Isis* XCII (2001) pp. 55–90, p. 60.
73. Francis Darwin and Dorothea F.M. Pertz 'On the artificial production of rhythm in plants' *Annals of botany* VI (1892) pp. 245–64; F. Darwin and D.F.M. Pertz 'On the effect of water currents on the assimilation of aquatic plants' *Proceedings of the Cambridge Philosophical Society* IX (1898) pp. 76–90; F. Darwin and D.F.M. Pertz 'On the injection of the intercellular spaces occurring in the leaves of *Elodea* during recovery from plasmolysis' *Proceedings of the Cambridge Philosophical Society* IX (1898) pp. 272–3; D.F.M. Pertz and F. Darwin 'Experiments on the periodic movement of plants' *Proceedings of the Cambridge Philosophical Society* X (1900) p. 259.
74. W. Bateson and A. Bateson 'On variations in the floral symmetry of certain plants having irregular corollas' *Journal of the Linnean Society* XXVIII (1891) pp. 386–424; W. Bateson 'Notes on hybrid Cinerarias produced by Mr Lynch and Miss Pertz' *Proceedings of the Cambridge Philosophical Society* IX (1898) pp. 308–9; W. Bateson and D.F.M. Pertz 'Notes on the inheritance of variation in the corolla of *Veronica Buxbaumii*' *Proceedings of the Cambridge Philosophical Society* X (1900) pp. 78–93.
75. William Bateson 'Hybridisation and cross-breeding as a method of scientific investigation' *Journal of the Royal Horticultural Society* XXIV (1900) pp. 59–66.
76. W. Bateson and E.R. Saunders 'Introduction' *Reports to the Evolution Committee of the Royal Society, Report* 1 1902 pp. 3–12.
77. Marsha L. Richmond 'Women in the early history of genetics: William Bateson and the Newnham College Mendelians, 1900–1910' *Isis* XCII (2001) pp. 55–90, pp. 60–3.
78. Marsha L. Richmond 'Women in the early history of genetics: William Bateson and the Newnham College Mendelians, 1900–1910' *Isis* XCII (2001) pp. 55–90, p. 73.
79. Cambridge Philosophical Society Archives, 2/1/6, minutes of Council meeting, 29 October 1900.
80. Cambridge Philosophical Society Archives, 2/1/7, minutes of Council meeting, 9 March 1914, 18 May 1914.
81. Marsha L. Richmond '"A lab of one's own": the Balfour biological laboratory for women at Cambridge University, 1884–1914' in Sally Gregory Kohlstedt (editor) *History of women in the sciences* Chicago: Chicago University Press 1999 pp. 235–68, pp. 265–6.
82. Cambridge Philosophical Society Archives, 2/1/7, minutes of Council meeting, 15 May 1922, 27 November 1922.
83. Cambridge Philosophical Society Archives, 2/1/8, minutes of Council meeting, 12 November 1928; 3/1/6, minutes of general meeting, 21 January 1929.

## Chapter 7

1. John (Cosmo Stuart) Rashleigh III, whose card I chose at random, came from a prominent Cornish family. He studied at Trinity College and qualified as a doctor in 1904; he later served as a justice of the peace and High Sheriff of Cornwall. In the 1940s, he rented out his family estate, Menabilly, to Daphne du Maurier, who

used it as inspiration for 'Manderley'. Cambridge Philosophical Society Archives, 12/2 and 12/1/10, anthropometrics.

2. For more on Galton as a pioneer, see, for example, his protégé Karl Pearson's biography of him: Karl Pearson *The life, letters, and labours of Francis Galton, volume II* Cambridge: Cambridge University Press 1924; Francis Galton 'Arithmetic by smell' *Psychological review* I (1894) pp. 61–2.
3. Anonymous 'A morning with the anthropometric detectives' *Pall Mall gazette* 16 November 1888 pp. 1–2; J. Venn 'Cambridge anthropometry' *The journal of the Anthropological Institute of Great Britain and Ireland* XVIII (1889) pp. 140–54, p. 141; Karl Pearson *The life, letters, and labours of Francis Galton, volume II* Cambridge: Cambridge University Press 1924 p. 270.
4. Francis Galton 'On the anthropometric laboratory at the International Health Exhibition' *The journal of the Anthropological Institute of Great Britain and Ireland* XIV (1885) pp. 205–21; Frans Lundgren 'The politics of participation: Francis Galton's anthropometric laboratory and the making of critical selves' *British journal for the history of science* XLVI (2011) pp. 445–66; Karl Pearson *The life, letters, and labours of Francis Galton, volume II* Cambridge: Cambridge University Press 1924 pp. 371–2.
5. Cambridge Philosophical Society Archives, 2/1/6, minutes of Council meeting, 7 February 1886, 1 March 1886.
6. See letter from Horace Darwin to Richard Glazebrook, Cambridge Philosophical Society Archives, 2/1/6, minutes of Council meeting, 10 May 1886; J. Venn 'Cambridge anthropometry' *The journal of the Anthropological Institute of Great Britain and Ireland* XVIII (1889) pp. 140–5, p. 141.
7. H. Darwin and R. Threlfall 'On Mr Galton's anthropometric apparatus at present in use in the Philosophical Library' *Proceedings of the Cambridge Philosophical Society* V (1886) p. 374.
8. Anonymous A *descriptive list of anthropometric apparatus, consisting of instruments for measuring and testing the chief physical characteristics of the human body. Designed under the direction of Mr Francis Galton and manufactured and sold by the Cambridge Scientific Instrument Company* Cambridge 1887.
9. Anonymous A *descriptive list of anthropometric apparatus, consisting of instruments for measuring and testing the chief physical characteristics of the human body. Designed under the direction of Mr Francis Galton and manufactured and sold by the Cambridge Scientific Instrument Company* Cambridge 1887 p. 3.
10. Anonymous A *descriptive list of anthropometric apparatus, consisting of instruments for measuring and testing the chief physical characteristics of the human body. Designed under the direction of Mr Francis Galton and manufactured and sold by the Cambridge Scientific Instrument Company* Cambridge 1887 p. 3.
11. Frans Lundgren 'The politics of participation: Francis Galton's anthropometric laboratory and the making of critical selves' *British journal for the history of science* XLVI (2011) pp. 445–66, pp. 455–6.
12. Anonymous 'A morning with the anthropometric detectives' *Pall Mall gazette* 16 November 1888 pp. 1–2. For a comprehensive overview of eugenics, see Alison

Bashford and Philippa Levine (editors) *The Oxford handbook of the history of eugenics* Oxford: Oxford University Press 2010.

13. Francis Galton 'On recent designs for anthropometric instruments' *The journal of the Anthropological Institute of Great Britain and Ireland* XVI (1887) pp. 2–8, p. 4.
14. Anonymous *A descriptive list of anthropometric apparatus, consisting of instruments for measuring and testing the chief physical characteristics of the human body. Designed under the direction of Mr Francis Galton and manufactured and sold by the Cambridge Scientific Instrument Company* Cambridge 1887 pp. 4–5.
15. Anonymous *A descriptive list of anthropometric apparatus, consisting of instruments for measuring and testing the chief physical characteristics of the human body. Designed under the direction of Mr Francis Galton and manufactured and sold by the Cambridge Scientific Instrument Company* Cambridge 1887 p. 5.
16. John Venn 'On the various notations adopted for expressing the common propositions of logic' *Proceedings of the Cambridge Philosophical Society* IV (1883) pp. 36–47; John Venn 'On the employment of geometrical diagrams for the sensible representation of logical propositions' *Proceedings of the Cambridge Philosophical Society* IV (1883) pp. 47–59; John Venn '*On the diagrammatic and mechanical representation of propositions and reasonings' The London, Edinburgh, and Dublin philosophical magazine and journal of science:* series 5 X (1880) pp. 1–18. The latter paper was published before those in the *Proceedings of the Cambridge Philosophical Society*, but it is believed that Venn's talks to the Cambridge Philosophical Society in December 1880 were his first on the topic.
17. Chris Pritchard 'Mistakes concerning a chance encounter between Francis Galton and John Venn' *BSHM bulletin: journal of the British Society for the History of Mathematics* 23 (2008) pp. 103–8; Nicholas Wright Gillham *A life of Sir Francis Galton: from African exploration to the birth of eugenics* Oxford: Oxford University Press 2001 p. 278; Philip Mirowski (editor) *Edgeworth on chance, economic hazard, and statistics* Rowman and Littlefield 1994.
18. J. Venn 'Cambridge anthropometry' *The journal of the Anthropological Institute of Great Britain and Ireland* XVIII (1889) pp. 140–54, pp. 142–3; J. Venn and Francis Galton 'Cambridge anthropometry' *Nature* XLI (13 March 1890) pp. 450–4, p. 451.
19. Karl Pearson *The life, letters, and labours of Francis Galton, volume II* Cambridge: Cambridge University Press 1924 p. 373.
20. Francis Galton 'Head growth in students at the University of Cambridge' *Nature* XXXVIII (3 May 1888) pp. 14–15.
21. F.M.T. 'Letters to the editor: head growth in students at the University of Cambridge' *Nature* XL (1 August 1889) pp. 317–18. The anonymous F.M.T. may have been Frederick Meadows Turner who trained as a physician in London and Cambridge and had an interest in statistics. He himself was measured several times in the Philosophical Library. See Anonymous 'Obituary: Frederick Meadows Turner, MD, BSc, DPH Medical Superintendent, South-Eastern Hospital, London County Council' *British medical journal* I (31 January 1931) p. 202; Cambridge Philosophical Society Archives, 12/1/3, item 2178 and item 3006.

22. Francis Galton 'Letters to the editor' *Nature* XL (1 August 1889) p. 318.
23. J. Venn 'Cambridge anthropometry' *The journal of the Anthropological Institute of Great Britain and Ireland* XVIII (1889) pp. 140–54, pp. 145–54. For more on Galton's work on percentiles, see Milo Keynes (editor) *Sir Francis Galton, FRS: the legacy of his ideas* Basingstoke: Macmillan 1991; Karl Pearson *The life, letters, and labours of Francis Galton, volume II* Cambridge: Cambridge University Press 1924, chapter 13; Francis Galton 'Anthropometric per-centiles' *Nature* XXXI (8 January 1885) pp. 223–5.
24. J. Venn 'Cambridge anthropometry' *The journal of the Anthropological Institute of Great Britain and Ireland* XVIII (1889) pp. 140–54, p. 147.
25. J. Venn and Francis Galton 'Cambridge anthropometry' *Nature* XLI (13 March 1890) pp. 450–4.
26. Joanne Woiak 'Karl Pearson (1857–1936)' *Oxford dictionary of national biography* Retrieved 1 June 2018 from http://www.oxforddnb.com/view/10.1093/ref:odnb/9780198614128.001.0001/odnb-9780198614128-e-35442; Karl Pearson 'On the correlation of intellectual ability with the size and shape of the head' *Proceedings of the Royal Society of London* LXIX (1902) pp. 333–42, p. 333.
27. W.R Macdonell 'On criminal anthropometry and the identification of criminals' *Biometrika* I (1902) pp. 177–227.
28. Alice Lee, Marie A. Lewenz, and Karl Pearson 'On the correlation of the mental and physical characters in man, part II' *Proceedings of the Royal Society of London* LXXI (1902–1903) pp. 106–14, p. 112.
29. Joanne Woiak 'Karl Pearson (1857–1936)' *Oxford dictionary of national biography* Retrieved 1 June 2018 from http://www.oxforddnb.com/view/10.1093/ref:odnb/9780198614128.001.0001/odnb-9780198614128-e-35442; Karl Pearson 'On the correlation of intellectual ability with the size and shape of the head' *Proceedings of the Royal Society of London* LXIX (1902) pp. 333–42, p. 333; J. Venn and Francis Galton 'Cambridge anthropometry' *Nature* XLI (13 March 1890) pp. 450–4, p. 453.
30. Anonymous 'A morning with the anthropometric detectives' *Pall Mall gazette* 16 November 1888 pp. 1–2; Karl Pearson *The life, letters, and labours of Francis Galton, volume II* Cambridge: Cambridge University Press 1924 p. 227.
31. Stephen Courtney 'Anthropometry and the biological sciences in late nineteenth-century Britain' Retrieved 11 May 2017 from http://anthropometryincontext.com/2017/05/01/blog-post-title/; Karl Pearson 'On the correlation of intellectual ability with the size and shape of the head' *Proceedings of the Royal Society of London* LXIX (1902) pp. 333–42, p. 336.
32. Anonymous 'The squeeze of 86' *Punch* (6 April 1889) p. 161.
33. Francis Galton, quoted in Karl Pearson *The life, letters, and labours of Francis Galton, volume II* Cambridge: Cambridge University Press 1924 p. 222.
34. Karl Pearson *The life, letters, and labours of Francis Galton, volume II* Cambridge: Cambridge University Press 1924 p. 222.
35. Hilary Perraton *A history of foreign students in Britain* Basingstoke: Palgrave Macmillan 2014. For a first-hand account of an Indian student's experiences in Cambridge, see Samuel Satthianadhan *Four years in an English University* Madras:

Lawrence Asylum Press 1890; Cambridge Philosophical Society archives, 12/2, anthropometrics. It is not known whether these cards were kept separate from the time of their creation, or whether they were separated out by a later user of the archive. They have recently been added to the main body of cards in chronological order.

36. Karl Pearson *The life, letters, and labours of Francis Galton, volume II* Cambridge: Cambridge University Press 1924 pp. 357–8.
37. Charles S. Myers 'The future of anthropometry' *The journal of the Anthropological Institute of Great Britain and Ireland* XXXIII (1903) pp. 36–40.
38. Donald A. MacKenzie *Statistics in Britain, 1865–1930* Edinburgh: Edinburgh University Press 1981 p. 104.
39. J.B.S. Haldane *Daedalus, or, science and the future* New York: E.P. Dutton and Company 1924 pp. 57–8.
40. Cambridge Philosophical Society archives, 12/2, anthropometrics, 1962.
41. Cambridge Philosophical Society archives, 12/2, anthropometrics, undated.
42. See, for example, the work of Deborah Oxley, who has studied similar data sets for Australian convicts and English child labourers, and is currently researching the anthropometric cards of the Cambridge Philosophical Society.
43. Anonymous 'Proceedings at the meetings held during the session 1922–1923' *Proceedings of the Cambridge Philosophical Society* XXI (1923) p. 795; Anonymous 'Proceedings at the meetings held during the session 1924–1925' *Proceedings of the Cambridge Philosophical Society* XXII (1925) p. 970; V.M. Quirke 'John Burdon Sanderson Haldane (1892–1964)' *Oxford dictionary of national biography* Retrieved 1 June 2018 from http://www.oxforddnb.com/view/10.1093/ref:odnb/9780198614128.001.0001/odnb-9780198614128-e-33641.
44. J.B.S. Haldane 'A mathematical theory of natural and artificial selection' *Transactions of the Cambridge Philosophical Society* XXIII (1931) pp. 19–42. This volume is dated 1931, but its papers spanned the years 1923 to 1928; individual papers were circulated shortly after they were accepted by the journal and so Haldane's article would have been available from 1924.
45. Lewis S. Feuer *Einstein and the generations of science* London and New York: Routledge 1982 p. 279.
46. Reginald Punnett 'Early days of genetics' *Heredity* IV (1950) pp. 1–10.
47. Richard M. Burian and Doris T. Zallan 'Genes' in Peter J. Bowler and John V. Pickstone (editors) *The Cambridge history of science, volume VI: the modern biological and earth sciences* Cambridge: Cambridge University Press 2009 pp. 432–50, p. 435; Donald A. MacKenzie *Statistics in Britain, 1865–1930* Edinburgh: Edinburgh University Press 1981 chapter 6.
48. Karl Pearson, 1902, quoted in Donald A. MacKenzie *Statistics in Britain, 1865–1930* Edinburgh: Edinburgh University Press 1981 p. 126.
49. Frank Yates (revised by Alan Yoshioka) 'George Udny Yule (1871–1951)' *Oxford dictionary of national biography* Retrieved 1 June 2018 from http://www.oxforddnb.com/view/10.1093/ref:odnb/9780198614128.001.0001/odnb-9780198614128-e-37086.

50. George Udny Yule 'Mendel's laws and their probable relations to intra-racial heredity' *The new phytologist* I (1902) pp. 195–6.
51. George Udny Yule 'Mendel's laws and their probable relations to intra-racial heredity' *The new phytologist* I (1902) pp. 222–38, p. 207.
52. For an analysis of why Yule's paper had limited impact, see James G. Tabery 'The "evolutionary synthesis" of George Udny Yule' *Journal of the history of biology* XXXVII (2004) pp. 73–101.
53. Hamish G. Spencer 'Ronald Aylmer Fisher (1890–1962)' *Oxford dictionary of national biography* Retrieved 1 June 2018 from http://www.oxforddnb.com/view/10.1093/ref:odnb/9780198614128.001.0001/odnb-9780198614128-e-33146; R.A. Fisher 'The correlation between relatives in the supposition of Mendelian inheritance' *Transactions of the Royal Society of Edinburgh* LII (1918) pp. 399–433.
54. R.A. Fisher 'The correlation between relatives in the supposition of Mendelian inheritance' *Transactions of the Royal Society of Edinburgh* LII (1918) pp. 399–433, p. 433.
55. J.B.S. Haldane 'A mathematical theory of natural and artificial selection' *Transactions of the Cambridge Philosophical Society* XXIII (1931) pp. 19–42, p. 19. There were ten papers in total, with nine published by the Society. The first paper was published in the final volume of the *Transactions*, the second was published in the first volume of the new journal *Biological reviews*, parts III–IX were published in the *Proceedings*, and the final part was published by the American journal *Genetics*.
56. J.B.S. Haldane 'A mathematical theory of natural and artificial selection' *Transactions of the Cambridge Philosophical Society* XXIII (1931) pp. 19–42, p. 19.
57. J.B.S. Haldane 'A mathematical theory of natural and artificial selection. Part II. The influence of partial self-fertilisation, inbreeding, assertive mating, and selective fertilisation on the composition of Mendelian populations, and on natural selection' *Biological reviews* I (1925) pp. 158–63; J.B.S. Haldane 'A mathematical theory of natural and artificial selection. Part III' *Proceedings of the Cambridge Philosophical Society* XXIII (1927) pp. 363–72; J.B.S. Haldane 'A mathematical theory of natural and artificial selection. Part IV' *Proceedings of the Cambridge Philosophical Society* XXIII (1927) pp. 607–15; J.B.S. Haldane 'A mathematical theory of natural and artificial selection. Part V' *Proceedings of the Cambridge Philosophical Society* XXIII (1927) pp. 838–44; J.B.S. Haldane 'A mathematical theory of natural and artificial selection. Part VI. Isolation' *Proceedings of the Cambridge Philosophical Society* XXVI (1930) pp. 220–30; J.B.S. Haldane 'A mathematical theory of natural and artificial selection. Part VII. Selection intensity as a function of mortality rate' *Proceedings of the Cambridge Philosophical Society* XXVII (1931) pp. 131–6; J.B.S. Haldane 'A mathematical theory of natural and artificial selection. Part VIII. Metastable population' *Proceedings of the Cambridge Philosophical Society* XXVII (1931) pp. 137–42; J.B.S. Haldane 'A mathematical theory of natural and artificial selection. Part IX. Rapid selection' *Proceedings of the Cambridge Philosophical Society* XXVIII (1932) pp. 244–8; J.B.S. Haldane 'A mathematical theory of natural and artificial selection. Part X. Some theorems on artificial selection' *Genetics* XIX (1934) pp. 412–29.

58. J.B.S. Haldane 'A mathematical theory of natural and artificial selection. Part IX. Rapid selection' *Proceedings of the Cambridge Philosophical Society* XXVIII (1932) pp. 244–8, p. 247.
59. J.B.S. Haldane *Daedalus, or, science and the future* New York: E.P. Dutton and Company 1924 p. 57.
60. Diane Paul 'Eugenics and the left' *Journal of the history of ideas* XLV (1984) pp. 567–90; Michael Freeden 'Eugenics and progressive thought' *The historical journal* XXII (1979) pp. 645–71.
61. J.B.S. Haldane *The causes of evolution* Ithaca: Cornell University Press 1966 p. 33.
62. Julian Huxley *Evolution: the modern synthesis* London: Allen and Unwin 1942 p. 13.
63. J.B.S. Haldane *The causes of evolution* Ithaca: Cornell University Press 1966 p. 215; A.W.F. Edwards 'Mathematising Darwin' *Behavioral ecology and sociobiology* LXV (2011) pp. 421–30; Jeffrey C. Schank and Charles Twardy 'Mathematical models' in Peter J. Bowler and John V. Pickstone (editors) *The Cambridge history of science, volume VI: the modern biological and earth sciences* Cambridge: Cambridge University Press 2009 pp. 416–31.
64. C.R. Marshall 'Note on the pharmacological action of cannabis resin' *Proceedings of the Cambridge Philosophical Society* IX (1898) pp. 149–50, p. 149.
65. Cambridge Philosophical Society archives, 2/1/7, minutes of Council meeting, 11 December 1922; H. Munro Fox 'The origin and development of *Biological reviews*' *Biological reviews* XL (1965) pp. 1–4. The first volume of the new biological journal was titled *Proceedings of the Cambridge Philosophical Society: biological sciences*, while the second was called *Biological reviews and biological proceedings*. From the mid-1930s, the name was shortened to *Biological reviews*—the title which is still used today.
66. Cambridge Philosophical Society archives, 2/1/8, minutes of Council meeting, 7 December 1925, 26 July 1926.
67. Anonymous 'Preface' *Transactions of the Cambridge Philosophical Society* I (1822) pp. iii–viii, p. v.

## Chapter 8

1. Cambridge Philosophical Society Archives, 2/1/6, minutes of Council meeting, 27 January 1896; 3/1/5, minutes of general meeting, 27 January 1896.
2. J.J. Thomson and E. Rutherford 'On the passage of electricity through gases exposed to Röntgen rays' *The London, Edinburgh, and Dublin philosophical magazine and journal of science: series 5* XLII (1896) pp. 392–407; Lawrence Badash 'Ernest Rutherford (1871–1937)' *Oxford dictionary of national biography* Retrieved 2 June 2018 from http://www.oxforddnb.com/view/10.1093/ref:odnb/9780198614128.001.0001/odnb-9780198614128-e-35891; Isobel Falconer 'Joseph John Thomson (1856–1940)' *Oxford dictionary of national biography* Retrieved 31 May 2018 from http://www.oxforddnb.com/view/10.1093/ref:odnb/9780198614128.001.0001/odnb-9780198614128-e-36506.
3. J.J. Thomson 'On the cathode rays' *Proceedings of the Cambridge Philosophical Society* IX (1898) pp. 243–4.

4. J.J. Thomson 'Cathode rays' *The electrician* XXXIX (1897) pp. 103–9. For a discussion of different accounts of the 'discovery' of the electron, see Isobel Falconer 'Corpuscles to electrons' in Jed Z. Buchwald and Andrew Warwick (editors) *Histories of the electron: the birth of microphysics* Cambridge: MIT Press 2001 pp. 77–100; Peter Achinstein 'Who really discovered the electron?' in Jed Z. Buchwald and Andrew Warwick (editors) *Histories of the electron: the birth of microphysics* Cambridge: MIT Press 2001 pp. 403–24.
5. Quoted in Graeme Gooday 'The questionable matter of electricity: the reception of J.J. Thomson's "corpuscle" amongst electrical theorists and technologists' in Jed Z. Buchwald and Andrew Warwick (editors) *Histories of the electron: the birth of microphysics* Cambridge: MIT Press 2001 pp. 101–34, p. 101.
6. *Proceedings of the Cambridge Philosophical Society* IX (1898); *Proceedings of the Cambridge Philosophical Society* X (1900); *Proceedings of the Cambridge Philosophical Society* XI (1902); *Proceedings of the Cambridge Philosophical Society* XII (1904); *Proceedings of the Cambridge Philosophical Society* XV (1910); *Proceedings of the Cambridge Philosophical Society* XVI (1912).
7. David Phillips 'William Lawrence Bragg (1890–1971)' *Oxford dictionary of national biography* Retrieved 2 June 2018 from http://www.oxforddnb.com/view/10.1093/ref:odnb/9780198614128.001.0001/odnb-9780198614128-e-30845; Talal Debs 'William Henry Bragg (1862–1942)' *Oxford dictionary of national biography* Retrieved 2 June 2018 from http://www.oxforddnb.com/view/10.1093/ref:odnb/9780198614128.001.0001/odnb-9780198614128-e-32031.
8. W.L. Bragg 'The diffraction of short electromagnetic waves by a crystal' *Proceedings of the Cambridge Philosophical Society* XVII (1914) pp. 43–57.
9. William Bragg, 21 November 1912, quoted in John Jenkin *William and Lawrence Bragg, father and son: the most extraordinary collaboration in science* Oxford: Oxford University Press 2008 p. 333.
10. William Bragg 'X-rays and crystals' *Nature* XC (28 November 1912) pp. 360–1.
11. C.T.R. Wilson 'On the cloud method of making visible ions and the tracks of ionising particles' *Nobel lectures, physics, 1922–41* (1965) p. 194.
12. C.T.R. Wilson 'On the formation of cloud in the absence of dust' *Proceedings of the Cambridge Philosophical Society* VIII (1895) p. 306.
13. C.T.R. Wilson 'On the action of uranium rays on the condensation of water vapour' *Proceedings of the Cambridge Philosophical Society* IX (1898) pp. 333–8; C.T.R. Wilson 'On the production of a cloud by the action of ultra-violet light on moist air' *Proceedings of the Cambridge Philosophical Society* IX (1898) pp. 392–3.
14. J.J. Thomson 'Nobel lecture' 11 December 1906 Retrieved 23 June 2017 from http://www.nobelprize.org/nobel_prizes/physics/laureates/1906/thomson-lecture.pdf. For more detailed discussion of the operation of the cloud chamber, see Peter Galison and Alexi Assmus 'Artificial clouds, real particles' in David Gooding, Trevor Pinch, and Simon Schaffer (editors) *The uses of experiment* Cambridge: Cambridge University Press 1989 pp. 225–75.
15. Malcolm Longair *Maxwell's enduring legacy: a scientific history of the Cavendish Laboratory* Cambridge: Cambridge University Press 2016 p. 158.

16. C.T.R. Wilson 'On a method of making visible the paths of ionising particles through a gas' *Proceedings of the Royal Society of London* LXXXV (1911) pp. 285–8.
17. P.M.S. Blackett 'Charles Thomson Rees Wilson, 1869–1959' in *Biographical memoirs of Fellows of the Royal Society* VI London: Royal Society 1960 pp. 269–95, p. 289.
18. Ernest Rutherford 'Professor C.T.R. Wilson (obituary)' *The Times* 16 November 1959 p. 16.
19. Christopher N.L. Brooke *A history of the University of Cambridge, volume IV: 1870–1990* Cambridge: Cambridge University Press 1993 p. 331.
20. M.G. Woods (later Mrs Waterhouse), quoted in Christopher N.L. Brooke *A history of the University of Cambridge, volume IV: 1870–1990* Cambridge: Cambridge University Press 1993 p. 332.
21. F.J.M. Stanton, quoted in Christopher N.L. Brooke *A history of the University of Cambridge, volume IV: 1870–1990* Cambridge: Cambridge University Press 1993 p. 333.
22. Christopher N.L. Brooke *A history of the University of Cambridge, volume IV: 1870–1990* Cambridge: Cambridge University Press 1993 p. 334.
23. Malcolm Longair *Maxwell's enduring legacy: a scientific history of the Cavendish Laboratory* Cambridge: Cambridge University Press 2016 p. 166.
24. Cambridge Philosophical Society Archives, 2/1/7, minutes of Council meeting, 9 November 1914, 8 February 1915, 22 November 1915, 7 February 1916, 21 February 1916, 5 February 1917, 27 October 1919.
25. Cambridge Philosophical Society Archives, 2/1/7, minutes of Council meeting, 8 February 1915.
26. Cambridge Philosophical Society Archives, 2/1/7, minutes of Council meeting, 7 February 1916, 5 February 1917.
27. Cambridge Philosophical Society Archives, 2/1/7, minutes of Council meeting, 4 February 1918.
28. Jon Agar *Science in the twentieth century and beyond* Cambridge: Polity Press 2012 p. 110–17.
29. Anonymous [H.H. Turner] 'From an Oxford note-book' *The Observatory* DXXIV (1918) p. 147.
30. Matthew Stanley '"An expedition to heal the wounds of war": the 1919 eclipse and Eddington as Quaker adventurer' *Isis* XCIV (2003) pp. 57–89, pp. 69–70.
31. Anonymous 'Proceedings at the meetings held during the session 1907–1908' *Proceedings of the Cambridge Philosophical Society* XIV (1908) p. 615.
32. Matthew Stanley '"An expedition to heal the wounds of war": the 1919 eclipse and Eddington as Quaker adventurer' *Isis* XCIV (2003) pp. 57–89.
33. C.W. Kilmister 'Arthur Stanley Eddington (1882–1944)' *Oxford dictionary of national biography* Retrieved 2 June 2018 from http://www.oxforddnb.com/view/10.1093/ref:odnb/9780198614128.001.0001/odnb-9780198614128-e-32967.
34. Albert Einstein 'Die Grundlage der allgemeinen Relativitätstheorie' *Annalen der physic* XLIX (1916) pp. 769–822. This paper was a consolidation of four papers that

Einstein had published towards the end of 1915, each dealing with an aspect of general relativity.

35. A.S. Eddington and Albert Einstein, quoted in Matthew Stanley '"An expedition to heal the wounds of war": the 1919 eclipse and Eddington as Quaker adventurer' *Isis* XCIV (2003) pp. 57–89, p. 69.
36. For a discussion of alternative ways of interpreting Eddington's images and the controversies they provoked, see John Earman and Clark Glymour 'Relativity and eclipses: the British eclipse expeditions of 1919 and their predecessors' *Historical studies in the physical sciences* XI (1980) pp. 49–85 or Harry Collins and Trevor Pinch *The Golem* Cambridge: Cambridge University Press 1998 chapter 2. For a discussion of the campaign to have the expedition and its results viewed in a favourable light, see Alistair Sponsel 'Constructing a "revolution in science": the campaign to promote a favourable reception for the 1919 solar eclipse experiments' *British journal for the history of science* XXXV (2002) pp. 439–67.
37. Arthur Eddington to Albert Einstein, 1 December 1919, quoted in John Earman and Clark Glymour 'The gravitational red shift as a test of general relativity: history and analysis' *Studies in the history and philosophy of science* XI (1980) pp. 175–214, p. 183 n. 22; Anonymous 'Proceedings at the meetings held during the session 1919–1920' *Proceedings of the Cambridge Philosophical Society* XX (1921) pp. 213–14.
38. Cambridge Philosophical Society Archives, 2/1/7, minutes of Council meeting, 24 January 1921.
39. Arthur Eddington to Albert Einstein, 1 December 1919, quoted in Matthew Stanley '"An expedition to heal the wounds of war": the 1919 eclipse and Eddington as Quaker adventurer' *Isis* XCIV (2003) pp. 57–89, p. 85.
40. Ernest Rutherford, quoted in Matthew Stanley '"An expedition to heal the wounds of war": the 1919 eclipse and Eddington as Quaker adventurer' *Isis* XCIV (2003) pp. 57–89, p. 58.
41. A. Rupert Hall *The Cambridge Philosophical Society: a history 1819–1969* Cambridge: Cambridge Philosophical Society 1969 p. 32.
42. Cambridge Philosophical Society Archives, 2/1/7, minutes of Council meeting, 10 March 1919.
43. Cambridge Philosophical Society Archives, 2/1/7, minutes of Council meeting, 2 June 1919, 6 October 1919, 24 November 1919, 22 November 1920.
44. A. Rupert Hall *The Cambridge Philosophical Society: a history 1819–1969* Cambridge: Cambridge Philosophical Society 1969 p. 37.
45. Cambridge Philosophical Society Archives, 2/1/7, minutes of Council meeting, 19 May 1913.
46. Cambridge Philosophical Society Archives, 2/1/7, minutes of Council meeting, 3 May 1920, 7 June 1920, 8 November 1920.
47. Cambridge Philosophical Society Archives, 2/1/7, minutes of Council meeting, 23 January 1922, 20 February 1922.
48. Cambridge Philosophical Society Archives, 2/1/7, minutes of Council meeting, 20 February 1922.

49. Cambridge Philosophical Society Archives, 2/1/7, minutes of Council meeting, 20 February 1922.
50. Cambridge Philosophical Society Archives, 2/1/7, minutes of Council meeting, 15 May 1922, 16 October 1922, 14 July 1924.
51. Cambridge Philosophical Society Archives, 2/1/7, minutes of Council meeting, 12 March 1923; 2/1/8, minutes of Council meeting, 26 November 1923, 14 July 1924, 2 February 1925, 15 February 1926.
52. Cambridge Philosophical Society Archives, 2/1/7, minutes of Council meeting, 16 February 1925, 27 January 1930.
53. Béla Bollobás 'Godfrey Harold Hardy (1877–1947)' *Oxford dictionary of national biography* Retrieved 2 June 2018 from http://www.oxforddnb.com/view/10.1093/ref:odnb/9780198614128.001.0001/odnb-9780198614128-e-33706; Anonymous 'Proceedings at the meetings held during the session 1901–1902' *Proceedings of the Cambridge Philosophical Society* XI (1902) p. 509.
54. Béla Bollobás (editor) *Littlewood's miscellany* Cambridge: Cambridge University Press 1986 p. 2.
55. G.H. Hardy *A mathematician's apology* Cambridge: Cambridge University Press 1948 pp. 24–5.
56. G.H. Hardy *A mathematician's apology* Cambridge: Cambridge University Press 1948 pp. 55–61.
57. Robert Kanigel 'Srinivasa Ramanujan (1887–1920)' *Oxford dictionary of national biography* Retrieved 2 June 2018 from http://www.oxforddnb.com/view/10.1093/ref:odnb/9780198614128.001.0001/odnb-9780198614128-e-51582.
58. Cambridge Philosophical Society Archives, 2/1/7, minutes of Council meeting, 11 February 1907, 24 February 1908, 4 May 1908, 25 January 1909.
59. Cambridge Philosophical Society Archives, 2/1/7, minutes of Council meeting, 4 February 1918.
60. Cambridge Philosophical Society Archives, 2/1/7, minutes of Council meeting, 6 October 1919, 15 May 1922; 2/1/8, minutes of Council meeting, 3 December 1923.
61. Cambridge Philosophical Society Archives, 2/1/7, minutes of Council meeting, 11 December 1922. As mentioned in Chapter 7, the first volume of the new biological journal was titled *Proceedings of the Cambridge Philosophical Society: biological sciences*, while the second was called *Biological reviews and biological proceedings*. From the mid-1930s, the name was shortened to *Biological reviews*—the title which is still used today.
62. Cambridge Philosophical Society Archives, 2/1/8, minutes of Council meeting, 7 December 1925, 26 July 1926.
63. Cambridge Philosophical Society Archives, 2/1/8, minutes of Council meeting, 18 May 1931, 9 November 1931.
64. Malcolm Longair *Maxwell's enduring legacy: a scientific history of the Cavendish Laboratory* Cambridge: Cambridge University Press 2016 p. 184.
65. Ernest Rutherford 'Capture and loss of electrons by α particles' *Proceedings of the Cambridge Philosophical Society* XXI (1923) pp. 504–10; Ernest Rutherford and

W.A. Wooster 'The natural x-ray spectrum of radium B' *Proceedings of the Cambridge Philosophical Society* XXII (1925) pp. 834–7.

66. G.P. Thomson 'A note on the nature of the carriers of the anode rays' *Proceedings of the Cambridge Philosophical Society* XX (1921) pp. 210–11; P.B. Moon (revised by Anita McConnell) 'George Paget Thomson (1892–1975)' *Oxford dictionary of national biography* Retrieved 2 June 2018 from http://www.oxforddnb.com/view/10.1093/ref:odnb/9780198614128.001.0001/odnb-9780198614128-e-31758.
67. J. Chadwick and C.D. Ellis 'A preliminary investigation of the intensity distribution in the $\beta$-ray spectra of radium B and C' *Proceedings of the Cambridge Philosophical Society* XXI (1923) pp. 274–80. Charles Ellis, who co-authored this paper, had met Chadwick in the Ruhleben internment camp in Germany, where both men had been held during World War I; Isobel Falconer 'James Chadwick (1891–1974)' *Oxford dictionary of national biography* Retrieved 2 June 2018 from http://www.oxforddnb.com/view/10.1093/ref:odnb/9780198614128.001.0001/odnb-9780198614128-e-30912; Jeffrey A. Hughes 'Charles Drummond Ellis (1895–1980)' *Oxford dictionary of national biography* Retrieved 2 June 2018 from http://www.oxforddnb.com/view/10.1093/ref:odnb/9780198614128.001.0001/odnb-9780198614128-e-31070.
68. J.R. Oppenheimer 'On the quantum theory of vibration-rotation bands' *Proceedings of the Cambridge Philosophical Society* XXIII (1927) pp. 327–35.
69. P.A.M. Dirac 'Dissociation under a temperature gradient' *Proceedings of the Cambridge Philosophical Society* XXII (1925) pp. 132–7.
70. A. Rupert Hall *The Cambridge Philosophical Society: a history 1819–1969* Cambridge: Cambridge Philosophical Society 1969 pp. 64–6.
71. Cambridge Philosophical Society Archives, 2/1/8, minutes of Council meeting, 3 December 1923, 18 February 1924; Niels Bohr 'On the application of the quantum theory to atomic structure, part 1: the fundamental postulates' Supplement to *Proceedings of the Cambridge Philosophical Society* XXII (1925) pp. 1–42.
72. Jon Agar *Science in the twentieth century and beyond* Cambridge: Polity Press 2012 pp. 213–15.
73. Cambridge Philosophical Society Archives, 2/1/8, minutes of Council meeting, 8 May 1933.
74. Cambridge Philosophical Society Archives, 2/1/8, minutes of Council meeting, 14 November 1938, 27 November 1939.
75. Cambridge Philosophical Society Archives, 2/1/8, minutes of Council meeting, 9 October 1939, 30 October 1939.
76. Cambridge Philosophical Society Archives, 2/1/8, minutes of Council meeting, 18 May 1942, 9 November 1942, 23 November 1942, 6 March 1944.
77. Christopher N.L. Brooke *A history of the University of Cambridge, volume IV: 1870–1990* Cambridge: Cambridge University Press 1993 pp. 506–7.
78. Malcolm Longair *Maxwell's enduring legacy: a scientific history of the Cavendish Laboratory* Cambridge: Cambridge University Press 2016 p. 261.
79. Sam Edwards 'Obituaries: Rudolph E. Peierls' *Physics today* XLIX (February 1996) pp. 74–5.

80. Anonymous 'Proceedings at the meetings held during the session 1935–1936' *Proceedings of the Cambridge Philosophical Society* XXXII (1936) p. 686; Anonymous 'Proceedings at the meetings held during the session 1936–1937' *Proceedings of the Cambridge Philosophical Society* XXXIII (1937) p. 588.
81. Ronald W. Clark *The birth of the bomb* London: Phoenix House 1961 p. 42; Richard Rhodes *The making of the atomic bomb* New York: Simon and Schuster 1988 p. 321.
82. R. Peierls 'Critical conditions in neutron multiplication' *Proceedings of the Cambridge Philosophical Society* XXXV (1939) pp. 610–15. Peierls had moved to Birmingham when he began this strand of research.
83. Rudolph Peierls quoted in Ronald W. Clark *The birth of the bomb* London: Phoenix House 1961 p. 43.
84. Richard Rhodes *The making of the atomic bomb* New York: Simon and Schuster 1986 p. 322.
85. Rudolph Peierls, quoted in Richard Rhodes *The making of the atomic bomb* New York: Simon and Schuster 1986 p. 323.
86. Otto Frisch, quoted in Richard Rhodes *The making of the atomic bomb* New York: Simon and Schuster 1986 p. 323.
87. Frisch–Peierls memorandum March 1940 Retrieved 23 June 2017 from http://web.stanford.edu/class/history5n/FPmemo.pdf.
88. K. Fuchs 'The conductivity of thin metallic films according to the electron theory of metals' *Proceedings of the Cambridge Philosophical Society* XXXIV (1938) pp. 100–8; K. Fuchs 'On the stability of nuclei against $\beta$-emission' *Proceedings of the Cambridge Philosophical Society* XXXV (1939) pp. 242–55.
89. Christoph Laucht *Elemental Germans: Klaus Fuchs, Rudolf Peierls and the making of British nuclear culture 1939–1959* Basingstoke: Palgrave Macmillan 2012.

## Chapter 9

1. Alexander G. Liu, Jack J. Matthews, Latha R. Menon, Duncan McIlroy, and Martin D. Brasier '*Haootia quadriformis* n. gen., n. sp., interpreted as a muscular cnidarian impression from the late Ediacaran period (approx. 560 Ma)' *Proceedings of the Royal Society B* CCLXXXI (2014) 20141202.
2. The quotation 'to keep alive the spirit of inquiry' is taken from the Cambridge Philosophical Society's first volume of *Transactions*: Anonymous 'Preface' *Transactions of the Cambridge Philosophical Society* I (1822) pp. iii–viii, p. v.
3. For a discussion of how we differentiate between the different kingdoms, see Susannah Gibson *Animal, vegetable, mineral?* Oxford: Oxford University Press 2015.
4. J.W. Salter 'On fossil remains in the Cambrian rocks of the Longmynd and North Wales' *Quarterly journal of the Geological Society* XII (1856) pp. 246–51; J.W. Salter 'On annelide-burrows and surface markings from the Cambrian rocks of the Longmynd' *Quarterly journal of the Geological Society* XIII (1857) pp. 199–207; Richard H.T. Callow, Duncan McIlroy, and Martin D. Brasier 'John Salter and the Ediacara Fauna of the Longmyndian Supergroup' *Ichnos* XVIII (2011) pp. 176–87.

5. Charles Darwin *On the origin of species by means of natural selection* London: John Murray 1859 p. 307.
6. L.R. Menon, D. McIlroy, A.G. Liu, and M.D. Brasier 'The dynamic influence of microbial mats on sediments: fluid escape and pseudofossil formation in the Ediacaran Longmyndian Supergroup, UK' *Journal of the Geological Society* CLXXIII (2016) pp. 177–85; Alexander G. Liu 'Reviewing the Ediacaran fossils of the Long Mynd, Shropshire' *Proceedings of the Shropshire Geological Society* XVI (2011) pp. 31–43.
7. Cambridge Philosophical Society Archives, minutes of Council meeting, 11 May 2009. The Henslow Fellowships have been awarded in conjunction with colleges with limited funds for such research, including Girton College, Downing College, Murray Edwards College, Selwyn College, Robinson College, Wolfson College, St Edmund's College, Darwin College, Hughes Hall, Fitzwilliam College, Lucy Cavendish College, and St Catharine's College.
8. The research briefly mentioned here has been conducted by Sarah Morgan, George Gordon, Arne Jungwirth, Bartomeu Monserrat, Tanya Hutter, and Glenn Masson.
9. Cambridge Philosophical Society Archives, 2/1/11, minutes of Council meeting, 28 April 1969, 23 July 1973.
10. Cambridge Philosophical Society Archives, minutes of Council meeting, 3 March 2014.
11. Cambridge Philosophical Society Archives, 2/1/7, minutes of Council meeting, 24 November 1919.

# FIGURE AND PLATE CREDITS

## Figure credits

| | |
|---|---|
| Figure 1 | Wellcome Collection. CC BY 4.0 |
| Figure 2 | Wellcome Collection. CC BY 4.0 |
| Figure 3 | © 2018. Sedgwick Museum of Earth Sciences, University of Cambridge. Reproduced with permission |
| Figure 4 | Wellcome Collection. CC BY 4.0 |
| Figure 5 | Wellcome Collection. CC BY 4.0 |
| Figure 6 | Reproduced by kind permission of the Syndics of Cambridge University Library (Maps.bb.53.79.1) |
| Figure 7 | http://www.antique-maps-online.co.uk |
| Figure 8 | From Harraden's *Picturesque Views of Cambridge*, 1800 |
| Figure 9 | © Cambridge Philosophical Society |
| Figure 10 | British Library, London, UK /© British Library Board. All Rights Reserved/ Bridgeman Images |
| Figure 11 | © Cambridge Philosophical Society |
| Figure 12 | Private Collection/© Look and Learn/Illustrated Papers Collection/ Bridgeman Images |
| Figure 13 | Jim Woodhouse, © Cambridge Philosophical Society |
| Figure 14 | Jim Woodhouse, © Cambridge Philosophical Society |
| Figure 15 | Herschel Family Archive |
| Figure 16 | Herschel Family Archive |
| Figure 17 | Universal History Archive/Getty Images |
| Figure 18 | Down House, Downe. FineArt/Alamy Stock Photo |
| Figure 19 | © Cambridge Philosophical Society |
| Figure 20 | © Cambridge Philosophical Society |
| Figure 21 | © Cambridge Philosophical Society |
| Figure 22 | Granger/Bridgeman Images |
| Figure 23 | Pictorial Press Ltd/Alamy Stock Photo |
| Figure 24 | RIBA Collections |
| Figure 25 | © Cambridge Philosophical Society |
| Figure 26 | The Mistress and Fellows, Girton College, Cambridge |
| Figure 27 | Courtesy of Marylebone Cricket Club |
| Figure 28 | Jim Woodhouse, © Cambridge Philosophical Society |
| Figure 29 | The Angus Library and Archive, Regent's Park College. Shelf mark: 378.4259 KEU STACK, p. 110 |

Figure 30 © Museum of Zoology, University of Cambridge/Chris Green (Biochemistry)
Figure 31 Reproduced by kind permission of the Museum of Zoology, University of Cambridge
Figure 32 Reproduced by kind permission of the Syndics of Cambridge University Library (P.VIII.1-10)
Figure 33 Copyright the Cavendish Laboratory, University of Cambridge
Figure 34 © Cambridge Philosophical Society
Figure 35 The Principal and Fellows, Newnham College, Cambridge
Figure 36 The Cartoon Collector/Print Collector/Getty Images
Figure 37 Copyright the Cavendish Laboratory, University of Cambridge
Figure 38 Reinhold Thiele/Thiele/Getty Images
Figure 39 The Principal and Fellows, Newnham College, Cambridge
Figure 40 The Principal and Fellows, Newnham College, Cambridge
Figure 41 Jim Woodhouse, © Cambridge Philosophical Society
Figure 42 Chronicle/Alamy Stock Photo
Figure 43 Whipple Museum of the History of Science, University of Cambridge
Figure 44 © Cambridge Philosophical Society
Figure 45 The Bodleian Library, University of Oxford, Soc. 190 d. 49, ill. Fac. P. 156 & 155
Figure 46 The Bodleian Library, University of Oxford, R. i 10, p. 45
Figure 47 SSPL/Getty Images
Figure 48 Oxford Science Archive/Print Collector/Getty Images
Figure 49 Chronicle/Alamy Stock Photo
Figure 50 © Cambridge Philosophical Society
Figure 51 Science History Images/Alamy Stock Photo
Figure 52 © Cambridge Philosophical Society
Figure 53 Dr Alex Liu, University of Cambridge
Figure 54 Photo by Dr Alex Liu, University of Cambridge; sketch by the late Professor Martin Brasier, University of Oxford
Figure 55 Popperfoto/Getty Images
Figure 56 © Cambridge Philosophical Society

## Plate credits

Plate 1 © 2018 Sedgwick Museum of Earth Sciences, University of Cambridge. Reproduced with permission
Plate 2 Reproduced by permission of the Geological Society of London
Plate 3 University of Cambridge, Institute of Astronomy
Plate 4 The Brenthurst Press, Johannesburg
Plate 5 SSPL/Getty Images
Plate 6 Courtesy of Marylebone Cricket Club

# BIBLIOGRAPHY

Peter Achinstein 'Who really discovered the electron?' in Jed Z. Buchwald and Andrew Warwick (editors) Histories of the electron: the birth of microphysics Cambridge: MIT Press 2001 pp. 403–24.

John Couch Adams 'On the meteoric shower of November, 1866' Proceedings of the Cambridge Philosophical Society II (1866) p. 60.

Jon Agar Science in the twentieth century and beyond Cambridge: Polity Press 2012.

George Biddell Airy 'On the use of silvered glass for the mirrors of reflecting telescopes' Transactions of the Cambridge Philosophical Society II (1827) pp. 105–18.

George Biddell Airy 'On an inequality of long period in the motions of the Earth and Venus' Philosophical transactions of the Royal Society CXXI (1832) pp. 67–124.

George Biddell Airy Account of the Northumberland equatoreal and dome attached to the Cambridge Observatory Cambridge: Cambridge University Press 1844.

S.J.M.M. Alberti 'Natural history and the philosophical societies of late Victorian Yorkshire' Archives of natural history XXX (2003) pp. 342–58.

R. Alcock 'The digestive processes of *Ammocœtes*' Proceedings of the Cambridge Philosophical Society VII (1892) pp. 252–6.

Anonymous 'Review of Transactions of the Geological Society, volume I' The Edinburgh review, or critical journal XIX (1812) pp. 207–29.

Anonymous Cambridge chronicle 5 November 1819.

Anonymous 'Review of meeting of Cambridge Philosophical Society' The Edinburgh philosophical journal III (1820) pp. 184–5.

Anonymous 'Review of meeting of Cambridge Philosophical Society' The London literary gazette and journal of belles lettres 1820 p. 172.

Anonymous 'Donations to the library' Transactions of the Cambridge Philosophical Society I (1822) pp. 461–3.

Anonymous 'Donations to the museum' Transactions of the Cambridge Philosophical Society I (1822) pp. 463–4.

Anonymous 'Preface' Transactions of the Cambridge Philosophical Society I (1822) pp. iii–viii.

Anonymous 'Regulations' Transactions of the Cambridge Philosophical Society I (1822) pp. xvii–xxiii.

Anonymous 'Review of Transactions of the Cambridge Philosophical Society, volume I, part II' The British critic: a new review XVIII (1822) pp. 386–95.

Anonymous The crimes of the clergy; or, the pillars of priestcraft shaken London 1823.

Anonymous 'Review of Transactions of the Cambridge Philosophical Society, volume I, part I' The Cambridge quarterly review and academical register I (1824) pp. 163–82.
Anonymous 'Donations to the library' Transactions of the Cambridge Philosophical Society II (1827) pp. 446–9.
Anonymous 'Donations to the museum' Transactions of the Cambridge Philosophical Society II (1827) pp. 450–1.
Anonymous 'Donations to the library' Transactions of the Cambridge Philosophical Society III (1830) pp. 445–7.
Anonymous 'Donations to the museum' Transactions of the Cambridge Philosophical Society III (1830) pp. 447–8.
Anonymous [David Brewster] 'Decline of science in England' Quarterly review XLIII (1830) p. 327.
Anonymous A new guide to the University and Town of Cambridge Cambridge 1831.
Anonymous [William Whewell] 'Review of Transactions of the Cambridge Philosophical Society, volume the third' The British critic IX (1831) pp. 71–90.
Anonymous Report of the third meeting of the British Association for the Advancement of Science, held at Cambridge in 1833 London: John Murray 1834.
Anonymous 'Geological Society of London' The Athenaeum Issue 421, 21 November 1835 p. 876.
Anonymous 'Cambridge Philosophical Society, 16 November 1835' The London, Edinburgh, and Dublin philosophical magazine and journal of science: series 3 VIII (1836) pp. 78–80.
Anonymous 'Review of a meeting of the Cambridge Philosophical Society, 22 February 1836' The London, Edinburgh, and Dublin philosophical magazine and journal of science: series 3 VIII (1836) p. 429.
Anonymous 'Cambridge Philosophical Society, 27 February 1837' The London, Edinburgh, and Dublin philosophical magazine and journal of science: series 3 X (1837) p. 316.
Anonymous 'Anemometers of Messrs. Whewell and Osler' Arcana of science and art (1838) pp. 279–81.
Anonymous 'University news: Cambridge' The British magazine and monthly register XVII (1840) p. 707.
Anonymous 'Fatal accident on the London and Brighton Railway four lives lost' The Times 4 October 1841.
Anonymous 'Report on the Inquest' The Times 5 October 1841.
Anonymous [Robert Chambers] Vestiges of the natural history of creation London: John Churchill 1844.
Anonymous 'Review of *Vestiges of creation*' The Examiner 9 November 1844 pp. 707–9.
Anonymous 'Review of *Vestiges of creation*' The Spectator 9 November 1844 p. 1072.
Anonymous The railway chronicle 26 April 1845 p. 488.
Anonymous [Robert Chambers] Explanations: a sequel to 'Vestiges of the natural history of creation' London: John Churchill 1845.

Anonymous [Adam Sedgwick] 'Review of *Vestiges of the natural history of creation*' Edinburgh review CLXV (1845) pp. 1–85.

Anonymous 'Cambridge Philosophical Society' The London, Edinburgh, and Dublin philosophical magazine and journal of science: series 3 XXIX (1846) pp. 65–6.

Anonymous Inverness courier 28 October 1846.

Anonymous Report of the fifteenth meeting of the British Association for the Advancement of Science, held at Cambridge in June 1845 London: John Murray 1846.

Anonymous 'A choral ode addressed to HRH Prince Albert' Cambridge chronicle 10 July 1847.

Anonymous 'Professor Airey's [sic] statement regarding Leverrier's planet' The British quarterly review VI (1847) pp. 1–40.

Anonymous Royal Cornwall gazette 17 September 1847.

Anonymous The Gazette Issue 21180, 11 February 1851 p. 357.

Anonymous Cambridge independent press 2 August 1851.

Anonymous The Cambridge Ray Club. Instituted March 11, 1837. A short account of the Club, with its laws, and a list of members Cambridge 1857.

Anonymous Cambridge herald and Huntingdonshire gazette 19 May 1860.

Anonymous 'Our national industries' Nature VI (6 June 1872) p. 97.

Anonymous 'A voice from Cambridge' Nature VIII (8 May 1873) p. 21.

Anonymous 'A voice from Cambridge II' Nature VIII (15 May 1873) p. 41.

Anonymous 'Report of the Council of the Cambridge Philosophical Society' Proceedings of the Cambridge Philosophical Society IV (1883) pp. 101–6.

Anonymous 'The woman of the future' Punch (10 May 1884) p. 225.

Anonymous A descriptive list of anthropometric apparatus, consisting of instruments for measuring and testing the chief physical characteristics of the human body. Designed under the direction of Mr Francis Galton and manufactured and sold by the Cambridge Scientific Instrument Company Cambridge 1887.

Anonymous 'A morning with the anthropometric detectives' Pall Mall gazette 16 November 1888 pp. 1–2.

Anonymous 'The squeeze of 86' Punch (6 April 1889) p. 161.

Anonymous [F.M.T.] 'Letters to the editor: head growth in students at the University of Cambridge' Nature XL (1 August 1889) pp. 317–18.

Anonymous 'Replies to queries' Yn lioar Manninagh I (1889) pp. 23–4.

Anonymous 'Proceedings at the meetings held during the session 1901–1902' Proceedings of the Cambridge Philosophical Society XI (1902) p. 509.

Anonymous 'Proceedings at the meetings held during the session 1907–1908' Proceedings of the Cambridge Philosophical Society XIV (1908) p. 615.

Anonymous A history of the Cavendish Laboratory, 1871–1910 London: Longman, Green and Co. 1910.

Anonymous [H.H. Turner] 'From an Oxford note-book' The Observatory DXXIV (1918) p. 147.

Anonymous 'Science teaching at Cambridge in the seventies' British medical journal II (9 October 1920) p. 572.

Anonymous 'Proceedings at the meetings held during the session 1919–1920' Proceedings of the Cambridge Philosophical Society XX (1921) pp. 213–14.
Anonymous 'Address to the President' The Eagle XLII (1922) pp. 161–70.
Anonymous 'Proceedings at the meetings held during the session 1922–1923' Proceedings of the Cambridge Philosophical Society XXI (1923) p. 795.
Anonymous 'Proceedings at the meetings held during the session 1924–1925' Proceedings of the Cambridge Philosophical Society XXII (1925) p. 970.
Anonymous 'Obituary: Frederick Meadows Turner, MD, BSc, DPH Medical Superintendent, South-Eastern Hospital, London County Council' British medical journal I (31 January 1931) p. 202.
Anonymous 'Proceedings at the meetings held during the session 1935–1936' Proceedings of the Cambridge Philosophical Society XXXII (1936) p. 686.
Anonymous 'Proceedings at the meetings held during the session 1936–1937' Proceedings of the Cambridge Philosophical Society XXXIII (1937) p. 588.
Anonymous An inventory of the historical monuments in the City of Cambridge London: Her Majesty's Stationery Office 1959.
Eric Ashby and Mary Anderson 'Introduction' in Adam Sedgwick A discourse on the studies of the University of Cambridge Leicester: Leicester University Press 1969 pp. (7)–(17).
Charles Babbage Reflections on the decline of science in England, and on some of its causes London: B. Fellowes 1830.
Anna Maria Babington (editor) Memorials, journal and botanical correspondence of Charles Cardale Babington Cambridge: Cambridge University Press 2013.
Lawrence Badash 'Ernest Rutherford (1871–1937)' Oxford dictionary of national biography Retrieved 2 June 2018 from http://www.oxforddnb.com/view/10.1093/ref:odnb/9780198614128.001.0001/odnb-9780198614128-e-35891.
Nora Barlow (editor) Darwin and Henslow: the growth of an idea, letters 1831–1860 London: Murray 1967.
Ruth Barton '"Huxley, Lubbock, and half a dozen others": professionals and gentlemen in the formation of the X Club, 1851–1864' Isis LXXXIX (1998) pp. 410–44.
Ruth Barton 'Lockyer's columns of controversy in *Nature*' Retrieved 9 October 2018 from http://www.nature.com/nature/history/full/nature06260.html.
Alison Bashford and Philippa Levine (editors) The Oxford handbook of the history of eugenics Oxford: Oxford University Press 2010.
A. Bateson and F. Darwin 'On the change in shape in turgescent pith' Proceedings of the Cambridge Philosophical Society VI (1889) pp. 358–9.
W. Bateson 'Notes on hybrid Cinerarias produced by Mr Lynch and Miss Pertz' Proceedings of the Cambridge Philosophical Society IX (1898) pp. 308–9.
William Bateson 'Hybridisation and cross-breeding as a method of scientific investigation' Journal of the Royal Horticultural Society XXIV (1900) pp. 59–66.
W. Bateson and A. Bateson 'On variations in the floral symmetry of certain plants having irregular corollas' Journal of the Linnean Society XXVIII (1891) pp. 386–424.

W. Bateson and A. Bateson 'On variations in the floral symmetry of certain flowers having irregular corollas' Proceedings of the Cambridge Philosophical Society VII (1892) p. 96.

W. Bateson and D.F.M. Pertz 'Notes on the inheritance of variation in the corolla of Veronica Buxbaumii' Proceedings of the Cambridge Philosophical Society X (1900) pp. 78–93.

W. Bateson and E.R. Saunders 'Introduction' Reports to the Evolution Committee of the Royal Society, Report 1 1902 pp. 3–12.

Harvey W. Becher 'Voluntary science in nineteenth-century Cambridge University to the 1850s' British journal for the history of science 19 (1986) pp. 80–1.

Harvey W. Becher 'George Peacock (1791–1858)' Oxford dictionary of national biography Retrieved 2 March 2018 from http://www.oxforddnb.com/view/10.1093/ref:odnb/9780198614128.001.0001/odnb-9780198614128-e-21673.

Harvey W. Becher 'Robert Woodhouse (1773–1827)' Oxford dictionary of national biography Retrieved 2 March 2018 from http://www.oxforddnb.com/view/10.1093/ref:odnb/9780198614128.001.0001/odnb-9780198614128-e-29926.

R.M. Beverley A letter to His Royal Highness the Duke of Gloucester London 1833.

Ronald M. Birse 'Philip Kelland (1808–1879)' Oxford dictionary of national biography Retrieved 31 May 2018 from http://www.oxforddnb.com/view/10.1093/ref:odnb/9780198614128.001.0001/odnb-9780198614128-e-15284.

P.M.S. Blackett 'Charles Thomson Rees Wilson, 1869–1959' Biographical memoirs of Fellows of the Royal Society VI (1960) pp. 269–95.

Niels Bohr 'On the application of the quantum theory to atomic structure, part 1: the fundamental postulates' Supplement to Proceedings of the Cambridge Philosophical Society XXII (1925) pp. 1–42.

Béla Bollobás (editor) Littlewood's miscellany Cambridge: Cambridge University Press 1986.

Béla Bollobás 'Godfrey Harold Hardy (1877–1947)' Oxford dictionary of national biography Retrieved 2 June 2018 from http://www.oxforddnb.com/view/10.1093/ref:odnb/9780198614128.001.0001/odnb-9780198614128-e-33706.

Henry J.H. Bond 'A statistical report of Addenbrooke's Hospital, for the year 1836' Transactions of the Cambridge Philosophical Society VI (1838) pp. 361–78.

Henry J.H. Bond 'A statistical report of Addenbrooke's Hospital, for the year 1837' Transactions of the Cambridge Philosophical Society VI (1838) pp. 565–75.

T.G. Bonney 'Science at Cambridge' Nature VIII (29 May 1873) p. 83.

William H. Bragg 'X-rays and crystals' Nature XC (28 November 1912) pp. 360–1.

W.L. Bragg 'The diffraction of short electromagnetic waves by a crystal' Proceedings of the Cambridge Philosophical Society XVII (1914) pp. 43–57.

M.G. Brock and M.C. Curthoys (editors) The history of the University of Oxford, volume VI: nineteenth-century Oxford, part 1 Oxford: Clarendon Press 1997.

Christopher N.L. Brooke A history of the University of Cambridge, volume IV: 1870–1990 Cambridge: Cambridge University Press 1993.

Janet Browne 'Dorothea Pertz (1859–1939)' Oxford dictionary of national biography Retrieved 1 June 2018 from http://www.oxforddnb.com/view/10.1093/ref:odnb/9780198614128.001.0001/odnb-9780198614128-e-58481.

Alexandrina Buchanan Robert Willis (1800–1875) and the foundation of architectural history Cambridge: Cambridge University Press 2013.

Jed Z. Buchwald and Sungook Hong 'Physics' in David Cahan (editor) From natural philosophy to the sciences Chicago: Chicago University Press 2003 pp. 163–95.

William Buckland Reliquiae diluvianae, or, observations on the organic remains contained in caves, fissures and diluvial gravel, and on other geological phenomena London: John Murray 1823.

Richard M. Burian and Doris T. Zallan 'Genes' in Peter J. Bowler and John V. Pickstone (editors) The Cambridge history of science, volume VI: the modern biological and earth sciences Cambridge: Cambridge University Press 2009 pp. 432–50.

R.W. Burkhardt The spirit of system: Lamarck and evolutionary biology Cambridge, Mass., and London: Harvard University Press 1995.

Lucilla Burn The Fitzwilliam Museum: a history London: Philip Wilson 2016.

J.P.T. Bury (editor) Romilly's Cambridge Diary, 1832–1842 Cambridge: Cambridgeshire Records Society 1967.

M.E. Bury and J.D. Pickles (editors) Romilly's Cambridge Diary, 1842–1847 Cambridge: Cambridgeshire Records Society 1994.

Günther Buttmann The shadow of the telescope: a biography of John Herschel Guildford: Lutterworth Press 1974.

George Gordon, Lord Byron 'Thoughts suggested by a college examination' Hours of idleness Newark 1807.

David Cahan (editor) From natural philosophy to the sciences Chicago: Chicago University Press 2003.

Richard H.T. Callow, Duncan McIlroy, and Martin D. Brasier 'John Salter and the Ediacara fauna of the Longmyndian Supergroup' Ichnos XVIII (2011) pp. 176–87.

Susan F. Cannon Science in culture: the early Victorian period New York: Dawson and Science History Publications 1978.

Walter F. Cannon 'John Herschel and the idea of science' Journal of the history of ideas XXII (1961) pp. 215–39.

Walter F. Cannon 'The impact of uniformitarianism: two letters from John Herschel to Charles Lyell, 1836–1837' Proceedings of the American Philosophical Society CV (1961) pp. 301–14.

Walter F. Cannon 'Scientists and broad churchmen: an early Victorian intellectual network' Journal of British studies IV (1964) pp. 65–88.

Nicolas Carlisle A memoir of the life of William Wyon London: W. Nichol 1837.

A.R. Catton 'On the synthesis of formic acid' Proceedings of the Cambridge Philosophical Society I (1863) p. 235.

Arthur Cayley 'A new theorem on the equilibrium of four forces acting on a solid body' Proceedings of the Cambridge Philosophical Society I (1863) p. 235.

J. Chadwick and C.D. Ellis 'A preliminary investigation of the intensity distribution in the $\beta$-ray spectra of radium B and C' Proceedings of the Cambridge Philosophical Society XXI (1923) pp. 274–80.

Elizabeth Chambers Patterson Mary Somerville and the cultivation of science, 1815–1840 The Hague: Nijhoff 1983.

Allan Chapman 'George Biddell Airy (1801–1892)' Oxford dictionary of national biography Retrieved 15 March 2018 from http://www.oxforddnb.com/view/10.1093/ref:odnb/9780198614128.001.0001/odnb-9780198614128-e-251.

John Willis Clark 'The foundation and early years of the Society' Proceedings of the Cambridge Philosophical Society VII (1892) pp. i–l.

John Willis Clark 'President's address' Proceedings of the Cambridge Philosophical Society VII (1892) pp. 2–4.

John Willis Clark and Thomas McKenny Hughes The life and letters of Adam Sedgwick, volume I Cambridge: Cambridge University Press 2009.

J.W. Clark (revised by H.G.C. Matthew) 'Connop Thirlwall (1797–1875)' Oxford dictionary of national biography Retrieved 15 March 2018 from http://www.oxforddnb.com/view/10.1093/ref:odnb/9780198614128.001.0001/odnb-9780198614128-e-27185.

J.W. Clark (revised by Michael Bevan) 'William Clark (1788–1869)' Oxford dictionary of national biography Retrieved 30 May 2018 from http://www.oxforddnb.com/view/10.1093/ref:odnb/9780198614128.001.0001/odnb-9780198614128-e-5478.

Ronald W. Clark The birth of the bomb London: Phoenix House 1961.

William Clark 'A case of human monstrosity, with a commentary' Transactions of the Cambridge Philosophical Society IV (1833) pp. 219–56.

Edward Daniel Clarke 'Observations upon the ores which contain cadmium, and upon the discovery of this metal in the Derbyshire silicates and other ores of zinc' Annals of philosophy XV (1820) pp. 272–6.

A.M. Clerke (revised by Anita McConnell) 'Thomas Catton (1758–1838)' Oxford dictionary of national biography Retrieved 2 March 2018 from http://www.oxforddnb.com/view/10.1093/ref:odnb/9780198614128.001.0001/odnb-9780198614128-e-4903.

William Cobbett 'The parson and the boy' Cobbett's weekly register XLVII (2 August 1823) pp. 256–319.

Samuel Taylor Coleridge 'General introduction, or, a preliminary treatise on method' in Edward Smedley, Hugh James Rose, and Henry John Rose (editors) Encyclopædia metropolitana, volume I London: J.J. Griffin 1849, pp. (1)–(27).

Harry Collins and Trevor Pinch The golem Cambridge: Cambridge University Press 1998.

Charles Henry Cooper Annals of Cambridge, volume V Cambridge: Warwick and Co. 1850–1856.

Thompson Cooper (revised by Julia Tompson) 'Bewick Bridge (1767–1833)' Oxford dictionary of national biography Retrieved 2 March 2018 from http://www.

oxforddnb.com/view/10.1093/ref:odnb/9780198614128.001.0001/odnb-9780198614128-e-3386.

Thompson Cooper (revised by M.C. Curthoys) 'Thomas Turton (1780–1864)' Oxford dictionary of national biography Retrieved 2 March 2018 from http://www.oxforddnb.com/view/10.1093/ref:odnb/9780198614128.001.0001/odnb-9780198614128-e-27895.

Stephen Courtney 'Anthropometry and the biological sciences in late 19th-century Britain' Retrieved 11 May 2017 from http://anthropometryincontext.com/2017/05/01/blog-post-title/.

Alex D.D. Craik Mr Hopkins' men: Cambridge reform and British mathematics in the 19th century London: Springer 2007.

Elizabeth Crawford The women's suffrage movement: a reference guide 1866–1928 London: University College London Press 1999.

Mary R.S. Creese Ladies in the laboratory? American and British women in science 1800–1900 London: Scarecrow 1998.

Mary R.S. Creese (revised by V.M. Quirke) 'Edith Rebecca Saunders (1865–1945)' Oxford dictionary of national biography Retrieved 1 June 2018 from http://www.oxforddnb.com/view/10.1093/ref:odnb/9780198614128.001.0001/odnb-9780198614128-e-37936/version/0.

Mary R.S. Creese 'Mary Somerville (1780–1872)' Oxford dictionary of national biography Retrieved 1 June 2018 from http://www.oxforddnb.com/view/10.1093/ref:odnb/9780198614128.001.0001/odnb-9780198614128-e-26024.

Tony Crilly 'The Cambridge mathematical journal and its descendants: the linchpin of a research community in the early and mid-Victorian age' Historia mathematica XXXI (2004) pp. 455–97.

Michael J. Crowe 'John Frederick William Herschel (1792–1871)' Oxford dictionary of national biography Retrieved 15 March 2018 from http://www.oxforddnb.com/view/10.1093/ref:odnb/9780198614128.001.0001/odnb-9780198614128-e-13101.

Andrew Cunningham 'How the *Principia* got its name; or, taking natural philosophy seriously' History of science XXIX (1991) pp. 377–92.

E. Dale 'On certain outgrowths (intumescences) on the green parts of *Hibiscus vitifolius* Linn.' Proceedings of the Cambridge Philosophical Society X (1900) pp. 192–210.

W.C.D. Dampier (revised by Frank A.J.L. James) 'George Liveing (1827–1924)' Oxford dictionary of national biography Retrieved 31 May 2018 from http://www.oxforddnb.com/view/10.1093/ref:odnb/9780198614128.001.0001/odnb-9780198614128-e-34559.

Charles Darwin Extracts from letters addressed to Professor Henslow Cambridge: Cambridge Philosophical Society 1835.

Charles Darwin Narrative of the surveying voyages of His Majesty's ships *Adventure* and *Beagle*, volume III: journal and remarks, 1832–1836 London: Henry Colburn 1839.

Charles Darwin On the origin of species by means of natural selection London: John Murray 1859.

Charles Darwin (edited by Nora Barlow) The works of Charles Darwin, volume I: diary of the voyage of H.M.S. *Beagle* New York: New York University Press 1986.

Francis Darwin and Dorothea F.M. Pertz 'On the artificial production of rhythm in plants' Annals of botany VI (1892) pp. 245–64.

F. Darwin and D.F.M. Pertz 'On the effect of water currents on the assimilation of aquatic plants' Proceedings of the Cambridge Philosophical Society IX (1898) pp. 76–90.

F. Darwin and D.F.M. Pertz 'On the injection of the intercellular spaces occurring in the leaves of *Elodea* during recovery from plasmolysis' Proceedings of the Cambridge Philosophical Society IX (1898) pp. 272–3.

H. Darwin and R. Threlfall 'On Mr Galton's anthropometric apparatus at present in use in the Philosophical Library' Proceedings of the Cambridge Philosophical Society V (1886) p. 374.

Lorraine Daston and H. Otto Sibum 'Scientific personae and their histories' Science in context XVI (2003) pp. 1–8.

Emily Davies 'The influence of university degrees on the education of women' Victoria magazine I (1863) pp. 260–71.

Augustus de Morgan 'On some points of the integral calculus' Proceedings of the Cambridge Philosophical Society I (1863) pp. 106–9.

Sophia Elizabeth de Morgan Memoir of Augustus De Morgan Cambridge: Cambridge University Press 2010.

Talal Debs 'William Henry Bragg (1862–1942)' Oxford dictionary of national biography Retrieved 2 June 2018 from http://www.oxforddnb.com/view/10.1093/ref:odnb/9780198614128.001.0001/odnb-9780198614128-e-32031.

Adrian Desmond Archetypes and ancestors: palaeontology in Victorian London, 1850–1875 London: Blond and Briggs 1982.

Adrian Desmond 'Thomas Henry Huxley (1825–1895)' Oxford dictionary of national biography Retrieved 31 May 2018 from http://www.oxforddnb.com/view/10.1093/ref:odnb/9780198614128.001.0001/odnb-9780198614128-e-14320.

P.A.M. Dirac 'Dissociation under a temperature gradient' Proceedings of the Cambridge Philosophical Society XXII (1925) pp. 132–7.

Brian Dolan Exploring European frontiers: British travellers in the age of Enlightenment Basingstoke: Macmillan 2000.

Richard Dunn and Rebecca Higgitt Ships, clocks, and stars: the quest for longitude Glasgow: Collins and London: Royal Museums Greenwich 2014.

John Earman and Clark Glymour 'Relativity and eclipses: the British eclipse expeditions of 1919 and their predecessors' Historical studies in the physical sciences XI (1980) pp. 49–85.

John Earman and Clark Glymour 'The gravitational red shift as a test of general relativity: history and analysis' Studies in the history and philosophy of science XI (1980) pp. 175–214.

A.W.F. Edwards 'Mathematising Darwin' Behavioral ecology and sociobiology LXV (2011) pp. 421–30.

Sam Edwards 'Obituaries: Rudolph E. Peierls' Physics today XLIX (February 1996) pp. 74–5.

Albert Einstein 'Die Grundlage der allgemeinen Relativitätstheorie' Annalen der physic XLIX (1916) pp. 769–822.

Albert Einstein 'Considerations concerning the fundaments of theoretical physics' Science XCI (1940) pp. 487–92.

Philip C. Enros 'The Analytical Society (1812–1813): precursor of the renewal of Cambridge mathematics' Historia mathematica 10 (1983) pp. 24–47.

F. Eves 'On some experiments on the liver ferment' Proceedings of the Cambridge Philosophical Society V (1886) pp. 182–3.

Isobel Falconer 'Corpuscles to electrons' in Jed Z. Buchwald and Andrew Warwick (editors) Histories of the electron: the birth of microphysics Cambridge: MIT Press 2001 pp. 77–100.

Isobel Falconer 'James Chadwick (1891–1974)' Oxford dictionary of national biography Retrieved 2 June 2018 from http://www.oxforddnb.com/view/10.1093/ref:odnb/9780198614128.001.0001/odnb-9780198614128-e-30912.

Isobel Falconer 'Joseph John Thomson (1856–1940)' Oxford dictionary of national biography Retrieved 31 May 2018 from http://www.oxforddnb.com/view/10.1093/ref:odnb/9780198614128.001.0001/odnb-9780198614128-e-36506.

William Farish A plan of a course of lectures on arts and manufactures, more particularly such as relate to chemistry Cambridge: J. Burgess 1796.

William Farish 'On isometrical perspective' Transactions of the Cambridge Philosophical Society I (1822) pp. 1–20.

Lewis S. Feuer Einstein and the generations of science London and New York: Routledge 1982.

R.A. Fisher 'The correlation between relatives in the supposition of Mendelian inheritance' *Transactions of the Royal Society of Edinburgh* LII (1918) pp. 399–433.

Walter Fletcher 'Letter to the editor' The Spectator 30 October 1841 p. 12.

Sophie Forgan 'Context, image and function: a preliminary enquiry into the architecture of scientific societies' British journal for the history of science XIX (1986) pp. 89–113.

Robert Fox 'John Dawson (bap. 1735–1820)' Oxford dictionary of national biography Retrieved 2 March 2018 from http://www.oxforddnb.com/view/10.1093/ref:odnb/9780198614128.001.0001/odnb-9780198614128-e-7350.

Michael Freeden 'Eugenics and progressive thought' The historical journal XXII (1979) pp. 645–71.

K. Fuchs 'The conductivity of thin metallic films according to the electron theory of metals' Proceedings of the Cambridge Philosophical Society XXXIV (1938) pp. 100–8.

K. Fuchs 'On the stability of nuclei against $\beta$-emission' Proceedings of the Cambridge Philosophical Society XXXV (1939) pp. 242–55.

Aileen Fyfe Steam-powered knowledge: William Chambers and the business of publishing, 1820–1860 Chicago: University of Chicago Press 2012.

Aileen Fyfe 'Peer review: not as old as you might think' Times higher education 25 June 2015 Retrieved 8 October 2018 from http://www.timeshighereducation.com/features/peer-review-not-old-you-might-think.

Peter Galison and Alexi Assmus 'Artificial clouds, real particles' in David Gooding, Trevor Pinch, and Simon Schaffer (editors) The uses of experiment Cambridge: Cambridge University Press 1989 pp. 225–75.

Francis Galton 'On the anthropometric laboratory at the International Health Exhibition' The journal of the Anthropological Institute of Great Britain and Ireland XIV (1885) pp. 205–21.

Francis Galton 'On recent designs for anthropometric instruments' The journal of the Anthropological Institute of Great Britain and Ireland XVI (1887) pp. 2–8.

Francis Galton 'Head growth in students at the University of Cambridge' Nature XXXVIII (3 May 1888) pp. 14–15.

Francis Galton 'Letters to the editor' Nature XL (1 August 1889) p. 318.

Francis Galton 'Arithmetic by smell' Psychological review I (1894) pp. 61–2.

John Gascoigne 'The universities and the scientific revolution: the case of Newton and Restoration Cambridge' History of science XXIII (1985) pp. 391–434.

John Gascoigne Cambridge in the age of the Enlightenment Cambridge: Cambridge University Press 1989.

Stephen Gaukroger The emergence of a scientific culture Oxford: Clarendon Press 2006.

Kostas Gavroglu 'John William Strutt, third Baron Rayleigh (1842–1919)' Oxford dictionary of national biography Retrieved 31 May 2018 from http://www.oxforddnb.com/view/10.1093/ref:odnb/9780198614128.001.0001/odnb-9780198614128-e-36359.

Gerald L. Geison Michael Foster and the Cambridge school of physiology: the scientific enterprise in late Victorian society Princeton: Princeton University Press 1978.

John Gibbins John Grote, Cambridge University and the development of Victorian thought Exeter: Imprint Academic 2007.

Susannah Gibson Animal, vegetable, mineral? Oxford: Oxford University Press 2015.

T.R. Glover Cambridge retrospect Cambridge: Cambridge University Press 1943.

Graeme Gooday 'The questionable matter of electricity: the reception of J.J. Thomson's "corpuscle" among electrical theorists and technologists' in Jed Z. Buchwald and Andrew Warwick (editors) Histories of the electron: the birth of microphysics Cambridge: MIT Press 2001 pp. 101–34.

George Cornelius Gorham Memoirs of John Martyn, F.R.S., and of Thomas Martyn, B.D., F.R.S., F.L.S., Professors of Botany in the University of Cambridge London: Hatchard 1830.

Paula Gould 'Women and the culture of university physics in late nineteenth-century Cambridge' British journal for the history of science XXX (1997) pp. 127–49.

Edward Grant A history of natural philosophy Cambridge: Cambridge University Press 2007.

I. Grattan-Guinness 'Mathematics and mathematical physics from Cambridge, 1815–1840: a survey of the achievements and of the French influences' in P.M. Harman (editor) Wranglers and physicists: studies on Cambridge physics in the nineteenth century Manchester: Manchester University Press 1985 pp. 84–111.

Mott T. Green 'Genesis and geology revisited: the order of nature and the nature of order in nineteenth-century Britain' in David C. Lindberg and Ronald L. Numbers (editors) When science and Christianity meet Chicago and London: University of Chicago Press 2003 pp. 139–60.

J.B.S. Haldane Daedalus, or, science and the future New York: E.P. Dutton and Company 1924.

J.B.S. Haldane 'A mathematical theory of natural and artificial selection. Part II. The influence of partial self-fertilisation, inbreeding, assertive mating, and selective fertilisation on the composition of Mendelian populations, and on natural selection' Biological reviews I (1925) pp. 158–63.

J.B.S. Haldane 'A mathematical theory of natural and artificial selection. Part III' Proceedings of the Cambridge Philosophical Society XXIII (1927) pp. 363–72.

J.B.S. Haldane 'A mathematical theory of natural and artificial selection. Part IV' Proceedings of the Cambridge Philosophical Society XXIII (1927) pp. 607–15.

J.B.S. Haldane 'A mathematical theory of natural and artificial selection. Part V' Proceedings of the Cambridge Philosophical Society XXIII (1927) pp. 838–44.

J.B.S. Haldane 'A mathematical theory of natural and artificial selection. Part VI. Isolation' Proceedings of the Cambridge Philosophical Society XXVI (1930) pp. 220–30.

J.B.S. Haldane 'A mathematical theory of natural and artificial selection' Transactions of the Cambridge Philosophical Society XXIII (1931) pp. 19–42.

J.B.S. Haldane 'A mathematical theory of natural and artificial selection. Part VII. Selection intensity as a function of mortality rate' Proceedings of the Cambridge Philosophical Society XXVII (1931) pp. 131–6.

J.B.S. Haldane 'A mathematical theory of natural and artificial selection. Part VIII. Metastable population' Proceedings of the Cambridge Philosophical Society XXVII (1931) pp. 137–42.

J.B.S. Haldane 'A mathematical theory of natural and artificial selection. Part IX. Rapid selection' Proceedings of the Cambridge Philosophical Society XXVIII (1932) pp. 244–8.

J.B.S. Haldane 'A mathematical theory of natural and artificial selection. Part X. Some theorems on artificial selection' Genetics XIX (1934) pp. 412–29.

J.B.S. Haldane The causes of evolution Ithaca: Cornell University Press 1966.

A. Rupert Hall The Cambridge Philosophical Society: a history 1819–1969 Cambridge: Cambridge Philosophical Society 1969.

Thomas Hamilton (revised by John D. Haigh) 'Samuel Lee (1783–1852)' Oxford dictionary of national biography Retrieved 2 March 2018 from http://www.oxforddnb.com/view/10.1093/ref:odnb/9780198614128.001.0001/odnb-9780198614128-e-16309.

G.H. Hardy A mathematician's apology Cambridge: Cambridge University Press 1948.

P.M. Harman (editor) Wranglers and physicists: studies on Cambridge physics in the nineteenth century Manchester: Manchester University Press 1985.

P.M. Harman (editor) The scientific letters and papers of James Clerk Maxwell, volume I: 1846–1862 Cambridge: Cambridge University Press 1990.

P.M. Harman 'James Clerk Maxwell (1831–1879)' Oxford dictionary of national biography Retrieved 31 May 2018 from http://www.oxforddnb.com/view/10.1093/ref:odnb/9780198614128.001.0001/odnb-9780198614128-e-5624.

Negley Harte and John North The world of University College London 1828–1978 London: University College London 1978.

John Herschel 'On certain remarkable instances of deviation from Newton's scale in the tints developed by crystals, with one axis of double refraction, on exposure to polarised light' Transactions of the Cambridge Philosophical Society I (1822) pp. 21–42.

John Herschel A preliminary discourse on the study of natural philosophy London: Longman, Rees, Orme, Brown and Green 1831.

John Herschel 'Address' Report of the fifteenth meeting of the British Association for the Advancement of Science, held at Cambridge in June 1845 London: John Murray 1846 pp. xxvii–xliv.

John Herschel (edited by David S. Evans, Terence J. Deeming, Betty Hall Evans, and Stephen Goldfarb) Herschel at the Cape: diaries and correspondence of Sir John Herschel, 1834–1838 Austin: University of Texas Press 1969.

Jeffrey A. Hughes 'Charles Drummond Ellis (1895–1980)' Oxford dictionary of national biography Retrieved 2 June 2018 from http://www.oxforddnb.com/view/10.1093/ref:odnb/9780198614128.001.0001/odnb-9780198614128-e-31070.

G.M. Humphry Reporter 19 October 1870 p. 26.

Bruce Hunt 'Doing science in a global empire: cable telegraphy and electrical physics in Victorian Britain' in Bernard Lightman (editor) Victorian science in context Chicago: University of Chicago Press 1997 pp. 312–33.

Roger Hutchins British university observatories, 1772–1939 Aldershot: Ashgate 2008.

Roger Hutchins 'John Couch Adams (1819–1892)' Oxford dictionary of national biography Retrieved 31 May 2018 from http://www.oxforddnb.com/view/10.1093/ref:odnb/9780198614128.001.0001/odnb-9780198614128-e-123.

Julian Huxley Evolution: the modern synthesis London: Allen and Unwin 1942.

Frank James 'Michael Faraday, the City Philosophical Society and the Society of Arts' Royal Society of Arts Journal CXL (1992) pp. 192–9.

John Jenkin William and Lawrence Bragg, father and son: the most extraordinary collaboration in science Oxford: Oxford University Press 2008.

J. Vernon Jensen 'The X Club: fraternity of Victorian scientists' British journal for the history of science V (1970) pp. 63–72.

Leonard Jenyns 'Cambridge Philosophical Society museum' in J.J. Smith (editor) The Cambridge portfolio London: J.W. Parker 1840 pp. 127–9.

Leonard Jenyns Memoir of the Rev. John Stevens Henslow, M.A., F.L.S., F.G.S., F.C.P.S.: late rector of Hitcham and Professor of Botany in the University of Cambridge London: John Van Voorst 1862.

Leonard Jenyns Chapters in my life Cambridge: Cambridge University Press 2011.

Walter Jerrold Michael Faraday: man of science London: S.W. Partridge & Co. 1892.

Adrian Johns 'Miscellaneous methods: authors, societies and journals in early modern England' British journal for the history of science XXXIII (2000) pp. 159–86.

Alice Johnson 'On the development of the pelvic girdle and skeleton of the hind limb in the chick' Proceedings of the Cambridge Philosophical Society IV (1883) pp. 328–31.

Robert Kanigel 'Srinivasa Ramanujan (1887–1920)' Oxford dictionary of national biography Retrieved 2 June 2018 from http://www.oxforddnb.com/view/10.1093/ref:odnb/9780198614128.001.0001/odnb-9780198614128-e-51582.

Milo Keynes (editor) Sir Francis Galton, FRS: the legacy of his ideas Basingstoke: Macmillan 1991.

C.W. Kilmister 'Arthur Stanley Eddington (1882–1944)' Oxford dictionary of national biography Retrieved 2 June 2018 from http://www.oxforddnb.com/view/10.1093/ref:odnb/9780198614128.001.0001/odnb-9780198614128-e-32967.

Dong-Won Kim Leadership and creativity: a history of the Cavendish Laboratory, 1871–1919 Dordrecht: Kluwer 2002.

W. Towler Kingsley 'Application of the microscope to photography' Journal of the Society of Arts, London I (1853) pp. 289–92.

W. Towler Kingsley 'Application of photography to the microscope' Proceedings of the Cambridge Philosophical Society I (1863) pp. 117–19.

E. Kitson Clark The history of 100 years of life of the Leeds Philosophical and Literary Society Leeds: Jowett and Sowry 1924.

H.G. Klaassen 'On the effect of temperature on the conductivity of solutions of sulphuric acid' Proceedings of the Cambridge Philosophical Society VII (1892) pp. 137–41.

Helge Kragh 'The vortex atom: a Victorian theory of everything' Centaurus XLVI (2002) pp. 32–114.

Sylvestre Lacroix (translated by Charles Babbage, John Herschel, and George Peacock) An elementary treatise on the differential and integral calculus Cambridge: Cambridge University Press 1816.

Dionysius Lardner (editor) The museum of science and art, volume I London: Walton and Maberly 1854.

Christoph Laucht Elemental Germans: Klaus Fuchs, Rudolf Peierls and the making of British nuclear culture 1939–1959 Basingstoke: Palgrave Macmillan 2012.

Alice Lee, Marie A. Lewenz, and Karl Pearson 'On the correlation of the mental and physical characters in man, part II' Proceedings of the Royal Society of London LXXI (1903) pp. 106–14.

Elisabeth Leedham Green 'The arrival of research degrees in Cambridge' Darwin College Research Report 2011 Retrieved 12 October 2018 from http://www.darwin.cam.ac.uk/drupal7/sites/default/files/Documents/publications/dcrr010.pdf.

J.M. Levine 'John Woodward (1665/1668–1728)' Oxford dictionary of national biography Retrieved 2 March 2018 from http://www.oxforddnb.com/view/10.1093/ref:odnb/9780198614128.001.0001/odnb-9780198614128-e-29946.

C.L.E. Lewis and S.J. Knell (editors) The making of the Geological Society of London London: Geological Society 2009.

Bernard Lightman 'Huxley and the Devonshire Commission' in Gowan Dawson and Bernard Lightman (editors) Victorian scientific naturalism Chicago: University of Chicago Press 2014 pp. 101–30.

Christopher F. Lindsey 'James Cumming (1777–1861)' Oxford dictionary of national biography Retrieved 2 March 2018 from http://www.oxforddnb.com/view/10.1093/ref:odnb/9780198614128.001.0001/odnb-9780198614128-e-6896.

Peter Linehan (editor) St John's College, Cambridge: a history Woodbridge: Boydell Press 2011.

Alexander G. Liu 'Reviewing the Ediacaran fossils of the Long Mynd, Shropshire' Proceedings of the Shropshire Geological Society XVI (2011) pp. 31–43.

Alexander G. Liu, Jack J. Matthews, Latha R. Menon, Duncan McIlroy, and Martin D. Brasier '*Haootia quadriformis* n. gen., n. sp., interpreted as a muscular cnidarian impression from the late Ediacaran period (approx. 560 Ma)' Proceedings of the Royal Society B CCLXXXI (2014) 20141202.

Malcolm Longair Maxwell's enduring legacy: a scientific history of the Cavendish Laboratory Cambridge: Cambridge University Press 2016.

William C. Lubenow The Cambridge Apostles, 1820–1914 Cambridge: Cambridge University Press 1998.

William C. Lubenow 'Only connect': learned societies in nineteenth-century Britain Woodbridge: Boydell Press 2015.

Frans Lundgren 'The politics of participation: Francis Galton's anthropometric laboratory and the making of critical selves' British journal for the history of science XLVI (2011) pp. 445–66.

Charles Lyell Life, letters and journals of Sir Charles Lyell, Bart, volume I London: John Murray 1881 p. 368.

K.M. Lyell (editor) Life, letters and journals of Sir Charles Lyell, Bart Cambridge: Cambridge University Press 2010.

W.R Macdonell 'On criminal anthropometry and the identification of criminals' Biometrika I (1902) pp. 177–227.

Donald A. MacKenzie Statistics in Britain, 1865–1930 Edinburgh: Edinburgh University Press 1981.

Roy M. MacLeod 'The X Club: a social network of science in late-Victorian England' Notes and records of the Royal Society of London XXIV (1970) pp. 305–22.

Roy M. MacLeod 'The support of Victorian science: the endowment of research movement in Great Britain, 1868–1900' Minerva IX (1971) pp. 197–230.

Roy MacLeod (editor) Days of judgement: science, examinations and the organization of knowledge in late Victorian England Driffield: Nafferton 1982.

Roy MacLeod and Russell Moseley 'Breaking the circle of the sciences: the Natural Sciences Tripos and the "examination revolution"' in Roy MacLeod (editor) Days of judgement: science, examinations and the organization of knowledge in late Victorian England Driffield: Nafferton 1982 pp. 189–212.

C.R. Marshall 'Note on the pharmacological action of cannabis resin' Proceedings of the Cambridge Philosophical Society IX (1898) pp. 149–50.

F. Martin 'Expansion produced by electric discharge' Proceedings of the Cambridge Philosophical Society IX (1898) pp. 11–17.

Theodore Martin The life of His Royal Highness the Prince Consort, volume II Cambridge: Cambridge University Press 2013.

James Clerk Maxwell 'On the transformation of surfaces by bending' Transactions of the Cambridge Philosophical Society IX (1856) pp. 445–70.

James Clerk Maxwell On the stability of the motion of Saturn's rings Cambridge: Macmillan 1859.

James Clerk Maxwell 'On physical lines of force, part I' The London, Edinburgh, and Dublin philosophical magazine and journal of science: series 4 XXI (1861) pp. 161–75.

James Clerk Maxwell 'On Faraday's lines of force' Proceedings of the Cambridge Philosophical Society I (1863) pp. 160–6.

James Clerk Maxwell 'On Faraday's lines of force' Transactions of the Cambridge Philosophical Society X (1864) pp. 27–83.

James Clerk Maxwell 'A dynamical theory of the electromagnetic field' Philosophical transactions of the Royal Society CLV (1865) pp. 459–512.

James Clerk Maxwell 'On the proof of the equations of motion of a connected system' Proceedings of the Cambridge Philosophical Society II (1876) pp. 292–4.

James Clerk Maxwell 'On a problem in the calculus of variations in which the solution is discontinuous' Proceedings of the Cambridge Philosophical Society II (1876) pp. 294–5.

James Clerk Maxwell 'Introductory lecture on experimental physics' in W.D. Niven (editor) The scientific papers of James Clerk Maxwell, volume II Cambridge: Cambridge University Press 1890 pp. 241–55.

Anita McConnell 'William Farish (1759–1837)' Oxford dictionary of national biography Retrieved 2 March 2018 from http://www.oxforddnb.com/view/10.1093/ref:odnb/9780198614128.001.0001/odnb-9780198614128-e-9162.

David McKitterick A history of Cambridge University Press, volume II Cambridge: Cambridge University Press 1998.

David McKitterick Cambridge University Library: a history: the eighteenth and nineteenth centuries Cambridge: Cambridge University Press 2009.

David McKitterick 'Joseph Power (1798–1868)' Oxford dictionary of national biography Retrieved 31 May 2018 from http://www.oxforddnb.com/view/10.1093/ref:odnb/9780198614128.001.0001/odnb-9780198614128-e-22666.

L.R. Menon, D. McIlroy, A.G. Liu, and M.D. Brasier 'The dynamic influence of microbial mats on sediments: fluid escape and pseudofossil formation in the Ediacaran Longmyndian Supergroup, UK' Journal of the Geological Society CLXXIII (2016) pp. 177–85.

Philip Mirowski (editor) Edgeworth on chance, economic hazard, and statistics Rowman and Littlefield 1994.

P.B. Moon (revised by Anita McConnell) 'George Paget Thomson (1892–1975)' Oxford dictionary of national biography Retrieved 2 June 2018 from http://www.oxforddnb.com/view/10.1093/ref:odnb/9780198614128.001.0001/odnb-9780198614128-e-31758.

Dennis Moralee A hundred years and more of Cambridge physics Cambridge: Cambridge University Physics Society 1995.

Jack Morrell 'William Vernon Harcourt (1789–1871)' Oxford dictionary of national biography Retrieved 15 March 2018 from http://www.oxforddnb.com/view/10.1093/ref:odnb/9780198614128.001.0001/odnb-9780198614128-e-12249.

Jack Morrell and Arnold Thackray Gentlemen of science: early years of the British Association for the Advancement of Science Oxford: Clarendon Press 1981.

H. Munro Fox 'The origin and development of Biological reviews' Biological reviews XL (1965) pp. 1–4.

Terence D. Murphy 'Medical knowledge and statistical methods in early nineteenth-century France' Medical history XXV (1981) 301–19.

Charles S. Myers 'The future of anthropometry' The journal of the Anthropological Institute of Great Britain and Ireland XXXIII (1903) pp. 36–40.

Kathryn A. Neeley Mary Somerville: science, illumination, and the female mind Cambridge: Cambridge University Press 2001.

Norma C. Neudoerffer 'The function of a nineteenth-century catalogue belonging to the Cambridge Philosophical Library' Transactions of the Cambridge Bibliographical Society IV (1967) pp. 293–301.

David M. Night 'Scientists and their publics: popularisation of science in the nineteenth century' in Mary Jo Nye (editor) The Cambridge history of science, volume V: the modern physical and mathematical sciences Cambridge: Cambridge University Press 2003 pp. 72–90.

Matthew O'Brien'On the symbolical equation of vibratory motion of an elastic medium, whether crystallized or uncrystallized' Transactions of the Cambridge Philosophical Society VIII (1849) pp. 508–23.

Robert Olby 'William Bateson (1861–1926)' Oxford dictionary of national biography Retrieved 1 June 2018 from http://www.oxforddnb.com/view/10.1093/ref:odnb/9780198614128.001.0001/odnb-9780198614128-e-30641.

J.R. Oppenheimer 'On the quantum theory of vibration-rotation bands' Proceedings of the Cambridge Philosophical Society XXIII (1927) pp. 327–35.

A.D. Orange Philosophers and provincials: the Yorkshire Philosophical Society from 1822 to 1844 York: Yorkshire Philosophical Society 1973.

William Otter The life and remains of the Reverend Edward Daniel Clarke, LL.D., Professor of Mineralogy in the University of Cambridge London: J.F. Dove 1824.

Richard Owen 'Description of an extinct lacertian reptile, *Rhynchosaurus articeps*, (Owen,) of which the bones and footprints characterize the Upper New Red Sandstone at Grinsill, near Shrewsbury' Transactions of the Cambridge Philosophical Society VII (1842) pp. 355–70.

Diane Paul 'Eugenics and the left' Journal of the history of ideas XLV (1984) pp. 567–90.

Karl Pearson 'On the correlation of intellectual ability with the size and shape of the head' Proceedings of the Royal Society of London LXIX (1902) pp. 333–42.

Karl Pearson The life, letters, and labours of Francis Galton, volume II Cambridge: Cambridge University Press 1924.

Morse Peckham 'Dr. Lardner's *Cabinet Cyclopaedia*' The papers of the Bibliographical Society of America XLV (1951) pp. 37–58.

R. Peierls 'Critical conditions in neutron multiplication' Proceedings of the Cambridge Philosophical Society XXXV (1939) pp. 610–15.

R. Peirson 'The theory of the long inequality of Uranus and Neptune' Transactions of the Cambridge Philosophical Society IX (1856) Appendix I pp. i–lxvii.

Hilary Perraton A history of foreign students in Britain Basingstoke: Palgrave Macmillan 2014.

D.F.M. Pertz and F. Darwin 'Experiments on the periodic movement of plants' Proceedings of the Cambridge Philosophical Society X (1900) p. 259.

David Phillips 'William Lawrence Bragg (1890–1971)' Oxford dictionary of national biography Retrieved 2 June 2018 from http://www.oxforddnb.com/view/10.1093/ref:odnb/9780198614128.001.0001/odnb-9780198614128-e-30845.

John D. Pickles 'John Hailstone (1759–1847)' Oxford dictionary of national biography Retrieved 2 March 2018 from http://www.oxforddnb.com/view/10.1093/ref:odnb/9780198614128.001.0001/odnb-9780198614128-e-11874.

Roy Porter 'John Woodward: "a droll sort of philosopher"' Geological magazine CXVI (September 1979) pp. 335–43.

Roy Porter 'The natural science tripos and the "Cambridge school of geology", 1850–1914' History of universities II (1982) pp. 193–216.

Roy Porter 'Science, provincial culture and public opinion in Enlightenment England' in Peter Borsay (editor) The eighteenth-century town London and New York: Longman Group 1990 pp. 243–67.

Joseph Power 'An enquiry into the causes which led to the fatal accident on the Brighton Railway (Oct. 2 1841), in which is developed a principle of motion of the greatest importance in guarding against the disastrous effects of collision under whatever circumstances it may occur' Transactions of the Cambridge Philosophical Society VII (1842) pp. 301–17.

Chris Pritchard 'Mistakes concerning a chance encounter between Francis Galton and John Venn' BSHM Bulletin: journal of the British Society for the History of Mathematics 23 (2008) pp. 103–8.

Reginald Punnett 'Early days of genetics' Heredity IV (1950) pp. 1–10.

V.M. Quirke 'John Burdon Sanderson Haldane (1892–1964)' Oxford dictionary of national biography Retrieved 1 June 2018 from http://www.oxforddnb.com/view/10.1093/ref:odnb/9780198614128.001.0001/odnb-9780198614128-e-33641.

William Strutt, Lord Rayleigh 'On the minimum aberration of a single lens for parallel rays' Proceedings of the Cambridge Philosophical Society III (1880) pp. 373–5.

William Strutt, Lord Rayleigh 'On a new arrangement for sensitive flames' Proceedings of the Cambridge Philosophical Society IV (1883) pp. 17–18.

William Strutt, Lord Rayleigh 'The use of telescopes on dark nights' Proceedings of the Cambridge Philosophical Society IV (1883) pp. 197–8.

William Strutt, Lord Rayleigh 'On a new form of gas battery' Proceedings of the Cambridge Philosophical Society IV (1883) p. 198.

William Strutt, Lord Rayleigh 'On the mean radius of coils of insulated wire' Proceedings of the Cambridge Philosophical Society IV (1883) pp. 321–4.

William Strutt, Lord Rayleigh 'On the invisibility of small objects in a bad light' Proceedings of the Cambridge Philosophical Society IV (1883) p. 324.

Richard Rhodes The making of the atomic bomb New York: Simon and Schuster 1988.

Marsha L. Richmond '"A lab of one's own": the Balfour biological laboratory for women at Cambridge University, 1884–1914' in Sally Gregory Kohlstedt (editor) History of women in the sciences Chicago: Chicago University Press 1999 pp. 235–68.

Marsha L. Richmond 'Women in the early history of genetics: William Bateson and the Newnham College Mendelians, 1900–1910' Isis XCII (2001) pp. 55–90.

Marsha L. Richmond 'Adam Sedgwick (1854–1913)' Oxford dictionary of national biography Retrieved 1 June 2018 from http://www.oxforddnb.com/view/10.1093/ref:odnb/9780198614128.001.0001/odnb-9780198614128-e-36003.

J.P.C. Roach (editor) A history of the county of Cambridge and the Isle of Ely, volume III: the City and University of Cambridge London: Victoria County History 1959.

Sydney C. Roberts (revised by Herbert H. Huxley) 'Terrot Glover (1869–1943)' Oxford dictionary of national biography Retrieved 1 June 2018 from http://www.oxforddnb.com/view/10.1093/ref:odnb/9780198614128.001.0001/odnb-9780198614128-e-33427.

Humphry Davy Rolleston The Cambridge medical school Cambridge: Cambridge University Press 1932.

Terrie M. Romano 'Michael Foster (1836–1907)' Oxford dictionary of national biography Retrieved 1 June 2018 from http://www.oxforddnb.com/view/10.1093/ref:odnb/9780198614128.001.0001/odnb-9780198614128-e-33218.

Hugh James Rose The tendency of prevalent opinions about knowledge considered Cambridge 1826.

J.S. Rowlinson Sir James Dewar, 1842–1923: a ruthless chemist London: Routledge 2012.

Katherina Rowold (editor) Gender and science: late nineteenth-century debates on the female mind and body Bristol: Thoemmes Press 1996.

M.J.S. Rudwick 'The early Geological Society in its international context' in C.L.E. Lewis and S.J. Knell (editors) The making of the Geological Society of London London: Geological Society 2009 pp. 145–54.

Andrea A. Rusnock Vital accounts: quantifying health and population in eighteenth-century England and France Cambridge: Cambridge University Press 2002.

Ernest Rutherford 'Capture and loss of electrons by $\alpha$ particles' Proceedings of the Cambridge Philosophical Society XXI (1923) pp. 504–10.
Ernest Rutherford 'Professor C.T.R. Wilson (obituary)' The Times 16 November 1959 p. 16.
Ernest Rutherford and W.A. Wooster 'The natural x-ray spectrum of radium B' Proceedings of the Cambridge Philosophical Society XXII (1925) pp. 834–7.
J.W. Salter 'On fossil remains in the Cambrian rocks of the Longmynd and North Wales' Quarterly journal of the Geological Society XII (1856) pp. 246–51.
J.W. Salter 'On annelide-burrows and surface markings from the Cambrian rocks of the Longmynd' Quarterly journal of the Geological Society XIII (1857) pp. 199–207.
J.W. Salter 'Diagram of the relations of the univalve to the bivalve, and of this to the brachiopod' Transactions of the Cambridge Philosophical Society XI (1871) p. 485–8.
J.W. Salter 'On the succession of plant life upon the Earth' Proceedings of the Cambridge Philosophical Society II (1876) pp. 125–8.
Samuel Satthianadhan Four years in an English university Madras: Lawrence Asylum Press 1890.
Edith Rebecca Saunders 'Mrs G.P. Bidder (Marion Greenwood)' Newnham College letter (1932) p. 65.
Simon Schaffer 'Scientific discoveries and the end of natural philosophy' Social studies of science XVI (1986) pp. 387–420.
Simon Schaffer 'Rayleigh and the establishment of electrical standards' European journal of physics XV (1994) pp. 277–85.
Jeffrey C. Schank and Charles Twardy 'Mathematical models' in Peter J. Bowler and John V. Pickstone (editors) The Cambridge history of science, volume VI: the modern biological and earth sciences Cambridge: Cambridge University Press 2009 pp. 416–31.
Robert E. Schofield 'History of scientific societies: needs and opportunities for research' History of science II (1963) pp. 70–83.
Peter Searby A history of the University of Cambridge, volume III: 1750–1870 Cambridge: Cambridge University Press 1997.
James A. Secord Controversy in Victorian geology: the Cambrian–Silurian dispute Princeton: Princeton University Press 1986.
James A. Secord 'Introduction' to Vestiges of the natural history of creation and other evolutionary writings Chicago and London: University of Chicago Press 1994.
James A. Secord Victorian sensation Chicago: University of Chicago Press 2000.
James A. Secord Visions of science Oxford: Oxford University Press 2014.
J.A. Secord 'Adam Sedgwick (1785–1873)' Oxford dictionary of national biography Retrieved 2 March 2018 from http://www.oxforddnb.com/view/10.1093/ref:odnb/9780198614128.001.0001/odnb-9780198614128-e-25011.
Adam Sedgwick 'On the geology of the Isle of Wight' The annals of philosophy III (1822) pp. 329–55.

Adam Sedgwick 'On the physical structure of those formations which are immediately associated with the primitive ridge of Devonshire and Cornwall' Transactions of the Cambridge Philosophical Society I (1822) pp. 89–146.

Adam Sedgwick A discourse on the studies of the University of Cambridge Cambridge: University of Cambridge Press 1833.

Adam Sedgwick A discourse on the studies of the University of Cambridge Fifth edition London: Parker 1850.

A.C. Seward (editor) Darwin and modern science Cambridge: Cambridge Philosophical Society and Cambridge University Press 1910.

Steven Shapin The scientific life Chicago: University of Chicago Press 2008.

J.L.A. Simmons Report to the Commissioners of Railways, by Mr Walker and Captain Simmons, R.E., on the fatal accident on the 24th day of May 1847, by the falling of the bridge over the River Dee, on the Chester and Holyhead Railway London 1849.

Crosbie Smith 'Geologists and mathematicians: the rise of physical geology' in P.M. Harman (editor) Wranglers and physicists: studies on Cambridge physics in the nineteenth century Manchester: Manchester University Press 1985 pp. 49–83.

Crosbie Smith 'Force energy and thermodynamics' in Mary Jo Nye (editor) The Cambridge history of science, volume V: the modern physical and mathematical sciences Cambridge: Cambridge University Press 2003 pp. 289–310.

Crosbie Smith 'William Hopkins (1793–1866)' Oxford dictionary of national biography Retrieved 31 May 2018 from http://www.oxforddnb.com/view/10.1093/ref:odnb/9780198614128.001.0001/odnb-9780198614128-e-13756.

Crosbie Smith 'William Thomson (1824–1907)' Oxford dictionary of national biography Retrieved 31 May 2018 from http://www.oxforddnb.com/view/10.1093/ref:odnb/9780198614128.001.0001/odnb-9780198614128-e-36507.

Jonathan Smith 'Alfred Newton: the scientific naturalist who wasn't' in Bernard Lightman and Michael S. Reidy (editors) The age of scientific naturalism Pittsburgh: University of Pittsburgh Press 2016 pp. 137–56.

Jonathan Smith and Christopher Stray (editors) Teaching and learning in nineteenth-century Cambridge Woodbridge: Boydell Press 2001.

Robert W. Smith 'The Cambridge network in action: the discovery of Neptune' Isis 80 (1989) pp. 395–422.

Mary Somerville 'On the magnetizing power of the more refrangible solar rays' Philosophical transactions of the Royal Society CXVI (1826) pp. 132–9.

Mary Somerville The mechanism of the heavens London: John Murray 1831.

Mary Fairfax Somerville Personal recollections from early life to old age of Mary Somerville London: John Murray 1874.

Colin Speakman Adam Sedgwick: geologist and dalesman Broad Oak: Broad Oak Press 1982.

Hamish G. Spencer 'Ronald Aylmer Fisher (1890–1962)' Oxford dictionary of national biography Retrieved 1 June 2018 from http://www.oxforddnb.com/view/10.1093/ref:odnb/9780198614128.001.0001/odnb-9780198614128-e-33146.

Alistair Sponsel 'Constructing a "revolution in science": the campaign to promote a favourable reception for the 1919 solar eclipse experiments' British journal for the history of science XXXV (2002) pp. 439–67.

Matthew Stanley '"An expedition to heal the wounds of war": the 1919 eclipse and Eddington as Quaker adventurer' Isis XCIV (2003) pp. 57–89.

Leslie Stephen (revised by I. Grattan-Guinness) 'Augustus De Morgan (1806–1871)' Oxford dictionary of national biography Retrieved 31 May 2018 from http://www.oxforddnb.com/view/10.1093/ref:odnb/9780198614128.001.0001/odnb-9780198614128-e-7470.

George Gabriel Stokes 'Discussion of a differential equation relating to the breaking of railway bridges' Transactions of the Cambridge Philosophical Society VIII (1849) pp. 707–35.

George Gabriel Stokes and Joseph Larmor (editors) Memoir and scientific correspondence of the late George Gabriel Stokes, Bart, volume I Cambridge: Cambridge University Press 2010.

H.P. Stokes The esquire bedells of the University of Cambridge from the 13th century to the 20th century Cambridge: Cambridge Antiquarian Society 1911.

Willie Sugg A history of Cambridgeshire cricket, 1700–1890 Cambridge 2008 http://www.cambscrickethistory.co.uk/new writing.shtml.

Doron Swade 'Charles Babbage (1791–1871)' Oxford dictionary of national biography Retrieved 2 March 2018 from http://www.oxforddnb.com/view/10.1093/ref:odnb/9780198614128.001.0001/odnb-9780198614128-e-962.

James G. Tabery 'The "evolutionary synthesis" of George Udny Yule' Journal of the history of biology XXXVII (2004) pp. 73–101.

E.M. Tansey 'George Eliot's support for physiology: the George Henry Lewes Trust 1879–1939' Notes and records of the Royal Society of London XLIV (1990) pp. 221–40.

Roger Taylor and Larry John Schaaf Impressed by light: British photographs from paper negatives, 1840–1860 New Haven: Yale University Press 2007.

Sedley Taylor 'Physical science at Cambridge' Nature II (12 May 1870) p. 28.

John C. Thackray 'David Ansted (1814–1880)' Oxford dictionary of national biography Retrieved 31 May 2018 from http://www.oxforddnb.com/view/10.1093/ref:odnb/9780198614128.001.0001/odnb-9780198614128-e-577.

G.P. Thomson 'A note on the nature of the carriers of the anode rays' Proceedings of the Cambridge Philosophical Society XX (1921) pp. 210–11.

J.J. Thomson 'Note on the rotation of the plane of polarisation of light by a moving medium' Proceedings of the Cambridge Philosophical Society V (1886) pp. 250–4.

J.J. Thomson 'Some experiments on the electric discharge in a uniform electric field, with some theoretical considerations about the passage of electricity through gases' Proceedings of the Cambridge Philosophical Society V (1886) pp. 391–409.

J.J. Thomson 'The application of the theory of transmission of alternating currents along a wire to the telephone' Proceedings of the Cambridge Philosophical Society VI (1889) pp. 321–5.

J.J. Thomson 'On the effect of pressure and temperature on the electric strength of gases' Proceedings of the Cambridge Philosophical Society VI (1889) pp. 325–33.

J.J. Thomson 'On the absorption of energy by the secondary of a transformer' Proceedings of the Cambridge Philosophical Society VII (1892) pp. 249.

J.J. Thomson 'A method of comparing the conductivities of badly conducting substances for rapidly alternating currents' Proceedings of the Cambridge Philosophical Society VIII (1895) pp. 258–69.

J.J. Thomson 'Cathode rays' The electrician XXXIX (1897) pp. 103–9.

J.J. Thomson 'On the cathode rays' Proceedings of the Cambridge Philosophical Society IX (1898) pp. 243–4.

J.J. Thomson 'Nobel lecture' 11 December 1906 Retrieved 23 June 2017 from http://www.nobelprize.org/nobel_prizes/physics/laureates/1906/thomson-lecture.pdf.

J.J. Thomson and J. Monckman 'The effect of surface tension on chemical action' Proceedings of the Cambridge Philosophical Society VI (1889) pp. 264–9.

J.J. Thomson and H.F. Newell 'Experiments on the magnetisation of iron rods' Proceedings of the Cambridge Philosophical Society VI (1889) pp. 84–90.

J.J. Thomson and E. Rutherford 'On the passage of electricity through gases exposed to Röntgen rays' The London, Edinburgh, and Dublin philosophical magazine and journal of science: series 5 XLII (1896) pp. 392–407.

John Timbs The year-book of facts in science and art London 1847.

Isaac Todhunter (editor) William Whewell, Master of Trinity College, Cambridge: an account of his writings, volume 1 London: Macmillan 1876.

Joseph Train An historical and statistical account of the Isle of Man, from the earliest times to the present date: with a view of its ancient laws, peculiar customs, and popular superstitions, volume I Douglas: Mary A. Quiggin 1845.

Geoffrey Tresise and Michael J. King 'History of ichnology: the misconceived footprints of rhynchosaurs' Ichnos XIX (2012) pp. 228–37.

Raleigh Trevelyan 'Paulina Jermyn Trevelyan, Lady Trevelyan (1816–1866)' Oxford Dictionary of national biography retrieved 27 March 2018 from http://www.oxforddnb.com/view/10.1093/ref:odnb/9780198614128.001.0001/odnb-9780198614128-e-45577.

Pamela Tudor-Craig 'Thomas Kerrich (1748–1828)' Oxford dictionary of national biography Retrieved 2 March 2018 from http://www.oxforddnb.com/view/10.1093/ref:odnb/9780198614128.001.0001/odnb-9780198614128-e-15471.

Raymond D. Tumbleson '"Reason and religion": the science of Anglicanism' Journal of the history of ideas LVII (1996) pp. 131–56.

David A. Valone 'Hugh James Rose's Anglican critique of Cambridge: science, antirationalism, and Coleridgean idealism in late Georgian England' Albion: a quarterly journal concerned with British studies XXXIII (2001) pp. 218–42.

John Venn 'On the diagrammatic and mechanical representation of propositions and reasonings' The London, Edinburgh, and Dublin philosophical magazine and journal of science: series 5 X (1880) pp. 1–18.

John Venn 'On the various notations adopted for expressing the common propositions of logic' Proceedings of the Cambridge Philosophical Society IV (1883) pp. 36–47.

John Venn 'On the employment of geometrical diagrams for the sensible representation of logical propositions' Proceedings of the Cambridge Philosophical Society IV (1883) pp. 47–59.

J. Venn 'Cambridge anthropometry' The journal of the Anthropological Institute of Great Britain and Ireland XVIII (1889) pp. 140–54.

J. Venn and Francis Galton 'Cambridge anthropometry' Nature XLI (13 March 1890) 450–4.

John Venn and J.A. Venn (editors) Alumni Cantabrigienses, part II, 1752–1900, volume vi Cambridge: Cambridge University Press 1954.

S.M Walters and E.A. Stow Darwin's mentor: John Stevens Henslow, 1796–1861 Cambridge: Cambridge University Press 2001.

S. Max Walters 'John Stevens Henslow (1796–1861)' Oxford dictionary of national biography Retrieved 2 March 2018 from http://www.oxforddnb.com/view/10.1093/ref:odnb/9780198614128.001.0001/odnb-9780198614128-e-12990.

Andrew Warwick Masters of theory: Cambridge and the rise of mathematical physics Chicago: University of Chicago Press 2003.

Prudence Waterhouse A Victorian Monument: the buildings of Girton College Cambridge: Girton College 1990.

Mark Weatherall Gentlemen, scientists and doctors: medicine at Cambridge 1800–1940 Woodbridge: Boydell Press 2000.

Mark W. Weatherall 'John Haviland (1785–1851)' Oxford dictionary of national biography Retrieved 2 March 2018 from http://www.oxforddnb.com/view/10.1093/ref:odnb/9780198614128.001.0001/odnb-9780198614128-e-12636.

Christopher Webster and John Elliott (editors) 'A church as it should be': the Cambridge Camden Society and its influence Stamford: Shaun Tyas 2000.

William Whewell 'On the position of the apsides of orbits of great eccentricity' Transactions of the Cambridge Philosophical Society I (1822) pp. 179–94.

William Whewell Astronomy and general physics considered with reference to natural theology London: William Pickering 1833.

William Whewell 'Mathematical exposition of some of the leading doctrines in Mr. Ricardo's "Principles of political economy and taxation"' Transactions of the Cambridge Philosophical Society IV (1833) pp. 155–98.

William Whewell 'Review of *On the connexion of the physical sciences*' The quarterly review LI (1834) pp. 54–68.

William Whewell 'On the results of observations made with a new anemometer' Transactions of the Cambridge Philosophical Society VI (1838) pp. 301–15.

William Whewell Indications of the creator London: John W. Parker 1845.

William Whewell 'On the fundamental antithesis of philosophy' Transactions of the Cambridge Philosophical Society VIII (1849) pp. 170–82.

William Whewell 'Second memoir on the fundamental antithesis of philosophy' Transactions of the Cambridge Philosophical Society VIII (1849) pp. 614–20.

William Whewell 'On the intrinsic equation of a curve' Transactions of the Cambridge Philosophical Society VIII (1849) pp. 659–71.
William Whewell 'On Hegel's criticism of Newton's *Principia*' Transactions of the Cambridge Philosophical Society VIII (1849) pp. 696–706.
William Whewell 'Criticism of Aristotle's account of induction' Transactions of the Cambridge Philosophical Society IX (1856) part I pp. 63–72.
William Whewell 'Mathematical exposition of some doctrines of political economy' Transactions of the Cambridge Philosophical Society IX (1856) part I pp. 128–49.
William Whewell 'Second memoir on the intrinsic equation of a curve' Transactions of the Cambridge Philosophical Society IX (1856) part I pp. 150–6.
William Whewell 'Mathematical exposition of certain doctrines of political economy, third memoir' Transactions of the Cambridge Philosophical Society IX (1856) part II pp. 1–7.
William Whewell 'Of the transformation of hypotheses in the history of science' Transactions of the Cambridge Philosophical Society IX (1856) part II pp. 139–46.
William Whewell 'On Plato's survey of the sciences' Transactions of the Cambridge Philosophical Society IX (1856) part IV pp. 582–9.
William Whewell 'On Plato's notion of dialectic' Transactions of the Cambridge Philosophical Society IX (1856) part IV pp. 590–7.
William Whewell 'Of the intellectual powers according to Plato' Transactions of the Cambridge Philosophical Society IX (1856) part IV pp. 598–604.
James F. White The Cambridge movement: the ecclesiologists and the Gothic revival Cambridge: Cambridge University Press 1962.
Walter White The journals of Walter White Cambridge: Cambridge University Press 2012.
C.T.R. Wilson 'On the formation of cloud in the absence of dust' Proceedings of the Cambridge Philosophical Society VIII (1895) p. 306.
C.T.R. Wilson 'On the action of uranium rays on the condensation of water vapour' Proceedings of the Cambridge Philosophical Society IX (1898) pp. 333–8.
C.T.R. Wilson 'On the production of a cloud by the action of ultra-violet light on moist air' Proceedings of the Cambridge Philosophical Society IX (1898) pp. 392–3.
C.T.R. Wilson 'On a method of making visible the paths of ionising particles through a gas' Proceedings of the Royal Society of London LXXXV (1911) pp. 285–8.
C.T.R. Wilson 'On the cloud method of making visible ions and the tracks of ionising particles' Nobel lectures, physics, 1922–41 (1965) p. 194.
David B. Wilson 'The educational matrix: physics education at early-Victorian Cambridge, Edinburgh and Glasgow Universities' in P.M. Harman (editor) Wranglers and physicists: studies on Cambridge physics in the nineteenth century Manchester: Manchester University Press 1985 pp. 12–48.
David B. Wilson 'George Gabriel Stokes (1819–1903)' Oxford dictionary of national biography Retrieved 31 May 2018 from http://www.oxforddnb.com/view/10.1093/ref:odnb/9780198614128.001.0001/odnb-9780198614128-e-36313.

Joanne Woiak 'Karl Pearson (1857–1936)' Oxford dictionary of national biography Retrieved 1 June 2018 from http://www.oxforddnb.com/view/10.1093/ref:odnb/9780198614128.001.0001/odnb-9780198614128-e-35442.

William Wordsworth The excursion London 1814.

John Martin Frederick Wright Alma mater, or, seven years at the University of Cambridge, volume II Cambridge: Cambridge University Press 2010.

Nicholas Wright Gillham A life of Sir Francis Galton: from African exploration to the birth of eugenics Oxford: Oxford University Press 2001.

Carla Yanni Nature's museums: Victorian science and the architecture of display London: Athlone Press 2005.

Frank Yates (revised by Alan Yoshioka) 'George Udny Yule (1871–1951)' Oxford dictionary of national biography Retrieved 1 June 2018 from http://www.oxforddnb.com/view/10.1093/ref:odnb/9780198614128.001.0001/odnb-9780198614128-e-37086.

Richard Yeo Defining science Cambridge: Cambridge University Press 1993.

Richard Yeo 'William Whewell (1794–1866)' Oxford dictionary of national biography Retrieved 15 March 2018 from http://www.oxforddnb.com/view/10.1093/ref:odnb/9780198614128.001.0001/odnb-9780198614128-e-29200.

George Udny Yule 'Mendel's laws and their probable relations to intra-racial heredity' The new phytologist I (1902) pp. 222–38.

William Zachs, Peter Isaac, Angus Fraser, and William Lister 'Murray family (per. 1768–1967)' Oxford dictionary of national biography Retrieved 15 March 2018 from http://www.oxforddnb.com/view/10.1093/ref:odnb/9780198614128.001.0001/odnb-9780198614128-e-64907.

## Archives/databases

Cambridge Philosophical Society Archives
Darwin Correspondence Project
Privy Council
Archives of the Sedgwick Museum of Earth Sciences

# INDEX